THE
PHYSICAL
WORLD

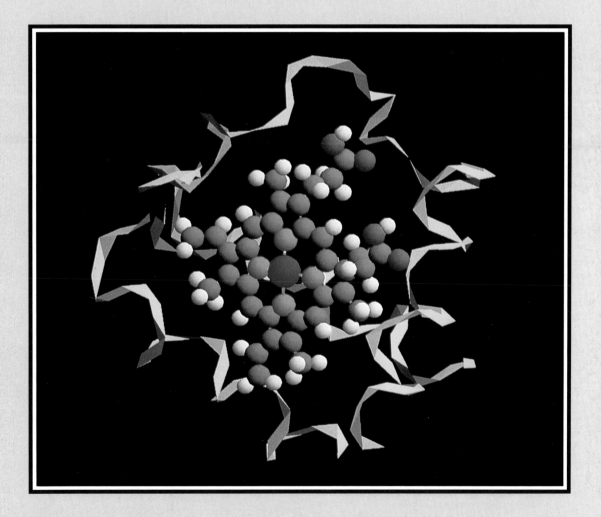

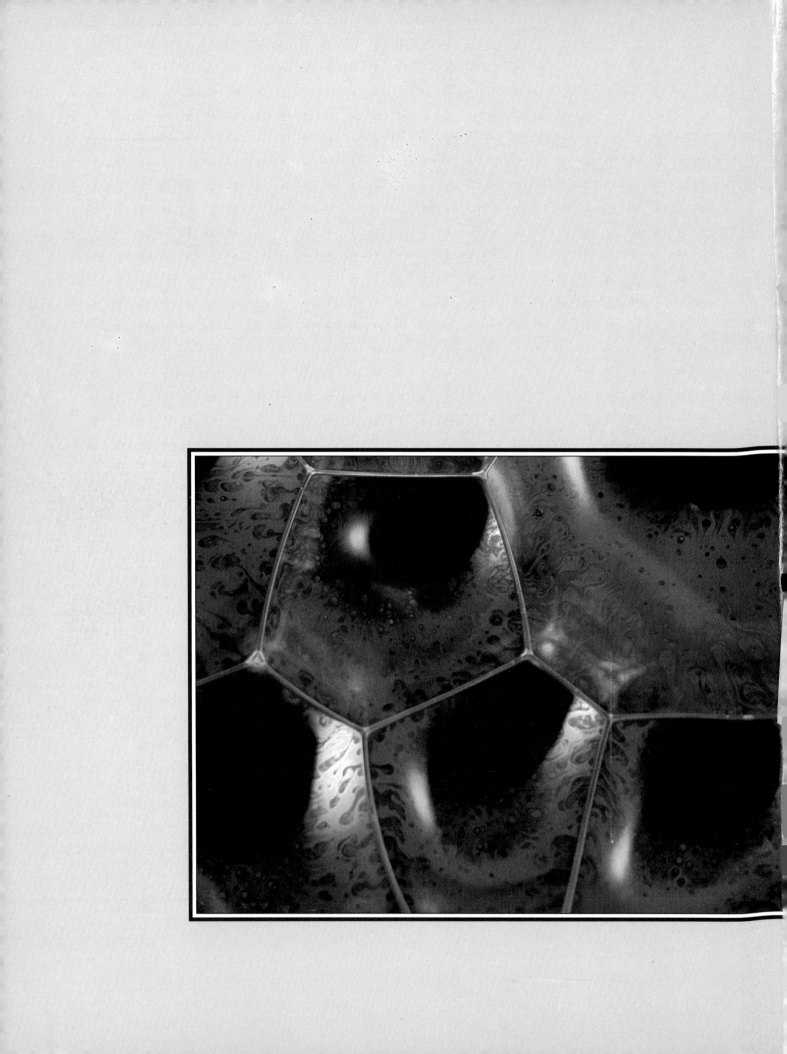

THE PHYSICAL WORLD

Edited by Martin Sherwood and Christine Sutton

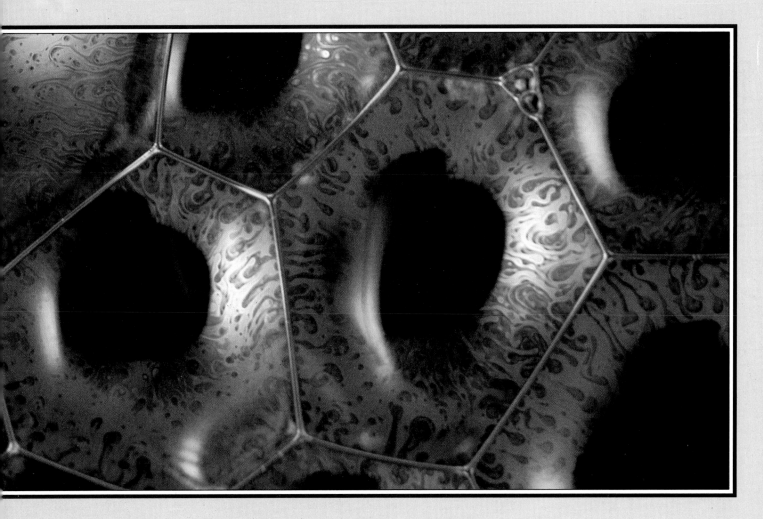

Oxford University Press

New York   Oxford

Editor Peter Furtado
Designers Ayala Kingsley,
Frankie MacMillan, Chris Munday,
Niki Overy
Picture Editor Mary Fane

Design Consultant John Ridgeway
Project Editor Lawrence Clarke

Advisors
Sir Alan Cottrell FRS, Master of
Jesus College, Cambridge
Steven Weinberg, University of
Austin, Texas

Contributors
David Bickerton (2, 28)
Dr M.G. Bowler (7, 8)
Nigel Davis (9, 10, 11, 12)
John Emsley   John Fry (4)
Helen Gasking (14)
Richard Jones (2)   Gill Lacey (6)
Jack Meadows (1)
Martin Sherwood (13, 14, 15, 16, 17,
18, 20, 21, 22, 23, 24, 31, 32)
Richard Stevenson (24)
Christine Sutton (25, 26, 27, 28, 29,
30)
Roger Upton (14)

Computer-generated molecular
graphics by Chemical Design Ltd,
Oxford, using Chem-X software.

AN EQUINOX BOOK

Copyright © Andromeda Oxford Ltd
1988, 1991

Published in the United States of
America by Oxford University Press
Inc, 200 Madison Avenue,
New York, NY 10016

First issued as an Oxford University
Press paperback, 1991

Oxford is a registered trademark of
Oxford University Press

**Library of Congress Cataloging-in-
Publication Data**
The Physical world
"An Equinox book"–t.p. verso.
Bibliography: p.
Includes index.
1. Physics.   2. Chemistry,
Physical and theoretical..
I. Sherwood, Martin.   II. Sutton,
Christine.
QC21.2.P475 1988 530   87-26333

ISBN 0-19-520632-0 (Cloth)
ISBN 0-19-520849-8 (Pbk)

Printed in Spain by Heraclio
Fournier SA, Vitoria

Introductory pictures (pages 1–8)
1 Computer graphic of hemoglobin
molecule (◆ page 88)
2–3 Colors in soap bubbles (◆ page
45)
4–5 Centrifugal force on a
fairground roundabout (◆ page 17)
7 Cosmic rays (◆ page 210)
8 Lighting over Tucson, Arizona
(◆ page 56)

Contents

Introduction

It is easy to forget how new a word physics is. The famous ninth edition of the *Encyclopaedia Britannica*, published a century ago, contains no entry for "physics" at all. There is an entry for "physical sciences" written by the Scottish physicist, James Clerk Maxwell. He listed in this most of the topics now included under physics – dynamics, heat, light, sound, electricity and magnetism – but treated them as distinct from each other. In addition, he mentioned chemistry, and observed that it occupied a high rank in the physical sciences. By contrast, the same edition contained a separate entry for chemistry stretching over 120 pages.

This difference between the treatment of physics and chemistry can be related to the question of how they are defined. Physics, according to one standard definition, deals with the general properties of matter and energy. It is easy for people to conceive of matter, but less so energy. It was not until the mid-19th century that the importance of energy was fully realized; so the present view of physics has only evolved over the past century. Chemistry was much easier to understand. The combination of different substances to form new compounds has been studied for centuries: a prime concern of alchemists was the possibility of transmuting one substance into another. Already by the end of the 18th century, the French scientist, Antoine Lavoisier, had begun to put chemistry onto a proper scientific footing.

The link between physics and chemistry

Scientists have always accepted that physics and chemistry must ultimately be interlinked. The chemical combination of substances must be related in some way to their general physical properties. But it proved difficult to be specific about the exact nature of this link. Since antiquity scientists have speculated that matter is made up of atoms. But modern ideas of the atom began in the late 17th century with Isaac Newton, the English physicist. "All things being considered," he wrote, it seems probable to me that God in the beginning formed matter in solid, massy, hard, impenetrable, moveable particles." Newton's influence was such that the idea of atomic matter has dominated the physical sciences ever since; but there have always been strong challengers. After all, what convincing evidence for atoms was there before the 20th century? Virtually all the observations then available could be interpreted in other terms. The demonstration of the atomic nature of matter has been the great triumph of 20th-century physics.

Even if 19th-century scientists accepted the idea of atoms – as most did – they did not necessarily find it very helpful.

Newton's description of atoms was acceptable to physicists, who were concerned with explanations of relatively simple systems, but it was hardly sufficient for chemists. They were interested in the way atoms combined to form molecules; an atom with the properties of a billiard ball helped them not at all. So, although 19th-century physicists and chemists agreed that matter must be atomic, in practice they went their own ways.

There was one exception to this rule – electrolysis. Experiments at the beginning of the 19th century revealed that the passage of electricity through solutions could release some of the dissolved substances. This discovery affected both physics and chemistry, and the scientists who explored it initially, such as the British scientists Humphry Davy and Michael Faraday, were masters of both branches of science. Later in the 19th century, electrolysis was mainly investigated by chemists. The link that exists today between physics and chemistry came from a different quarter, though it had some connection with the earlier work. Faraday argued that the next step, after studying the passage of electricity through liquids, was to investigate its passage through gases. Experiments along these lines led, at the end of the 19th century, to J.J. Thomson's discovery of the electron.

Physics and chemistry in the present century

Scientists quickly realized that electrons were produced by the disruption of atoms, and this meant, in turn, that atoms must have a complex internal structure. At the same time, it was found that naturally-occurring materials, such as uranium, were also breaking down, a process that came to be called, rather misleadingly, "radioactivity". The discovery affected both physicists and chemists. Radioactivity provided insight into the structure of the atom – a topic at the heart of modern physics. But the transmutation by radioactivity of one element into another was a question of concern to chemists: after all, this had been the original motivation of the alchemists. Both chemists and physicists participated in the study of radioactivity. The overlap in their activities was such that the New Zealander Ernest Rutherford, while being perhaps the greatest experimental physicist of the 20th century, was awarded a Nobel Prize in chemistry.

These investigations led in the 1920s to a theoretical explanation of atomic structure, variously called quantum or wave mechanics. Initially, quantum mechanics was applied strictly to atoms, but it soon became clear that it could also, in principle, explain why atoms join together to form molecules. Chemists, such as Linus Pauling in the United States, successfully applied quantum mechanics to explore the nature of interatomic bonds. Since this represents the application of a physical theory to the basic problem of chemistry, it might be supposed that chemistry now became simply a branch of physics. In practice, it has not worked out that way. Physicists soon concentrated their attention on the internal structure of the atom, and especially the nucleus, leaving bonding between atoms to chemists. They, in turn, soon found that the application of quantum mechanics to systems of atoms was far from simple. Chemists therefore have had to work out their own approaches to this problem and have consequently moved down a rather different line of research from that followed by most physicists.

Two disciplines working together

Despite this, links between physicists and chemists have become much closer during the 20th century. In "mission-oriented" research, the important requirement is to solve specific problems regardless of which scientific disciplines need to be invoked. The need in recent years to investigate new energy sources is an example. Fuel cells, which produce electrical current directly from the conversion of conventional fuels, demand both physical and chemical knowledge. So does the design and maintenance of nuclear reactors. The breakdown of radioactive material in a reactor depends on the careful control of the nuclear particles circulating there. Many years of experiment by physicists lie at the back of such control. Equally, the spent fuel from a reactor has to be processed to extract its radioactive contents. This is an operation that requires considerable chemical skill. A similar mix of knowledge is required to tackle another area of current concern – atmospheric pollution. How the atmosphere operates is a problem in physics, but its exact mode of operation depends on the chemical composition of the atmosphere, which can be affected by pollutants.

These are examples of research that might be labeled "applied", but there are similar areas of joint physical and chemical endeavor in pure research. One such is the study of water. Its peculiar properties have long interested scientists. Nowadays, attempts to examine these use a combination of physical and chemical methods. But the best example of all is the work over the past few decades on the basic nature of life. Two strands led to this – the growth of biophysics and of biochemistry. Here physical, chemical and biological knowledge must be applied simultaneously.

The Greek view of matter
The debate on whether matter was continuous or made up of discrete elements began with the earliest known Greek thinker, Thales (c.624- c.547 BC), who asserted that all matter was made of water. By "water" he meant some kind of fluid with no distinctive shape or color. Subsequently, Anaximenes (c.570 BC) suggested that this basic substance was actually air. Again, by "air" he meant not just the material making up our atmosphere, but an immaterial substance which breathed life into the universe. These early views led to the popular Greek picture of matter described first by Empedocles (c.500-c.430 BC), where there were four elements – earth, water, air and fire. All these proposals implied that matter is continuous.

The opposing view appeared later, beginning with the little-known Leucippos (c.474 BC) and fully expounded by his pupil Democritos (c.460-c.400 BC). This saw matter as consisting of solid "atoms" (the word means "indivisible") with empty space between them. The idea of empty space was, in its way, as great an innovation as atoms; for continuous matter left no gaps. Both views flourished in ancient Greece, but a belief in continuous matter was much commoner. The debate restarted in 17th-century Europe, still on the basis of the early Greek speculations, but this time it finally led to an acceptance of atomic matter (◆ *pages 12-13).*

The earliest efforts to understand the nature of the physical world around us began several thousand years ago. By the time of the ancient Greeks, over 2,000 years ago, these attempts at explanation had become both complex and sophisticated. They were characterized by the desire to find a single explanation which could be applied to all happenings in the physical world. For example, the description of the world that received most support supposed the existence of four primary chemical elements – earth, water, air and fire. This list may look odd to us but we should see it as something like the modern division of substances into solids, liquids and gases (◆ pages 29-38). These four elements were considered to have particular places where they were naturally at rest. The earth, preferentially accumulated at, or below, the Earth's surface; the water came next, lying on top of the Earth's surface; air formed a layer of atmosphere above the surface; and, finally, a layer of fire surrounded the atmosphere. This layering of the elements was invoked to explain how things moved on Earth. A stone thrown into the air fell back to the Earth's surface because that was its natural resting-place; flames leapt upwards in order to reach their natural home at the top of the atmosphere, and so on.

Greek philosophers set the scene for later studies of the material world by distinguishing between different types of theories of matter. The Greeks pointed out that two explanations are feasible. The first supposes that matter is continuous; so that it is always possible to chop up a lump of material into smaller and smaller pieces. The other theory supposes that matter consists of many small indivisible particles clumped together; so that chopping up a lump of matter must stop once it has reached the size of these particles.

The four humors

The chemical elements could combine to create new substances – in particular, they formed the "humors". Each individual human being contained a mixture of four humors, made up from the four elements, and the balance of these humors determined the individual's nature. This theory is still invoked today when we say someone is in a "good humor". Indeed, some of the Greek technical terms are still used: "melancholy" is simply the term for "black bile", one of the four humors. So the chemical elements of the ancient Greeks were involved in determining motion, a fundamental part of physics, and in determining human characteristics, an area now referred to as physiology and biochemistry. The Classical world did not distinguish between physics and chemistry, but saw all of what we would now call "science" as an integrated whole, known as natural philosophy; by the end of the period, however, a distinction between the two areas of study was beginning to emerge as practical studies in alchemy developed that field into a separate area of knowledge.

▼ *Much ancient study was devoted to the movements of the Sun, Moon and planets. Monuments such as Stonehenge in southern Britain were used as observatories. Here a partial eclipse of the Moon is seen above Stonehenge.*

The Chinese search for the elixir of life led to the discovery of gunpowder

◄ *An alchemist in Iran, where the study continues, with stress on its spiritual rather than its scientific aspect. Even in earlier times, alchemy was as much a philosophical investigation as a chemical attempt to transmute one element into another.*

Early Chinese physics and chemistry

The early Chinese view of the world differed in important respects from the Greek. The Chinese saw the world as a living organism, whereas the Greeks saw it in mechanical terms. In some ways this made little difference. For example, the Greeks concluded that all matter was made of four elements; the Chinese supposed there were five – water, earth, metal, wood and fire. The Chinese, like most Greeks, believed matter to be continuous. Perhaps their picture of the world as an organism prevented them from thinking of the alternative atomic theory, unlike the Greeks.

The Chinese led the world for many centuries in practical physics and chemistry. Their knowledge of magnetism advanced rapidly. They learnt at an early date how to magnetize iron by first heating it, and then letting it cool whilst held in a north-south direction (♦ page 51). They realized, 700 years before Western scientists, that magnetic north and south do not coincide with terrestrial north and south. In chemistry, too, practical knowledge was ahead. Thus, experiments seeking for the elixir of life led instead to the discovery that a mixture of saltpetre, charcoal and sulfur formed the potent explosive known as gunpowder (♦ page 181).

Why then, with this practical lead, did modern physics and chemistry not originate in China? Factors that have been suggested include the limitations of Chinese mathematics, the nature of the society, and even the structure of the language.

▲ *A reconstruction of Galileo's pendulum clock. The development of accurate clocks enabled scientific measurement, and allowed him to develop the study of forces and motion, initiating modern physics (♦ page 15).*

The division between physics and chemistry

One of the great problems in discussions of motion was to try and explain how the Sun, Moon and planets moved across the sky. This question had been enthusiastically attacked by the ancient Greeks, and their work was followed up by the Arabs, but in both cases on the assumption that all these bodies moved round a stationary Earth. The concentration on astronomical motions reduced interest in the link between physics and chemistry. The Greeks and Arabs believed that the heavens were made of a fifth element – labelled the "aether" – which had nothing in common with the terrestrial elements. Consequently, motions in the heavens could not be explained in terms of motions on the Earth; so study of these motions held little of consequence for the relationship between physics and chemistry.

At the same time, a form of chemistry arose which also diverted attention away from the link with physics. Called alchemy, it emphasized practical activity along with a diffuse theory, typically expressed in symbolic terms. Though alchemy first appeared in the late classical world notably in Alexandria, now in Egypt, it flourished particularly amongst the Arabs. A major aim was to transmute one metal into another, especially to turn "baser" metals into gold. Alchemists thought this could be done by finding an appropriate substance – often called the "philosopher's stone" – which would induce the change.

Over the centuries, Arabic studies led to a number of practical developments in physics and chemistry, but retained much the same theoretical framework as the Greeks. From AD1100 onwards, scholars in western Europe began to translate and study the Greek texts preserved by the Arabs, along with the developments made by the Arabs themselves. As the Arab world became gradually less interested in science, the Western world caught up and, by the 16th century, had reached the point where it could advance beyond either the Greeks or Arabs. The first breakthrough was in astronomy. A Polish cleric, Nicolaus Copernicus (1473-1543), worked out how the motions of the heavens could be explained if the Earth moved round the Sun, rather than *vice versa*. His initiative led over the next 150 years to an explanation of planetary motions that is still basically accepted today. This explanation showed that motions in the heavens and on the Earth were not basically different, as had been previously supposed. It also overthrew the old idea of a connection between the chemical elements and the nature of motion. A division between physics and chemistry therefore remained unbridged, as physics remained linked to astronomy and chemistry to alchemy. The English scientist Isaac Newton (1642-1726), for example, was not only one of the greatest mathematicians and physicists of all time, he was also an enthusiastic alchemist. Yet he seems to have made little connection between these activities.

One step in the 17th century which held some hope for renewing links between physics and chemistry was the fresh interest in an atomic theory. The idea that all matter was made up of tiny, invisible particles called "atoms" originated with the ancient Greeks, but has always been less popular than the belief in four elements. It was now revived, with the suggestion that the various materials in the world might all be formed from atoms grouping together in various ways. This sounds a very modern explanation, but it was not very useful in the 17th century. Atoms could not be studied, or their properties determined, with the equipment then available. So physics and chemistry continued to develop along their own lines.

By the mid-20th century, theoretical physics and chemistry were approaching very similar questions from slightly different angles

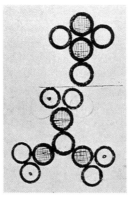

◀▲ *John Dalton was the first chemist to show molecules as compounds of elements arranged in a particular manner. His formulae for organic acids (1810-15) are shown here.*

▶ *A modern computer graphic illustration of part of the DNA molecule, which contains the genetic code.*

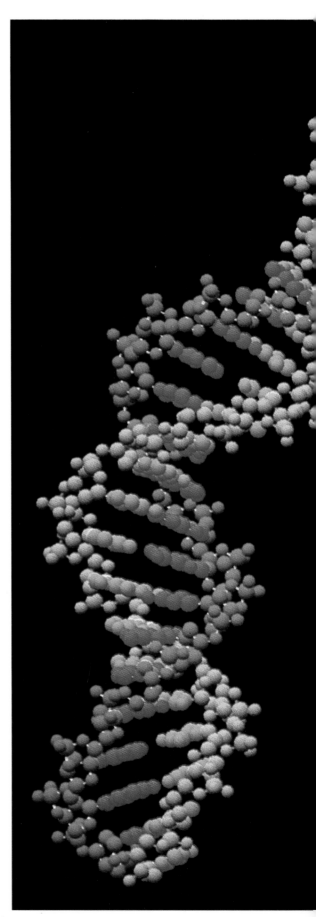

The 19th century

Up to the 18th century, physics had progressed more rapidly than chemistry, but now chemistry moved ahead. The theories of alchemy were rejected, but its concern in practical experiments was pursued vigorously. One area of particular concern was the analysis of gases. It became clear that the old element "air" actually consisted of a mixture of gases; other gases, not present in the atmosphere led to two major developments. In the first place, the Frenchman, Antoine Lavoisier (1743-1794) introduced the modern definition of a chemical element and the modern idea of elements combining to form a variety of chemical compounds (◆ pages 77-92). Secondly, John Dalton (1766-1844) in England and Amadeo Avogadro (1776-1856) in Italy showed that elements combined in simple proportions by weight, as would be expected if matter was made up of atoms.

This concept of chemical compounds as a series of atoms linked together led to one of the basic scientific advances of the 19th century. Each atom was assigned a certain number of bonds – now called "valence" bonds – by which it could attach itself to other atoms. The results of chemical analysis could be interpreted in terms of valences, and the theory also formed the basis for the synthesis of new compounds. Knowledge of chemical bonds improved throughout the century. For example, the carbon atom was assigned four valence bonds (◆ page 94). From studying the properties of carbon compounds, chemists worked out where in space these bonds pointed relative to each other. The spatial picture they derived was found to explain quite unrelated physical observations. It was also known that some properties of light were changed when it was passed through certain organic compounds. The chemists' explanation of carbon-atom bonding proved capable of explaining why the light was changed. In these instances chemistry provided a better insight into the nature of matter than physics could.

To most 19th-century physicists, atoms were little more than tiny billiard balls. Chemists recognized that atoms must be more complex than that, but could not, themselves, provide a better description. It

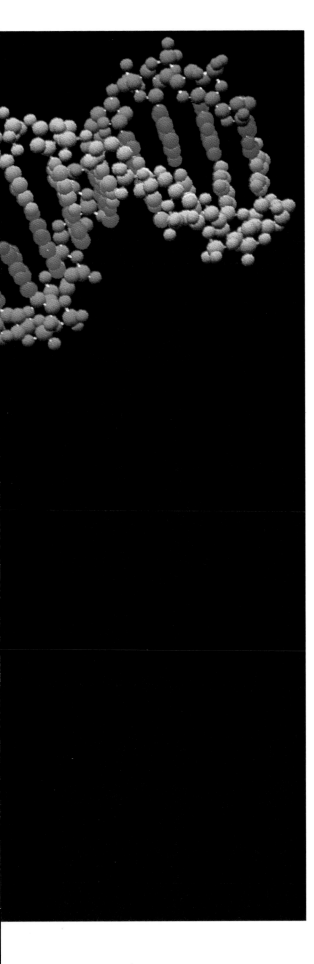

was the physicists who made the important breakthrough. Again, it came from the study of gases – in this case, from examining the passage of electricity through rarified gases. Experiments by the British physicist J. J. Thomson (1856-1940) showed that electrical "cathode rays" in gases seemed to consist of sub-atomic particles, which gave some insight into the nature of atoms. Thomson discovered that atoms contained particles – which he labeled "electrons" – with a low mass and a negative electrical charge (◆ page 73). Not long afterwards, the New Zealander Ernest Rutherford (1871-1937) deduced that atoms consisted of a cloud of electrons circling round a much more massive positively-charged nucleus (◆ page 187).

These were startling developments, but it was the next step that had the most impact on chemists – the explanation, "quantum mechanics" began with Niels Bohr (1885-1962) just before World War I, but reached a stage where it was useful in the 1920s. Quantum mechanics showed how electrons in different atoms could interact, so linking the atoms together. Now the valence bonds of the chemists could be explained in terms of the physicists' atom (◆ page 195).

Physics, chemistry and industry

By the 1920s the theoretical link between physics and chemistry was firmly established. But the practical applications of the two subjects continued on separate paths. A recognizable chemical industry had first appeared at the end of the 18th century. It remained small-scale for many years, and was mainly concerned with the production of simple chemicals, such as household soda ($NaOH$). In the latter part of the 19th century, attention turned to the production of organic compounds (containing carbon). The successful synthesis of new artificial dyestuffs led to a rapid growth of the chemical industry, which has continued ever since. An industry based on research in physics came later than in chemistry; but, by the end of the 19th century, earlier studies of electricity and magnetism had led to thriving industries in electrical engineering and communications. These physics-based industries had little in common with the chemical industry, and the gap was not bridged by any major developments in the first half of the 20th century.

The position has changed drastically in recent decades. Science, industry and defense have become intermeshed in a variety of ways, several of which involve joint activity in physics and chemistry. A good example concerns the Earth's upper atmosphere. This is a region of considerable importance, both for space activities and for military purposes. How it can be used depends on the properties of the gases present, and determining these has led to co-operative investigations of the region by physicists and chemists. However, the most revealing example of interdependence is molecular biology. The nature of biological materials has long been studied by applying various physical and chemical techniques, the most important being their interaction with X-rays. Results initially came slowly because of the complexity of biological compounds. But researchers, mainly in Britain and the United States, gradually pieced together information about the nature of biological molecules. The most significant advance was made in 1953, when Francis Crick (b.1916) and James Watson (b.1928) were able to describe the structure of the basic genetic material, DNA. From that work has come the new "biotechnology" industry. Today, the ancient Greeks' belief that these three branches of science are linked has been vindicated, but in a way far beyond their envisaging.

Physics **Chemistry**

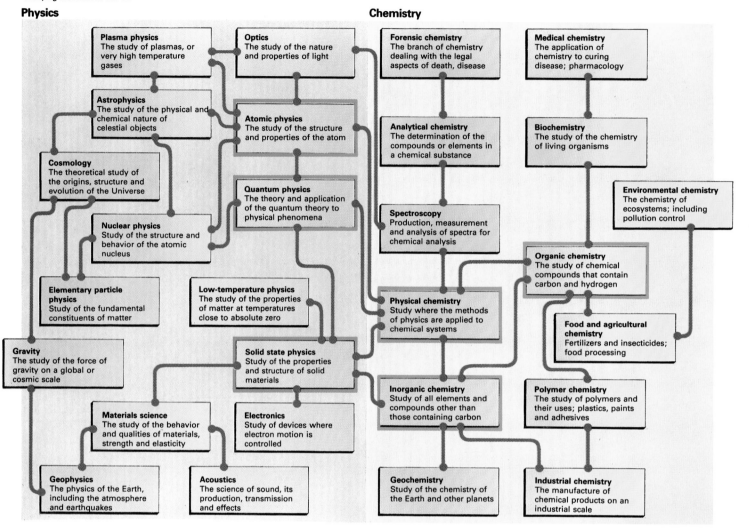

The range of physics and chemistry
Modern physicists and chemists can apply their skills to almost any area of science or technology. This is not too surprising. Questions involving physics and chemistry are basic to almost any attempt at understanding the world around us. So there are scientists who study the physics and chemistry of stars and planets, while others examine the physics and chemistry of plants and animals. The list is endless.

Physics has traditionally been divided into such categories as sound, heat, light, and so on. These divisions hardly suggest the complexity of modern physics, but do hint at the opportunities for applying physics. For example, the design of musical instruments now requires a detailed knowledge of sound. So does the design of music centers, and these also use the products of the huge new microelectronics industry, which is based on electromagnetism and solid-state physics. Physicists in this industry are concerned with applications varying from computers to biosensors (to detect the physical characteristics of living organisms). Electromagnetism figures in most modern forms of communication, and physicists are concerned with improvements to telephones, radio and television. Lasers have been developed for purposes ranging from

communication at one end to medicine at the other (where they are controlled by medical physicists). Lasers also appear in one of the most publicized employment areas of modern physics – the attempts to gain new sources of energy from atoms, as via fusion.

Chemistry, too, has its traditional divisions – into physical, inorganic and organic – but, as in physics, the boundaries are blurred nowadays, just as the boundaries between physics and chemistry themselves are increasingly doubtful. Chemists, like physicists, are often concerned with sources of energy. The oil industry, for example, employs chemists on tasks ranging from the discovery of oil to its use in internal combustion engines.

The pollution caused by such engines is monitored by other chemists, for environmental chemistry has expanded greatly in recent decades. Pollution studies often involve looking for small amounts of chemical, a problem shared by forensic scientists as they try to help the police. Much of this work consists of analysis – finding what substances are there – but many chemists are more concerned with the synthesis of new compounds. Vast amounts of time and money are spent on this in the pharmaceutical industry. Finally, physicists and chemists must think of the future of their subjects: so many are employed in some area of teaching.

▲ Together physics and chemistry provide a framework of interlinked subject areas that are used to explain matter, energy and the Universe. Physics has the wider span, encompassing the smallest subatomic particle at one extreme, and the infinity of the known Universe at the other. Chemistry, however, may limit itself to the level of atoms and molecules but these are the building blocks of all matter. In some areas, in the center of the diagram, physicists and chemists may be studying the same phenomena, but approaching them from different angles or asking different questions. Most of the disciplines in the boxes of this diagram emerged only in the past 50 years.

Forces, Energy and Motion

*Why do objects move?...Newton's laws of motion...
Friction...Energy at work...Conversion of energy...
Oscillating systems...PERSPECTIVE...Vectors, velocity
and acceleration...Circular motion...Gravity...Newton
and the apple...The tides...The physics of pool...
Defining work...Resonance*

Imagine a ball being hit by a stick like a golf club. The impact producing the movement is obvious, and the ball eventually stops rolling. Ancient Greek philosophers were puzzled by such situations because they could see no reason for the ball to continue moving after contact with the stick has been broken. Aristotle (384-322 BC) believed the medium through which the ball moves transmits thrust to the ball.

Eventually the Italian scientist Galileo Galilei (1564-1642) concluded that the problem was being considered from the wrong viewpoint. He argued that constant motion in a straight line is as unexceptional a condition as being stationary, but the continual presence of friction (◆ page 19) on moving objects conceals this. Without friction the ball would roll in a straight line forever, unless its direction is changed by hitting another object. It is therefore only *changes* in motion that deserve particular consideration.

Velocity and acceleration

Physicists distinguish between the concepts of speed and velocity. Speed indicates the distance covered by a body in a given period of time, irrespective of the direction it is moving. It may be measured in meters per second, for example. Velocity, on the other hand, is a so-called "vector" quantity: that is, a quantity that requires direction as well as magnitude. Two ships that travel equal distances in equal times have the same speed, but they have the same velocity only if they move in the same direction. Because directions are involved, adding velocities and other vectors requires special techniques. These involve drawing parallelograms in which each line represents the distance covered and the direction of each vector.

Acceleration (which is another vector quantity) is defined as the change in velocity per second, measured in meters/second^2 (m/s^2). A satellite in circular orbit will be traveling with constant speed, but its direction is continually changing. As a result, its velocity is similarly changing, and so it must have an acceleration. This acceleration is towards the center of the orbit, and is caused by gravity (◆ page 18).

▼ Motion is no more unusual than standing still; it is changes in motion that involve an external influence. When a horse slows down abruptly, the rider tends to continue in the same state of motion, and tumbles over the top.

Conservation of angular momentum explains why a skater pulls in her arms when she spins

Galileo also considered the motion of falling bodies, and showed that any two objects in free fall at the same place above the Earth's surface have the same acceleration. He deduced the basic relationships of dynamics, showing that the velocity of a uniformly accelerating body increases in proportion to the time, while the distance traveled is proportional to the time squared. Why all falling bodies should have the same acceleration was an unanswered question.

When the English scientist Isaac Newton (1642-1727) came to consider this problem, he set down three "laws of motion" as a foundation upon which to build his revolutionary theory of gravitation.

Law 1 stated that "a body will continue at rest, or in uniform motion in a straight line unless acted upon by a resultant force". Newton introduced the idea of "mass", or inertia, as a measure of a body's reluctance to start or stop moving.

In his second law ("the rate of change of momentum of a body is proportional to the resultant force on the body, and takes place in the direction of that force"), Newton attempted to describe the change in motion that a body would experience under the action of a resultant force. He introduced the quantity "momentum", the product of mass times velocity. In cases where the mass of the body is constant, this second law is stated simply as "force equals mass times acceleration".

Law 3 states that "if a body A experiences a force due to the action of a body B, then body B will experience an equal force due to body A, but in the opposite direction." Newton illustrated his third law through the example of a horse pulling a stone tied by a rope. While the stone experiences a force forwards, the horse experiences a force backwards. The tension in the rope acts equally to move the stone and to impede the movement of the horse.

A consequence of Newton's second and third laws is that when two objects collide with no external forces acting upon them, the total momentum before the collision is equal to the total momentum after the collision. This is the "conservation of linear momentum", and is of great value in analyzing collisions or interactions on any scale. For example, when a gun fires, the momentum of its recoil is equal and opposite to the momentum of the bullet, adding to a total momentum of zero – the same as before firing.

▲ *Once hit, an ice hockey puck shoots in a straight line, demonstrating Newton's first law of motion. According to his second law, the heavier an object, the greater the force needed to set it moving, as anyone knows who has tried to push or pull (right) a truck. Newton's third law equates action (here the upward pull of the athlete's muscles) with reaction (the downward force of the car's weight).*

▶ *These people flying round a roundabout do not travel in a straight line because they feel a centripetal force, acting toward the center of their circular path. This force is the net result of the weight of the chair and body, acting downward, and the tension in the wires.*

Circular motion

An object such as a seat on a fairground roundabout, traveling in a circle, can appear to be moving uniformly. However, its velocity is continually changing. To understand why, recall that velocity is a vector quantity, with a direction as well as a magnitude. At any point in time the velocity of the seat is in fact in the direction of the tangent to the circle at the roundabout's position. As the seat moves, this direction, and hence the velocity, changes. According to Newton's first law the seat must therefore be subject to a force and, indeed, this force is applied continually to the seat via the chain that holds it to the roundabout. If the chain were to break and the force it provides were thus suddenly interrupted, the seat would fly away in a straight line, as Newton's first law dictates.

Any force that produces circular motion of this kind is called a "centripetal force". It acts towards the center of the circle, and therefore at right angles to the motion round the circle. The size of the force is equal to the mass of the object multiplied by the square of the speed and divided by the radius of the circle. Here, the speed is the magnitude of the velocity.

Any object moving on a curved path or rotating on its own axis has an "angular speed". This is the angle the object travels through, with respect to the center of its motion, during a unit of time. An object traveling uniformly in a circle, like the roundabout seat, has a constant angular speed, although its velocity is changing all the time.

Objects with angular speed have "angular momentum", directly analogous to the "linear momentum" of objects moving in straight lines. Angular momentum is equal to mass multiplied by linear speed multiplied by the radius of the motion. In any system, the total angular momentum must be conserved if the system does not experience a turning force, or torque. So if, for instance, the radius decreases, the velocity increases provided the mass remains the same. This is why, for example, a figure skater spins slower when she stretches out her arms horizontally and faster when she pulls them in.

The concept of gravity enabled scientists to describe the orbits of the planets, the rhythms of the tides, falling objects and many other phenomena

Gravity

Gravity is the most obvious of nature's forces (♦ page 213). It keeps us on the ground, and it controls the behavior of the Universe. The structure and motion of the planets, stars and galaxies are all determined by gravity.

Newton was the first to realize that all bodies with mass attract each other. He showed that the force of attraction between two bodies is proportional to the product of their masses times a constant, and inversely proportional to the square of their distance apart.

The constant here is called the universal gravitational constant. It is usually denoted by G and is equal to $6\cdot673 \times 10^{-11}$ newton meters ^{2} per kilogram ^{2}. In proclaiming this a universal constant, Newton was assuming that the heavenly bodies – the Moon and the stars – obey the same rules as objects here on Earth. This was a revolutionary advance. From the time of the Greek philosopher Aristotle (384-322 BC), people had believed that earthly and heavenly objects obeyed different laws (♦ page 11). After Newton, however, physics could take the Universe as its laboratory; and his point of view remained unchallenged until the final years of the 19th century (♦ page 46).

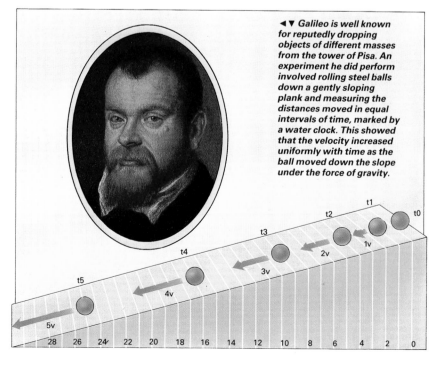

◄▼ *Galileo is well known for reputedly dropping objects of different masses from the tower of Pisa. An experiment he did perform involved rolling steel balls down a gently sloping plank and measuring the distances moved in equal intervals of time, marked by a water clock. This showed that the velocity increased uniformly with time as the ball moved down the slope under the force of gravity.*

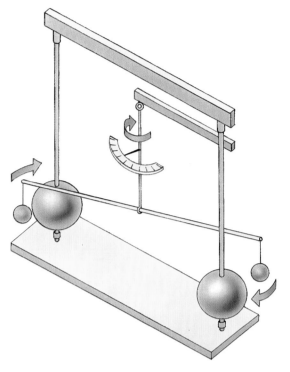

◄ **Free-fall parachutists experience a force due to air resistance that is equal and opposite to the force due to gravity. Thus, in accordance with Newton's first law of motion, they fall at a constant velocity.**

▼ **Fishing boats lie stranded on the sands around a harbor at low tide, as the seas respond to the changing gravitational pull of the Moon across the Earth's diameter.**

▲ **The English physicist Henry Cavendish (1731-1810) made the first measurements of the gravitational constant, using a "torsion balance". Two small balls were attached to the ends of a bar suspended at its center by a wire. Large balls held at either end, but on opposite sides of the bar, attracted the small balls through the gravitational force between them, and made the bar twist.**

"God said let Newton be, and all was light"

Isaac Newton was born in January 1643 in Woolsthorpe, Lincolnshire. As a schoolboy he was fascinated by mechanical devices and he went up to Cambridge University in 1660, graduating in 1665. When bubonic plague reached Cambridge in 1665 he returned to his mother's farm. The enforced rest left him free to develop his ideas on the law of gravitation which he published 20 years later, in his book "Principia Mathematica". At the same time he started a series of optical experiments and discovered, among other things, that white light is a mixture of colors (♦ page 42).

Newton was absent-minded and sensitive to criticism. He conducted an international dispute with the German mathematician Gottfried Wilhelm Leibniz (1646-1716) as to who had first discovered calculus. Nearer to home, he quarreled for years with the British physicist Robert Hooke (1635-1703). Hooke claimed that Newton had stolen some of his ideas and put them in the "Principia". Newton was finally forced to include a short passage acknowledging that Hooke and others had reached certain conclusions which he was now explaining in greater detail. These quarrels infuriated Newton, and contributed to his nervous breakdown in 1692.

Much of Newton's life was spent in trying to manufacture gold and in speculating on theology, yet he was honored and respected as few scientists have been before or since.

Gravity and the tides

The Earth and the Moon rotate about their common center of mass (the point where an outsider would consider all the mass of the system to be concentrated). Because the mass of the Earth is so much greater than that of the Moon, the center of mass is much closer to the Earth than to the Moon.

Newton showed that bodies move in straight lines at constant speed unless a force acts upon them. Thus there must be a force that keeps the Earth orbiting around the center of mass of the Earth-Moon system. This force, which is centripetal, is provided by the gravitational attraction of the Moon, and it is just the right size to keep the center of the Earth orbiting about the center of mass.

The Moon's gravitational force decreases as the distance from the Moon increases. For points on the Earth closer to the Moon than the Earth's center, the gravitational force is larger than required for the orbital motion. Here the Earth is stretched towards the Moon. The seas, being free to move, bulge towards the Moon. For points farther from the Moon than the Earth's center, the gravitational force is weaker than required and the seas bulge out away from the Moon. The Earth spins on its axis, rotating under these bulges which sweep over the surface of the Earth, causing two high tides each day.

The gravitational pull of the Sun also causes tides, but the Sun is so much farther from the Earth than the Moon that its gravitational pull changes less across the Earth's diameter. The tides are largest (spring tides) when the Sun, Moon and Earth reinforce each other, and weakest (neap tides) when the three bodies are 90° out of line and the tidal effects of the Sun and Moon tend to cancel.

◄ Frictional forces oppose the motion of objects sliding over each other. The downward force of the climber's weight is counterbalanced in part by the friction between the soles of his boots and the rock face. The soles are made of a soft rubber compound designed to "stick" to the rock, and they allow the climber to scale the vertical cliff without slipping.

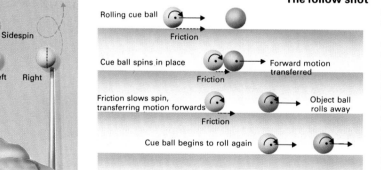

Topspin

Backspin

Sidespin

Left   Right

The follow shot

Rolling cue ball

Friction

Cue ball spins in place

Forward motion transferred

Friction

Friction slows spin, transferring motion forwards

Object ball rolls away

Friction

Cue ball begins to roll again

◄ In a game of pool a cue ball hit slightly above center (far left) is given "top spin", rotating in the direction of its motion; cueing below the center results in "back spin". Positioning the cue to left or right imparts "side spin", which allows the cue ball to swerve in the correct conditions. In detail, shots depend on the interplay between the motion of a ball and the friction between the ball and the table (left).

A trick shot

► In this pool shot, the aim is to pocket all six balls. A skilled player would hit the cue ball above left of center, toward the two ball. The net force (see inset) is such that the two ball hits the five ball and bounces into the pocket. The three ball ricochets off the cushion toward the opposite pocket, swerving slightly to the right due to friction with the two ball. The net force on the five ball sends it into the top pocket, while the one and four balls are pocketed at the same time. The top spin given to the cue ball allows it to travel on, curving due to side spin, so that it ricochets off three cushions, eventually knocking the six ball into the bottom pocket.

Net force

3

Friction

1

2

5

4

The physics of pool

The laws of motion are often described in terms of the interactions of "billiard balls", on the assumption that in a two-dimensional plane the momentum and angular displacement of bodies after collision can be calculated simply from their previous velocity and the angle of impact. It is convenient to think of billiard balls as behaving in this manner but in practice their behavior is more complex, being affected by friction.

When a ball moves across a snooker or pool table it has two types of motion. The first is a forward "translational" motion, the second is a rotation about the ball's center. For pure rolling there is a relationship between these two. In other cases skidding occurs at the table surface. This happens, for example, when a ball is hit centrally by a cue. Initially the ball moves off without rotating and slides across the table. However, friction between the ball and the table causes the ball both to slow down and to start rotating. When the rotational motion matches the translational motion pure rolling takes over, and the friction decreases correspondingly.

To eliminate this initial skidding the ball must be set moving with the correct amount of initial rotation. This is achieved by striking it slightly above the center. The cushions on the table are set rather higher than the center of the balls for similar reasons. When a rolling ball hits a stationary one, forward movement of the cue ball is transmitted to the object ball. The object ball moves off skidding, because it has been hit centrally. If the balls are smooth there is no significant friction between them and no rotation is transmitted in the impact. The cue ball is left instantaneously stationary, but still rotating. The frictional force which slows this rotation also gives the cue ball forward motion (and if strong enough, it may cause the cue ball to follow the object ball into the pocket!).

If the cue ball is still skidding as it makes the collision, the player has some control over the outcome. For example, if the cue ball is not rotating at all and is simply sliding across the table, it will stop dead after collision with the object ball. If, however, it is hit below its center its rotation will be in opposition to its forward motion, and friction will cause it to move backwards after the collision.

If the collision with the object ball is oblique rather than head on, the cue ball does not lose all its translational motion, but moves off in a different direction at reduced speed. The frictional force resisting skidding is now no longer aligned with the direction of movement. As a result, the ball swerves while skidding continues, before eventually moving in a straight line once pure rolling starts. This gives the player some control over the final direction of the cue ball, in anticipation of the next shot.

Similarly the player may swerve the cue ball around an obstacle. By cueing to the right or left of center, the spin produced is across the direction of forward motion. This resulting sideways frictional force at the surface allows the ball to swerve as long as skidding is taking place. These techniques all require that the cue ball has not started to roll; for a typical, firmly struck shot the ball must not have traveled more than about one meter.

Newton was conscious of two types of force. First there are those that involve contact of some kind including friction, tension and compression. Second, there are forces that are able to act across a distance, such as magnetic (◆ page 49) or electrostatic forces (◆ page 53) and the force that concerned Newton, gravity. Subsequently, scientists began to interpret forces in terms of the interaction between particles, such as the collisions of air molecules at a surface causing air pressure (◆ page 29), or the interatomic forces allowing a wire to withstand tension (◆ page 31). The concept of a "field" was introduced to explain forces acting at a distance. Today all the apparently different types of force may be accounted for by four fundamental forces (◆ page 213).

The interplay of forces underlies many physical features of the everyday world. Whenever two surfaces slide over each other, for example, friction has to be considered, even if its effects may be dismissed as negligible. In many circumstances it may be desirable to reduce it as much as possible (by lubrication in engines for example), yet without friction we would not be able to walk, or even stand.

The laws of friction may be demonstrated simply by investigating the force required to pull a block of metal across a horizontal metal surface. The frictional force always acts in the direction that opposes the motion of the block, and can have whatever value is necessary to prevent motion, from zero up to a maximum when sliding occurs. This limiting maximum value depends on the perpendicular force between the block and the surface, but not on the area of contact between the two. It also depends on the nature of the two sliding surfaces. Once the block starts to slip the frictional force usually decreases slightly.

Looking in detail at the surfaces in contact shows that no metal is perfectly smooth. There are only a few points of contact between the block and the surface. Here the local pressure is very high, and interatomic forces (◆ page 29) tend to bond the two together. For sliding to take place these local joints have to be broken, and this gives rise to the frictional force. As one set of joints is broken others form, in a continuous process. The number of local points of contact does not noticeably rise when the apparent area of contact increases, but does so when there is a larger normal force.

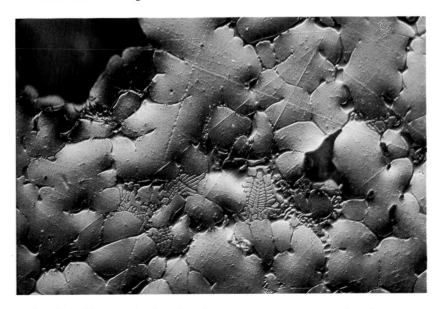

▲ *Even the highly polished surface of aluminum alloy appears rough through a microscope.*

The conservation of energy
A hydroelectric power station taps the store of potential energy that is held in a water reservoir. As the water is released, the potential energy is converted to kinetic energy when the water runs downhill.

Waste heat/Solid

There is a continual interplay between different types of energy. One of the simplest examples is provided by a ball confined to a hollow. If the ball is released at the top of one side of the hollow, it rushes down to the bottom and up the other side, slowly coming to a halt before rushing back down into the hollow and up the first side again. If there were no friction between the ball and the surface, this oscillating movement could continue for ever, but in practice the ball rises up the sides less and less each time until it eventually comes to rest in the base of the hollow.

What exactly is happening to the ball? It gains *kinetic energy* – energy of motion – as it falls into the hollow. The kinetic energy is gained as the ball falls downwards through the Earth's gravitational field. It is lost again as the ball moves upwards, against the gravitational field. The work done by the ball against gravity is defined as the force on the ball (due to gravity) multiplied by the vertical distance moved (that is, the difference between the heights of the top of the slope and the base of the hollow).

The change in energy of the ball is related to the work done – in one sense, an object's energy is its capacity to do work. But this is not the end of the story because once the ball comes to a stop – its kinetic energy is zero – it immediately falls back down the slope. In going up the slope it has gained another kind of energy, known as *gravitational potential energy*. It is a simple matter to show that the potential energy gained equals the kinetic energy lost, while when the ball is at the bottom of the hollow once again, the kinetic energy gained equals the potential energy lost. The total amount of energy remains the same;

At some level below the reservoir, the water drives round the blades of turbines and the linear kinetic energy of the water converts to the rotational energy of the turbine. The process is not totally efficient, because the water is not brought to a complete standstill, but continues to flow.

▲ *If a ball is released at the top of a hollow, it will roll back and forth, climbing the slope on the opposite side each time, gradually losing height and finally coming to rest at the lowest point. It is continually exchanging potential energy (due to height) for kinetic (due to motion) and vice versa. Gradually the ball loses its energy and comes to rest. Its energy is not destroyed, but rather lost to the system, turned into heat and noise by the action of friction with the surface.*

▶ *To a physicist, work takes place whenever a force moves something, or, in other words, when energy is changed to a different form. The greater the distance moved, the more the work done. James Joule was one of the first to appreciate the relationship between heat and mechanical work. The unit of one joule is equivalent to lifting a bag of sugar from one shelf to another in a cupboard; the act of shutting a door might use another five joules.*

Once the electricity supply reaches the consumer, the electrical energy is converted to other forms, in particular heat, light and sound — all pervasive at a pop concert. In the home, conversion to mechanical energy occurs in devices from washing machines to lawn-mowers. In cooking, the energy from electricity can fuel chemical changes, as when cakes rise.

one form of energy simply converts into the other, a change that occurs whenever work is done.

The transformation of energy from one kind to another is basic to the machines used in daily life, from simple devices like a can opener to the complex workings of a hydroelectric power station. Even the human body is a machine, continuously converting energy from one kind to another. The body transforms the energy contained in food, for example, to be stored as chemical energy in muscles, before being released as kinetic energy, in a runner, or converting to potential energy in the case of a high-jumper. None of these machines, from the body to a power station, is 100 percent efficient at converting one type of energy to another. In all cases, there are losses.

The principle of the conservation of energy is a fundamental physical law that applies to all kinds of energy: energy cannot be created or destroyed. There are many kinds of energy, but in any process, the total amount of energy always remains the same. As Einstein showed in his theory of relativity (◗ page 46), even mass is a form of "frozen" energy, which can be released in nuclear reactions. Electrical, chemical, and nuclear energy are all familiar in our daily lives, as are the forms of energy known better as heat, light and sound. Nuclear energy is used to heat water to drive turbines to produce electricity to heat and light homes; chemical energy released when petrol burns propels many kinds of vehicle. Ultimately most of the energy that is used on Earth derives from the Sun – from the heat that drives the climatic systems, and the light that makes plants grow through photosynthesis.

Defining work

The British scientist James Prescott Joule (1816-1889) was one of the first to appreciate that mechanical work can produce heat. He performed a series of experiments to show the heating effect of work done against friction, including his famous paddle-wheel experiment. For this, Joule used an arrangement of paddles on a central axle, which passed between fixed vanes attached to the walls of a vessel filled with water. As the paddles rotated on the axle, the water became warmed through frictional effects, thus converting the mechanical work done in rotating the paddles into heat, which could be measured through the temperature rise. A system of weights and pulleys allowed Joule to calculate the work done, and so equate work and heat quantitatively.

The modern unit of work done, and therefore of energy, is named in Joule's honor. One joule is the work done in applying a force of one newton through a distance of one meter. On Earth, the gravitational force on a mass of 1kg is 9.8 newtons, so a joule is roughly the energy used (or work done) in lifting 1kg through 0.1m. In terms of heat, the energy required to raise the temperature of 1gm of water through 1°C is equal to 4.18 joules. Electrical energy, on the other hand, is usually measured in terms of power, or the rate at which energy is flowing. In this respect the unit of power, the watt, is defined as the energy flow of one joule per second.

In the turbine house some energy is lost by the turbines in doing work against friction as the shafts rotate. This "lost" energy is converted to heat; other losses include the energy of the sounds produced. The turbines drive generators which convert the kinetic energy of the rotating shafts into electrical energy.

ROTATIONAL KINETIC ENERGY

The rotation of a turbine shaft in a power station causes a large electromagnet — the rotor — to rotate within a fixed coil, the stator. The movements of the electromagnet induce electric currents to flow in the stator, thereby converting kinetic energy to electrical energy. The electromagnet is moved rather than the pickup coil because it requires relatively low electric currents to create the magnetic field. The currents induced in the outer coil are much greater. At this stage losses are about 2 percent.

The electrical energy created by the generator is in the form of alternating current. Large currents at relatively low voltages from the generator are converted to lower currents at higher voltages for transmission. This conversion takes place in transformers, which are very efficient.

ELECTRICAL ENERGY

Electricity is transmitted by a grid system which links the power stations to the industrial and domestic consumers. Overhead transmission lines carry the electricity supply across long distances at high voltages so as to reduce losses that might be caused by electrical resistance in the wires, which dissipate energy as heat.

24

See also
Studying the Material World 9-14
Sound 25-8
Molecules and Matter 29-38
Electricity 53-60
Fundamental Forces 213-18

Resonance

All objects have their own natural frequency of vibration, and when an object is vibrated at this frequency it readily absorbs energy and vibrates through large amplitudes. This condition is known as resonance. It is made use of in musical instruments, in which vibrations are set up deliberately to produce pleasing sounds (♦ page 27). But resonance can also be a hazard, as unwanted vibrations can destroy an object. Thus soldiers may be required to break march across certain types of bridge, and it is said that some opera singers can shatter glasses by setting them in resonance with a particular note.

Resonance is not restricted to mechanical systems. In electronics, a resonant circuit is one in which the frequency response of a capacitor and inductor (♦ page 68) are matched in such a way that the circuit can pass large alternating currents. Such circuits are used in the transmission of radio waves. In atomic and nuclear physics, resonance occurs when electrons or the nuclei of atoms absorb radiation with a frequency corresponding to a particular transition, as for example in nuclear magnetic resonance (♦ page 201).

Oscillating systems

From the motion of the atoms within a molecule to the vibrations of a large engineering structure such as a bridge, oscillations are of great importance. Examples of oscillations such as a mass on a spring, or a pendulum swinging, approximate to "simple harmonic motion". This is an important class of oscillations where the resultant force acting on the moving mass or bob is always proportional to the displacement from the rest position, and directed towards it. Simple harmonic motion (SHM) is important not only because it is common, but because more complex oscillations can be broken down and analyzed in terms of it.

In an oscillating system such as a mass on a spring, there is a continual interchange between the elastic energy stored in the spring and the kinetic energy associated with the movement of the mass. In ideal SHM the period of oscillation is constant regardless of the amplitude of vibration, but it is affected by the elasticity of the spring and the size of the mass. In practical situations energy is lost and so the amplitude decreases. In many cases the motion is deliberately "damped" so that the vibrations die away rapidly. For example, the wheel of a car could oscillate dangerously on the end of the coil spring unless damped by the action of the shock absorber.

Oscillating motion

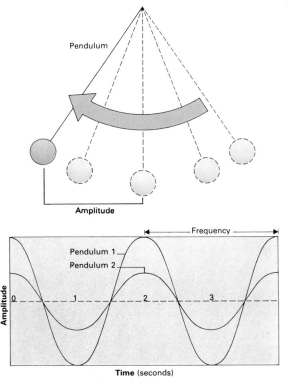

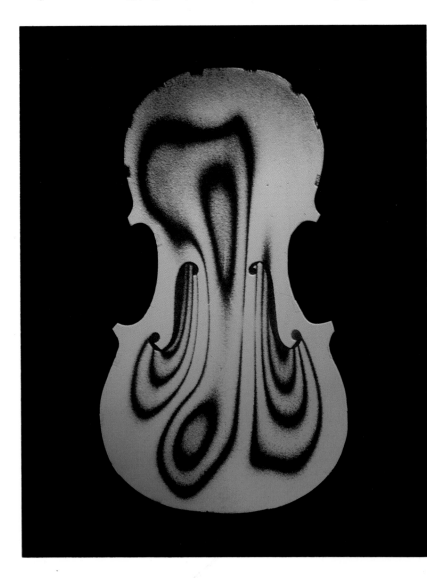

◄ In a violin, the vibrations of the strings pass via the bridge to the body of the instrument. The body has its own modes of vibration – made visible here by interference effects – which resonate with vibrations of the strings. The frequency of these modes is usually constrained to match the frequencies of the strings and gives the violin's tone.

▲ The swing of a pendulum bob typifies simple harmonic motion – a regular oscillatory motion that occurs in many physical systems. The angle to the vertical varies between a maximum value (the amplitude) on either side over a definite time period. The time period (frequency) varies only with the length of the string.

Sound

Sound waves...Frequency and wavelength...Diffraction and reflection...PERSPECTIVE...Loudness and intensity ...Pipes and strings...Sonic booms and the Doppler effect

Some 2,000 years ago the Roman architect Vitruvius (active in the 1st century BC) described the propagation of sounds through the air as like the motion of ripples across the surface of a pond. Vitruvius was largely ignored and it was not until 1,700 years later that the Italian scientist Galileo Galilei (1564-1642) decided for himself that sound is a wave motion, "produced by the vibration of a sonorous body".

A sound wave is a pressure wave and consists of alternating regions of compression and rarefaction. Therefore, unlike a light wave (◗ page 65), a sound wave needs a material to travel through.

Sound waves are the most familiar example of "longitudinal" waves: waves that vibrate and travel in the same direction. Light, on the other hand, is a "transverse" wave motion, vibrating at right angles to the direction of travel. The basic characteristics of a sound wave are its "amplitude", its "frequency" and its velocity. The amplitude refers to the size of the pressure variations; the frequency to the number of variations – waves – per second.

The velocity of sound depends on the substance through which it is traveling. Sound moves faster through liquids than gases. In sea water, for instance, the speed of sound is nearly 1,500 meters per second, four times the speed in air, which is a little less than 350 meters per second. In steel, sound travels at 5,000 meters per second. The speed also depends on temperature: the higher the temperature, the greater the velocity. The frequency of a sound wave is related to the "pitch" of the sound: higher notes correspond to higher frequencies, that is more waves per second, or hertz (Hz). Audible frequencies lie in the range 20–20,000 Hz. The inaudible sounds over this higher frequency are referred to as "ultrasonic".

▲ Experiments to show that sound waves need a medium such as air to travel through were carried out in the 18th century. Air was pumped from a chamber containing a bell. Without air, the bell no longer made a sound.

Propagation of a sound wave

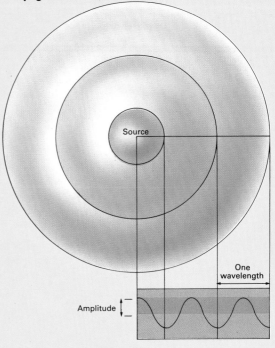

Source

One wavelength

Amplitude

▲ Sound waves spread out like ripples on a pond, but the ripples are variations in pressure that spread in three dimensions. "Crests" correspond to regions of increased pressure; "troughs" occur where the pressure is lower. Wavelength is the distance between crests; frequency the number of crests that pass a point each second.

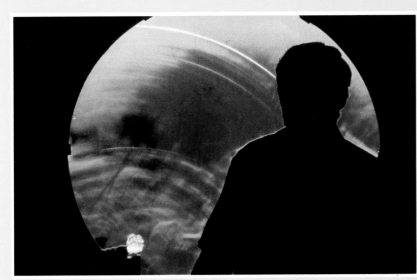

◀ Special photography shows a sound wave from a spark.

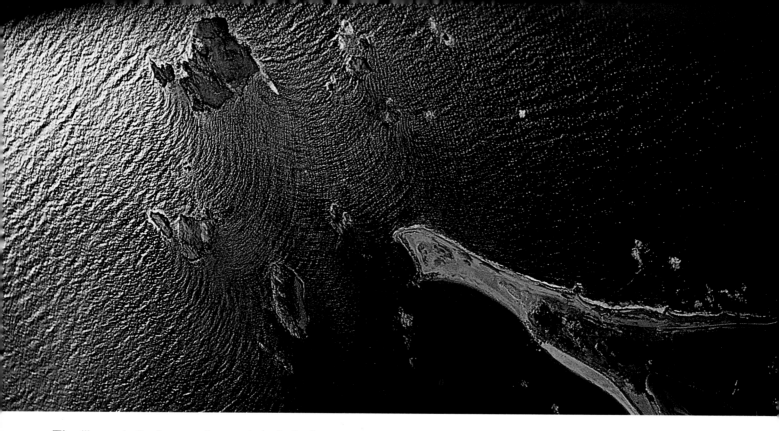

The "intensity" of a sound wave is technically given by the square of its amplitude, and it is related to the preceived loudness, albeit in a complicated way. The amplitude of a sound wave represents the pressure change involved, and the smallest pressure variations that can be heard are in the region of 0·00002 pascals (Pa). Human ears are sensitive to a variation in intensity of a factor of a million million.

Echoes and diffraction of sound

Sound waves demonstrate all the characteristic properties of waves. For example they reflect, refract and diffract just as light waves do. The reflection of sound is a common phenomenon, best known as the familiar echo. In a concert hall echoes can be a nuisance if the hall and its wall coverings are not properly designed, but in other circumstances echoes are vitally important. By timing the reflections of transmitted high-frequency sound waves given off by a sonar device, members of a ship's crew can tell how close their vessel is to the sea bed. And the fact that sound waves are reflected at the boundary between different substances has made ultrasonic sound useful in medical imaging, particularly for an object such as the fetus in a watery environment such as the womb. The refraction, or bending, of sound waves is most apparent at night when sounds often seem louder than during the day. This is because sound can travel further at night, being bent (refracted) back towards the ground by the atmosphere. Refraction occurs when a wave moves into a medium in which its velocity changes. Sound moves faster through warm air, and at night the air near the ground is cooler than the air above it. Sound waves traveling upwards into the warmer air are bent back towards the ground, carrying the sound far along the surface.

Although sound waves propagate basically in straight lines, sound can travel round corners – a wave phenomenon known as diffraction. The amount that the wave's path is bent depends on the frequency, lower frequencies being diffracted more than higher ones. Thus a conversation overheard round the corner of an open door, appears in mumbled, low tones. Similarly low noises, like drum beats, can be better heard around buildings than high noises like whistles; this is why a distant band often seems to consist only of drums.

▼ Two waves of the same frequency can cancel or reinforce, depending on their relative phase – the matching between peaks and troughs. Waves of different frequency (below) add together to give a complex waveform of varying amplitude.

▲ Reflection, interference and diffraction can be seen in this aerial photograph of waves in the sea. As the waves pass through a narrow gap, they spread out (diffract), and the interference of two waveforms is manifested in cross-patterned areas.

Cancellation
+
↓

Reinforcement
+
↓

Complex wave
↓

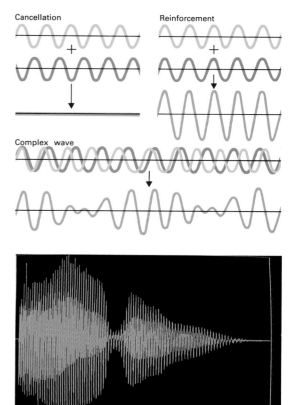

▲ The waveform of "baby", spoken by a speech synthesizer.

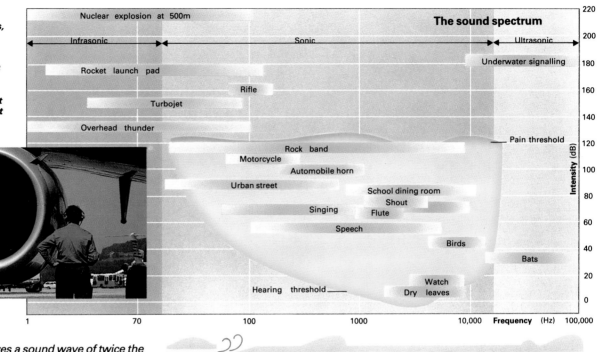

► ▼ *The human ear hears only a range of frequencies, being most sensitive to those around 5,000Hz. Sound levels above about 120dB relative to a zero dB level of 10^{-12}W/m² are painful, so the ears of people working close to jet engines, for example, must be protected (below).*

The sound spectrum

Measuring loudness

The human ear perceives a sound wave of twice the intensity of another as rather less than twice as loud. Moreover, the ear responds to such a large variation in intensity that it is useful to define a scale that somehow compresses this huge range. The scale used is the "sound intensity level" scale. Its basic unit is the "bel", named after the Scottish-American inventor Alexander Graham Bell (1847-1922). However, the "decibel" (dB) – one-tenth of a bel – is more convenient to use.

The scale's zero point is defined as the threshold of hearing, at an intensity of 10^{-12} watts/sq m. Other sounds are normally measured relative to this level. The scale is logarithmic, to approximate . to the actual response of the human ear. Thus a 10dB sound is 10 times as intense as one of 0dB, while a 20dB sound is 100 times, and 30dB 1000 times, as intense as the 0dB sound.

Pipes and strings

Most musical instruments produce sounds by setting a string vibrating or by initiating vibrations in a column of air. The basic process is to make the string or the air column vibrate at its own natural frequency, in other words to "resonate".

A stretched string, fixed at both ends and plucked at the center, will vibrate, the whole string moving from one side of its resting position to the other and back again. The vibration has a characteristic frequency which depends on the tension in the string, its weight and length. The shorter the string, the higher the frequency. The vibrating string sets the surrounding air molecules oscillating, generating a sound wave of the same frequency.

In a wind instrument such as a flute the musician sets air enclosed in a pipe in vibration. Air passes over a reed at the entrance to the pipe which causes eddies that generate vibrations in the column of air in the pipe. The frequency of the note produced depends on the length of the pipe, and whether it is closed at one end. The characteristic sound or "timbre" depends on the overtones that occur.

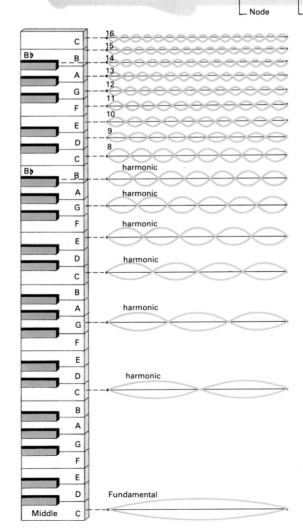

▲ ◄ *When a musician blows into a wind instrument such as a recorder (above), the air vibrates, setting up a "standing wave" in the pipe. This wave does not move along the tube, but consists of a stationary pattern of air moving by varying amounts. Positions where there is no movement are called nodes, while movement is greatest at the antinodes, for example at the ends of the pipe. In the simplest standing wave, one wavelength fits within the tube; this corresponds to the fundamental frequency of this note. Notes of higher fundamental frequency are made by shortening the tube – removing fingers covering holes along the tube. But each note contains overtones. These are weaker waves of higher frequency which also have antinodes at the open ends. Similar standing waves are set up when strings are plucked or struck, as in a piano (left). Here in the fundamental mode the ends of the string are held fixed, while the center vibrates. The profile of the vibrating string maps out half a wave pattern. The keyboard shows how notes of higher frequency correspond to the overtones, or harmonics, of the fundamental middle C.*

The Doppler effect

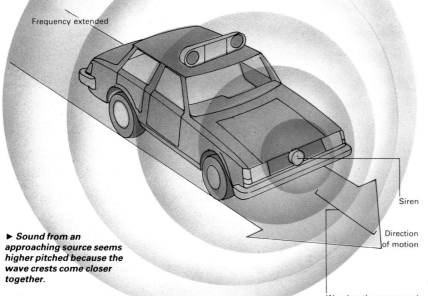

Frequency extended

Siren

Direction of motion

Wavelength compressed

▶ **Sound from an approaching source seems higher pitched because the wave crests come closer together.**

▼ **The shock wave due to a supersonic dart.**

The Doppler effect

A familiar wave phenomenon of sound is the change in pitch of the noise from a passing siren. This is an example of the Doppler effect, also observed for light waves. As the source of the sound moves closer to the listener, each successive compression is emitted closer to the previous one. The wave arriving at the listener is thus itself gradually squeezed together, so that its frequency appears higher as the source approaches. As soon as the source has passed, successive compressions are emitted at increasing intervals as the source moves away. The pitch of the sound drops.

Sonic booms

Sometimes the source of a sound travels faster than the waves it produces. A familiar example is the supersonic jet aircraft, which travels faster than the velocity of sound in the atmosphere. In such cases, the successive compressions arrive at the listener almost at the same time, and add together to produce a very loud noise. This "sonic boom" thus occurs continuously, and moves in the wake of the moving sound source, providing the source is moving faster than sound.

Molecules and Matter

Liquids, solids and gases...Oscillating molecules...
Forces between molecules...Latent heat...Melting and
boiling points...Viscosity...Thermodynamics...
PERSPECTIVE*...Pressure...Surface tension...Brownian*
motion...Stress and strain...Boltzmann...Boyle and the
expansion of gases...Phase diagrams...The critical
point...Amorphous solids and liquid crystals

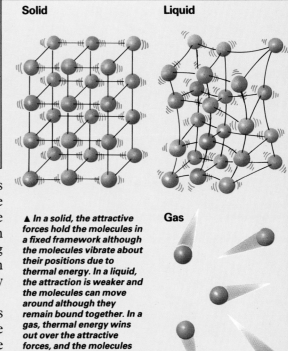

Solid   **Liquid**

Gas

▲ *In a solid, the attractive
forces hold the molecules in
a fixed framework although
the molecules vibrate about
their positions due to
thermal energy. In a liquid,
the attraction is weaker and
the molecules can move
around although they
remain bound together. In a
gas, thermal energy wins
out over the attractive
forces, and the molecules
are free to move
individually, spreading
through large volumes.*

The matter of the everyday world exists in one of three familiar states or "phases" – solid, liquid, or gaseous. Solids have a fixed shape, are usually rather dense, and are very difficult to compress. Liquids are also rather dense and difficult to compress, but they differ from solids in having no fixed shape and are able to flow with varying degrees of difficulty. Gases usually have much smaller densities than solids or liquids, are easily compressible, and flow even more easily than liquids.

A characterisitic feature of solids is that they often occur as crystals. Ice and gemstones are familiar examples of crystals, while modern electronics depend crucially upon crystalline silicon (◗ page 104). X-rays reveal that crystals are composed of a regular three-dimensional array of atoms spaced apart by a few tenths of a nano-meter. These atoms are bound in place, but vibrate; these vibrations grow by increasing amounts as the crystal is heated. In gases, by contrast, the molecules are not fixed in position. They move about randomly in space with speeds that increase as the gas is heated.

Liquids also show some regular structure, but only across a few molecules and over very short intervals of time. The key difference between liquids and solids is that in a liquid some molecules are missing from their places. This leaves empty spaces into which other molecules may jump every so often. Most of the energy of the molecule goes in vibrating about a fixed position, as in a solid, but it is the movement of molecules from one place to another within the body of the liquid that gives it its properties of flow and viscosity.

Pressure

When a force acts upon an object, its effect depends on both how the force is distributed and what the substance is made of. For example, snow shoes spread a person's weight over a large area, so the wearer does not sink so easily into soft snow. But if the person wears shoes with spike heels, much of the same force is now concentrated into the small area of the heels, which now sink easily into grass. The difference lies in the pressure, which is defined as the component of force perpendicular to the area divided by the size of the area. So the same force exerts a larger pressure over a smaller area, and vice versa.

The effect of pressure on a material depends on the microscopic structure of the substance. Increasing the pressure squashes the molecules closer to each other. In a solid, the rigid structure means that very little change in volume occurs and the pressure is transmitted through the structure. In a liquid, the molecules move more freely, so the pressure acts in all directions as the molecules push against each other. That is why water will shoot out sideways through a hole in the bottom of a tank although the weight of the water is acting downwards. The same is true of a gas, but in this case the molecules are so far apart that an increase in pressure causes a decrease in volume as the molecules are squashed closer together.

◀ *Experiments on air pressure became possible in the 17th century after the invention of the air pump by Otto von Guericke (1602-1686), a mayor of Magdeburg in Germany. Von Guericke himself performed a famous experiment demonstrating the pressure of the atmosphere, in which he showed how difficult it was to pull two hemispheres apart once the air had been pumped out from within them.*

Some materials, such as concrete, are able to resist compressive forces, but are very weak under tensile stress

Forces are required to hold molecules together. The fact that a single substance can exist in a solid, liquid or gaseous state reveals something about these forces. They must attract and repel other molecules, and be of short range. Without attractive forces the molecules would not coalesce to form liquids or solids; everything would be gaseous. Without repulsive forces matter would shrink to an infinitely dense point. The forces must decrease rapidly with increasing separation between molecules because physicists are able to describe the behavior of gases like air in everyday situations without reference to these forces; it is as if the molecules bounce apart from each other like billiard balls, even though they are separated on average by only a few nanometers. However, the attractive force must always be of longer range than the repulsive force if the molecules are to coalesce into a solid or liquid.

The force between a pair of electrically-neutral molecules, such as nitrogen, helium or water, decreases so rapidly with separation that only the "binding energy" between adjacent molecules is significant. Many properties of solids and liquids depend on the intermolecular forces and binding energy, and therefore many diverse physical phenomena are related to each other. The melting temperature, critical temperature, latent heat and surface tension are a few such related properties. The total binding energy of an assembly of molecules in the solid or liquid state is equal to the number of pairs of nearest neighbors, multiplied by the binding energy of a pair of molecules at their equilibrium spacing. At very low temperatures, this total binding energy is equal to the "latent heat of sublimation" – the energy needed to dissociate the solid into its separate molecules.

Some mechanical properties are also directly related to the inter-molecular binding energy. The "elastic moduli" measure how hard it is to change the separation between molecules in a material by stretching, twisting or compressing it. Thermal expansion occurs because the attractive force is of longer range than the repulsive force. At a temperature above absolute zero the molecules vibrate about their lattice positions, and as the temperature is raised, the average separation between the molecules increases. The length of a piece of the material in bulk is governed by the average separation between molecules, so as the temperature rises, the material expands.

▲ The balance between a repulsive force and an attractive force – to give a net result of zero – gives the average separation between atoms and molecules in gas or liquid.

▲ ► As temperature rises, the average separation between atoms and molecules increases, causing thermal expansion. In bridges, this is allowed for by expansion joints.

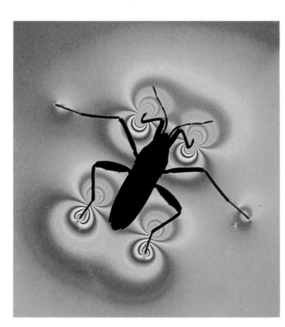

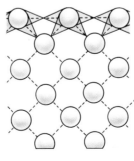

◄ ▲ Molecules at the surface of a liquid feel a net force pulling inward. This is surface tension. It provides a cohesive force between the surface molecules, which is sufficient to prevent the legs of a ripple bug from breaking through (left). The high surface tension in water is vital to many physiological processes.

Surface tension
Within a liquid, the attractive forces between molecules pull in all directions, so the net effect on a single molecule is zero. But at the surface there is an imbalance. A molecule there is pulled more towards the body of the liquid than in the opposite direction. This effect is known as surface tension. In a drop of liquid, the intermolecular forces are tending to pull the surface towards the center. The result is a spherical drop.

Surface tension can make a liquid climb "uphill" as when water climbs up a fine glass "capillary" tube. This happens because the attractive forces between the glass molecules and the water molecules are greater than those between the water molecules themselves. The surface of the water is pulled upwards, more so at the edges of the tube, creating a concave "meniscus". In other cases, such as with glass tubes and mercury, a convex meniscus forms and the liquid drops down the tube. This is because the forces between the glass molecules and the mercury molecules are weaker than those between the mercury molecules.

Young's modulus

If a length of copper wire is suspended from a support and a weight hung from the end, the force acting on the wire increases its length slightly. Provided the weight is not too great, the wire returns to its original length when the weight is removed. This is elastic behavior, and it is characterized by the modulus of elasticity, or "Young's modulus", in this example of a wire under tensile, or stretching, stress. Young's modulus is equal to the tensile stress (force per unit of area) divided by the change in length (also called the tensile "strain"), and its value for a material depends on the strength of intermolecular forces.

There is, however, a limit to this elastic behavior beyond which permanent stretching of the wire occurs when the load is removed. The wire in this case stretches irreversibly, as layers of atoms slide permanently over each other. Quite large increases in length are possible before a "ductile" material like copper finally breaks. A "brittle" material such as glass will fracture almost immediately after the elastic stage has been passed.

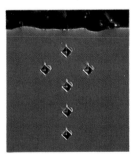

▲ **Hardness of materials depends on the forces between their atoms and molecules. Here diamonds are fired at the surface of a metal to test its hardness.**

▶ **A pole vaulter uses the elastic properties of the pole to help gain height, as the bent pole springs back and flings the athlete through the air.**

The founder of "statistical mechanics", Ludwig Boltzmann, committed suicide, depressed by the failure of others to appreciate his work

The temperature of a substance – its degree of "hotness" – reflects the energy of the molecules it contains. The higher the temperature, the greater the *average* energy of the molecules. Not every molecule has identical energy – at any particular temperature, a range of energies is possible although not equally likely. The exact distribution of energy among the molecules depends on the temperature, according to a law due to the Austrian physicist Ludwig Boltzmann (1844-1906). Boltzmann developed a "statistical" theory for the behavior of matter, based on the average motions of the many atoms and molecules within a substance. "Boltzmann's law" refers to a system of particles in thermal equilibrium, and it states how the average number of particles with a certain energy varies with absolute temperature, and rises exponentially with rising temperature.

Boltzmann's law underlies many features in the behavior of materials at varying temperatures. Many chemical reactions, in both inorganic and biological systems, proceed much more rapidly as the temperature increases (◗ page 94). For example, a change of one or two degrees in the processing temperature leads to a large change in the time required to develop a film (◗ page 167). The reason is related to Boltzmann's law. The probability for a molecule to have sufficient energy to react chemically with another molecule increases very rapidly with temperature.

When a crystal is heated, the most energetic molecules break free from their positions in the lattice and migrate through the crystal. Other molecules move into the "holes" left behind. This phenomenon of "diffusion" is of major importance to the electronics industry in the manufacture of large-scale integrated circuits (◗ page 59). As the temperature rises, the number of holes in the crystal increases exponentially, according to Boltzmann's law, but while the holes are relatively far apart the substance still behaves as a crystalline solid. However, when the number of holes becomes very large there is a

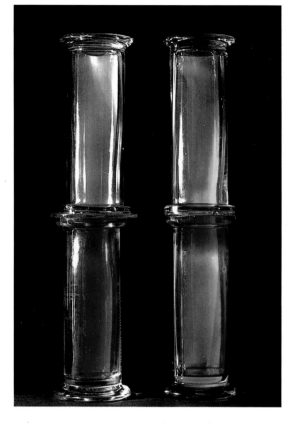

▲ *In a gas at room temperature, the molecules are moving around with a speed of nearly 500 m/s. This movement gives rise to diffusion, as the molecules spread out to fill any volume the gas enters. In this classic demonstration of molecular diffusion, bromine, the brown gas, and air (left) can be seen to mix once a plate keeping them apart has been removed (right).*

Ludwig Boltzmann

Boltzmann was born in Vienna in 1844. This was the era in which the theory of thermodynamics began to emerge (◗ page 38). At the same time, the kinetic theory of gases was also being developed showing how properties such as pressure could be understood in terms of the overall behavior of many atoms. Boltzmann's great achievement was to discover the links between these two apparently different theories, combining the thermodynamic properties of bulk matter and the microscopic world of kinetic theory. Using a statistical treatment of the average mechanical behavior of individual atoms, he deduced the thermodynamic properties and founded the theory of "statistical mechanics". His work bridged the classical theories of the 19th century and the quantum theories (◗ page 195) of the 20th. Yet this was at a time when atoms were not accepted by all scientists. He committed suicide at the age of 62, depressed by the failure of his fellows to appreciate his work. His tombstone is inscribed with the equation that encapsulated his statistical interpretation of "entropy" (◗ page 38).

▶ *Boltzmann was the father of statistical mechanics, which is used to study the average behavior of large collections of atoms. His work was based on foundations laid in particular by the Scottish physicist James Clerk Maxwell (1831-1879), who first worked out the distribution of velocities for gases at different temperatures.*

Brownian motion

Molecules are too small for their movement at high speeds to be seen directly. However, in 1827 the British botanist Robert Brown (1773-1858) first observed with a microscope the abrupt, random movements of very small solid particles (pollen grains) immersed in a liquid. These random jumps result from the impacts of molecules in the liquid on the particles, as required by Boltzmann's law.

The French physicist Jean Perrin (1870-1942) obtained further proof of Boltzmann's law early this century. He suspended in water microscopic particles of resin (having a density only slightly higher than water), and counted them at different heights using a microscope. The variation in their number with height was exactly what Boltzmann predicted. In the same experiment, he measured "Avagadro's number". Amedeo Avagadro (1776-1856), was an Italian physicist who first put forward the notion that equal volumes of gases, at the same temperature and pressure, contain the same number of molecules. Avagadro's number is the number of atoms in 12g of carbon-12, or 6.02×10^{23}.

Measuring the Sun's temperature

The French astronomer Audouin Dollfus (b.1924) used Boltzmann's law in 1953 to measure the Sun's corona. The Sun emits a red line from highly-ionized iron atoms. This line should be extremely narrow, but in the Sun's corona it is considerably broadened. Dollfus interpreted the broadening as due to wavelength shifting of the light emitted by molecules moving towards or away from an observer on Earth. He calculated the distribution of molecular speeds from these data, and found that it fitted well with the predictions of Boltzmann's law for a coronal temperature of 2.1 million degrees K.

good chance of adjacent lattice sites being vacant. The forces holding the nearby molecules in place are then greatly reduced and these molecules start to move about inside the crystal. This molecular mobility is manifest as "melting". As more heat energy is added to the melting solid it releases more molecules from their lattice sites. The temperature of the substance, meanwhile, remains constant until it is completely liquid. The heat energy required to melt a substance completely is called the "latent heat of melting".

Physicists can also explain the evaporation of a liquid in terms of the Boltzmann distribution of energies of the molecules. Much more energy is required for a molecule to break away from its neighbors and leave the liquid completely than for the molecule to change from one set of neighbors to another and move about in the liquid. At room temperature, for example, only a tiny fraction of water molecules have enough energy to evaporate. Nonetheless, a bowl of water will completely evaporate away over the course of a few days because the water slowly absorbs heat from its surroundings, and ultimately all the molecules will have acquired sufficient energy to escape. When a liquid is heated to higher temperature, however, a much larger fraction of molecules has the required energy, and it evaporates much faster.

"Sublimation", the evaporation of a solid directly into a gas, is most commonly and spectacularly seen at the theater when Cardice (solid carbon dioxide refrigerated at low temperatures) is thrown onto the stage to produce clouds of vapor looking like mist or fog. The energy required for a molecule in a solid to break away from its neighbors and evaporate is even greater than for a molecule in a liquid, and so we are not usually aware of ice, for example, subliming to water vapor. Even so, washing hung outside in sub-zero temperatures will eventually dry, because some of the molecules still have enough energy to escape.

◀ Mist gathers above a lake as the Sun rises in the early morning. Evaporation occurs when molecules in a liquid break away totally to form a gas. According to Boltzmann's law, the probability for a molecule to have enough energy to do this increases exponentially with temperature – puddles soon evaporate after a downpour on a hot day.

▼ Sublimation occurs when molecules have sufficient energy to escape directly from a solid to form a gas. This effect is seen in a theater, when solid carbon dioxide, or Cardice, which has been kept refrigerated, is thrown onto the stage to produce fog-like clouds of vapor as it sublimes. Sublimation requires more energy than evaporation.

Ice floats on water because water becomes less dense when it freezes

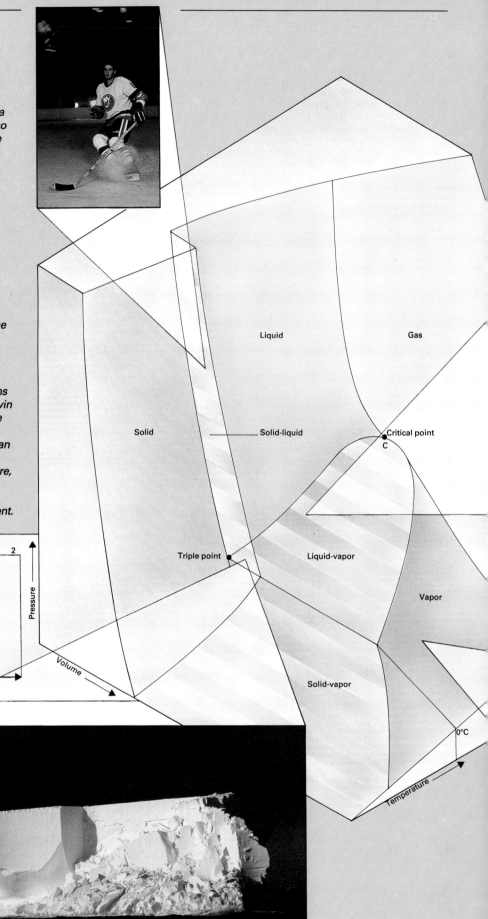

Pressure, volume and temperature

Whether a substance exists as a solid, a liquid or a gas depends not only on the temperature, but also on the pressure exerted on the substance and the volume it occupies. One of the first to study the relationship between these quantities was the Anglo-Irish physicist and chemist Robert Boyle (1627-1691). He showed that the product of pressure and volume for a fixed mass of gas at fixed temperature is approximately constant: double the pressure and the volume is halved.

Others extended this work by varying the temperature, and found that the pressure falls in proportion to decreases in temperature. Their results gave rise to the concept of an "absolute zero" of temperature, corresponding to a (hypothetical) zero pressure. Measurements indicated that this should occur at 273°C below the freezing point of water at atmospheric pressure. Thus -273°C became the starting point of the "absolute" scale of temperature, which has the same size of degree as the Celsius scale. Temperatures on this scale are referred to in terms of degrees (K), after the British physicist Lord Kelvin (1824-1907) who did much important work on the theory of heat and temperature.

The early work on gases by Boyle and others can be summarized in a single relationship: pressure times volume equals a constant times temperature, where the temperature is measured on the absolute scale. Both the "gas constant" and the volume are proportional to the mass of gas present.

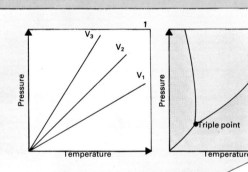

▲ ▶ The "pressure law" relates the pressure of an ideal gas (with negligible intermolecular forces) to its temperature for fixed volumes, V (1). If extended to low temperatures the lines for different volumes all meet at zero pressure and the absolute zero of temperature. A real gas (2) changes to liquid and solid phases, however, as the temperature falls and the force between molecules is no longer negligible compared with their kinetic energy. In water, shown here, increasing pressure can revert solid (ice) to liquid, and this partly explains the slipperiness of ice. Water becomes less dense on freezing, so ice floats on water.

◄ *The relationship between the phases of matter – solid, liquid, gaseous – at the different variables of temperature, pressure and volume, can be plotted on a single, three-dimensional "phase diagram". The diagram for water is shown here. The boundaries between the phases are plotted. Water is unusual in expanding when it freezes, and this is shown on the diagram by the notch in the face between solid and liquid. The point C is the critical point, the highest temperature at which the substance can exist as a liquid. It represents the substance's highest boiling point; below this, the boiling point varies with pressure. A gas below the critical point is known as a vapor.*

Phase diagrams

The values of pressure, volume and temperature related by the "equation of state" described opposite lie on a curved surface in a three-dimensional space depicted in a "phase diagram". Such a diagram is based on three axes which represent pressure, volume and temperature.

The surfaces corresponding to the simple equation of state occur on the phase diagram only in the region where the substance behaves as an "ideal" gas. Elsewhere different relationships hold between pressure, volume and temperature. Only for certain ranges of these quantities can a substance exist in a particular phase such as a solid, liquid or gas. Over other ranges, two phases, such as solid and liquid, coexist in equilibrium; and for one particular value of temperature and pressure all three phases are in equilibrium. This condition is shown as the horizontal line where the liquid-gaseous and the solid-gaseous phase boundaries meet.

Adding heat energy causes some of the solid to melt and some of the liquid to evaporate, so that the volume increases, but both the temperature and pressure remain constant. These values of temperature and pressure define the "triple point", which for water corresponds to a temperature of $-0.1°C$ at a pressure of somewhat less than one hundredth of atmospheric. Ice and water are in thermal equilibrium at atmospheric pressure at a slightly higher temperature – the melting point of ice.

347°C

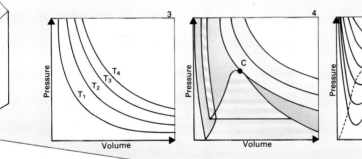

▲ ► *The pressure and volume of an ideal gas at constant temperature (T) are related by the smooth curves of Boyle's law (3). But this applies to a real substance such as water only at high temperatures (4). A better description of real gases comes from Van der Waals' theory, which takes into account the forces between molecules (5), although this does not describe the transitions to liquid and solid phases. In real substances, the temperature for boiling varies with pressure; so tea brewed at low pressure on a mountain boils at lower temperatures. Pressure changes in the fluid flow around a propeller can make bubbles of gas form.*

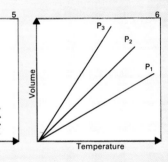

The behavior of real gases

Two hundred years after Boyle, the Irish physicist Thomas Andrews (1813-1885) made extensive measurements on carbon dioxide and drew up phase diagrams which reveal the difference in behavior of carbon dioxide from that of an ideal gas at high pressure and low temperature. These differences led the Dutch physicist Johannes van der Waals (1837-1923) to describe how real gases behave. He argued that the molecules of a gas take up space, so that the equivalent volume of an ideal gas is a little smaller than the measured volume of a real gas. And the measured pressure of a gas is smaller than the ideal gas pressure because of the net attraction between the molecules. Van der Waals' equation describes the behavior of real gases well over quite a wide range of pressures and temperatures. It fails only at high pressures and low temperatures where the separation of the molecules is similar to that in a liquid or solid.

Many different physical phenomena can be explained in terms of forces between molecules in a substance

Explaining the properties of matter

The molecules in a gas move about with very high speeds. At room temperature and atmospheric pressure, for example, the average molecular speed of the nitrogen and oxygen molecules in air is about 450 meters per second. However, the average distance traveled by a molecule before it collides with another molecule is very small (less than one ten-millionth of a meter). The diffusion of molecules from one region to another therefore involves many millions of molecular collisions. This process of diffusion explains the thermal conductivity of gases and also their viscosity. In the absence of convection currents or radiation, heat is transferred from a hot region of a gas to a cooler region by molecular collisions. Molecules in the hot region travel faster, and when they collide they give up some of their excess energy, thereby heating up the cooler regions of gas. Viscous forces can also be understood in terms of molecular collisions and diffusion.

Theory predicts that thermal conductivity and viscosity of gases should not depend on pressure, but should increase with the square root of the absolute temperature. By contrast, the viscosity of a liquid decreases with temperature, indicating that the mechanism of diffusion in a liquid is quite different from that in a gas. Indeed, diffusion in liquids is very similar to that in solids, and occurs because molecules jump into adjacent, vacant lattice sites. At higher temperatures the number of such holes increases strongly and so does the rate of diffusion. If it is easier for a molecule to move in one direction than another there will be a net rate of diffusion in this direction.

The viscosity of a liquid increases rapidly with pressure, in contrast to the behavior of a gas. This is because the effect of the pressure is to squeeze the holes and make it much more difficult for a molecule to force its way into an adjacent vacant position. This effect is of great importance in engineering. Many sliding mechanisms operate successfully only because the lubricating oil is not squeezed out. Heavily loaded gear teeth may enmesh with a contact pressure of several tonnes per square centimeter, at which pressure the viscosity of a typical lubricant may have increased a million-fold.

Although the description of a substance as being in the solid, liquid or gaseous phase is convenient, it can be misleading for it applies strictly to the properties of ideal substances. Such common substances as ice, pitch and lead flow like very viscous liquids when large forces and pressures are applied to them; water shows rigidity, a property of solids, if one attempts to change its shape too rapidly; gases moving in bulk close to the speed of sound can sustain sharp changes in density and pressure over quite small distances ... and alloys, plastics and glasses are all much more complicated to classify.

A simple understanding of matter is possible only under the rather special conditions of gases at high temperatures and low pressures (when intermolecular forces are negligible) or crystalline solids at very low temperature (when imperfections and diffusion can be ignored). The liquid state, in particular, is difficult to understand. Physicists can describe some features, such as evaporation and superheating, by thinking of the liquid as a very dense gas. Other features, such as the tensile strength and viscosity of a liquid, can be understood only by considering the liquid as an imperfect solid. There is, however, a link between these two, which is apparent in the behavior of a substance at pressures and temperatures where separation between molecules remains close to that of the liquid phase.

Amorphous solids and liquid crystals
Most materials are naturally crystalline in their solid form. Some materials have no regular structure; they are "amorphous". In these solids, the atoms do not form in a regular pattern.

Some amorphous materials, such as rubber, consist of molecules in the form of long chains (polymers) which have become tangled together (♦ page 93). In other instances, the solid is like a "supercooled" liquid in which the irregular pattern of atoms of the liquid state has become "frozen in". Glass is perhaps the most ubiquitous amorphous solid. It is made from a mixture of soda and lime with sand, all of which fuse together in a liquid at temperatures of around 1500°C. The glassy state forms as the liquid cools and rapidly becomes viscous, preventing crystals from forming as it solidifies.

Liquid crystals are, by contrast, liquids with a structure like a solid. They are liquids in which a high degree of ordering can occur, for example when an applied electric field organizes the normally random arrangement of the molecules. This can alter the optical properties of the liquid crystal, as in the displays on electronic watches.

► Ice in a glacier flows slowly downhill, like a very viscous liquid, influenced by tremendous forces.

◄ Glass is perhaps the most familiar amorphous solid, used for centuries in windows for example. Its irregular atomic structure, like that of a "frozen" liquid, characterizes other "glassy" materials.

▼ An array of rod-shaped liquid crystals seen in polarized light, showing one of the several possible regular arrangements of the crystals.

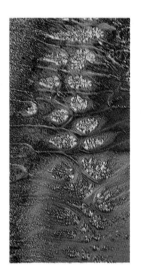

▼ Measurements of the viscosity of liquids such as lubricating oils are crucial to industry. Here, instruments for measuring viscosity are being calibrated.

See also
Forces, Energy and Motion 15-24
Sound 25-8
Light 39-48
Explosives 181-6
The Quantum World 195-204

Thermodynamics

The pressure of a gas, the viscosity of a liquid and even the rate of a chemical reaction can change with temperature. Although these changes occur in widely differing systems, there is a generalized framework that can be applied to them. This is thermodynamics.

"Classical" thermodynamics was developed around 1850 by the Scottish physicist William Thomson (1824-1907) and the German Rudolf Clausius (1832-1888). These men built upon work by Frenchman Sadi Carnot (1796-1832), who in 1824 published a treatise on heat engines – engines that use heat to perform work. He proved that no engine could be more efficient than his idealized engine, operating a reversible cycle between two temperatures, and that the efficiency depends on the temperatures between which the engine operates.

Carnot's insight became enshrined in two laws of thermodynamics. What is now known as the first law is a statement of the conservation of energy (◀ page 22), with heat taken into account. The first law showed that heat supplied to a system goes both in doing work and in changing the internal energy of the system. The second law is that heat cannot flow from a colder to a hotter body, without some other changes occurring. The second law reflects a basic lack of symmetry in the physical world: processes that can occur spontaneously in one direction will not occur equally well in reverse. If a partition between compartments containing two gases is removed, the gases will eventually mix; but they will not "unmix" again. The "arrow" that imposes this kind of direction is known as "entropy". Entropy, like energy, is a property of a system that changes when heat is supplied. In controlled conditions, the change in entropy is equal to the heat supplied divided by the temperature. Understanding of matter based on atoms leads to a deeper insight into entropy as a measure of "order" in a system. When the gases mix they become less ordered: entropy increases. And this reveals a more general statement of the second law – only those processes occur naturally in which entropy increases. Overall, energy is conserved (the first law) but entropy rises (the second law).

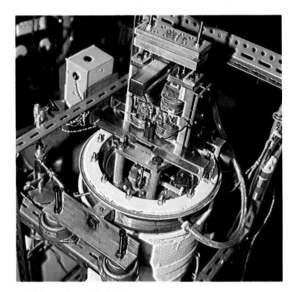

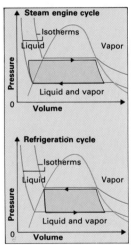

▲ *This methane-fueled generator is an example of a Stirling engine, the most efficient and cleanest form of engine yet devised.*

▶ *A steam engine works by allowing pressurized steam to expand and push a piston as the temperature falls. The resulting mix of liquid and steam is condensed to liquid and the pressure increased before reheating in the boiler. A refrigerator works on a similar cycle operating in reverse.*

▼ *The steam engine revolutionized work, from industry to agriculture, throughout the 19th century.*

Steam engine cycle

Isotherms
Liquid — Vapor
Pressure
Liquid and vapor
0 — Volume

Refrigeration cycle

Isotherms
Liquid — Vapor
Pressure
Liquid and vapor
0 — Volume

Light

Light rays...Lenses and mirrors...Reflection and refraction...Prisms...Colors...The wave nature of light... Polarization...Light and particles...PERSPECTIVE... Measuring the speed of light...White light...Einstein and special relativity...Flying clocks around the Earth

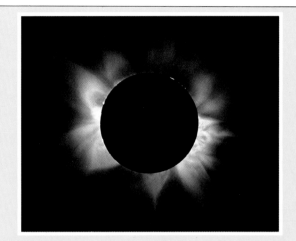

▲ *A total eclipse of the Sun in 1980*

Simple observations reveal some of the more obvious characteristics of light. The outlines of shadows in strong sunlight show that light travels in straight lines, at least on a macroscopic scale, and sunshine filtering into a room through small openings appears to form well-defined beams. This gives rise to the idea of "rays" of light, a concept that is useful in appreciating some of the basic properties of light, as well as the operation of many optical instruments. But what is light?

By the early 18th century, scientists had found that they could explain many optical effects in terms of the general properties of waves (◆ page 52). As with sound waves (◀ page 26) and ripples on the surface of water, light can be seen to undergo reflection, refraction, interference and diffraction. However, what exactly constitutes a "light wave" was not answered until a century later, with the work of the British physicist James Clerk Maxwell (1831-1879). Maxwell drew together many observations concerning electricity and magnetism and incorporated them in a single theory which predicted the existence of "electromagnetic waves" (◆ page 61). According to this theory, these waves travel through a vacuum at a velocity given by two constants related to electric and magnetic units, and this velocity is the same as the velocity of light. This was a revelation: it showed that light is an electromagnetic wave with a particular range of wavelengths. It forms part of a huge spectrum that ranges from gamma rays to radiowaves.

For many purposes the wave theory of light is adequate. But when it comes to explaining the absorption and emission of light on the atomic scale, the wave description is not tenable. Light must then be described as packets of energy, or "particles" of light, called photons, which interact individually with electrons in atoms. The discovery of the dual nature to light, demonstrating that it behaves both as waves and particles, brought about the development of quantum theory (◆ pages 195-204) and revolutionized physics in the 20th century.

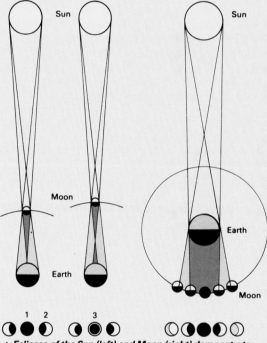

▲ *Eclipses of the Sun (left) and Moon (right) demonstrate how light travels in straight lines, casting shadows over great distances. On a much smaller scale, light can bend round corners, however, when it diffracts ◆ page 42).*

Measuring the velocity of light

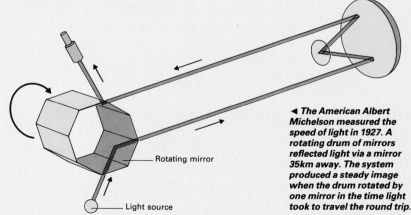

◀ *The American Albert Michelson measured the speed of light in 1927. A rotating drum of mirrors reflected light via a mirror 35km away. The system produced a steady image when the drum rotated by one mirror in the time light took to travel the round trip.*

Rotating mirror

Light source

Measuring the velocity of light

The velocity of light in a vacuum, 299,792·5km/s, is one of the fundamental constants of physics, usually denoted by the letter c. As Einstein showed in his special theory of relativity (◆ pages 46-47), this is a universal "speed limit". Nothing can travel faster than the speed of light in free space.

The Danish astronomer Olaf Roemer (1644-1710) determined the velocity of light in 1676. Roemer measured the times at which one of the moons of Jupiter emerged from the shadow of the planet. He found that the end of such an eclipse occurred later when the Earth was further from Jupiter. A knowledge of the orbits then gave a value for the velocity of light. The result was lower than the correct value by about 25 percent, but it confirmed that light travels much faster than sound.

The refraction of light as it crosses the boundary between two substances depends on the relative speeds of light in the two materials

▶▼ **Light rays "bend" – change direction – when they pass from one material to another. This is the process of refraction and it is related to the difference in the velocity of light in differing materials. It is put to use in lenses which can converge light rays (left), bringing light parallel to the axis of the lens together at a single point, the focus F. Lenses can also make parallel rays diverge (right), as if from a single point.**

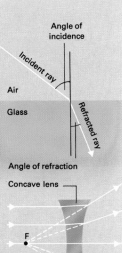

Angle of incidence

Incident ray

Air

Glass

Refracted ray

Angle of refraction

Concave lens

Parallel light — Convex lens

F

F

Focal length

Focal length

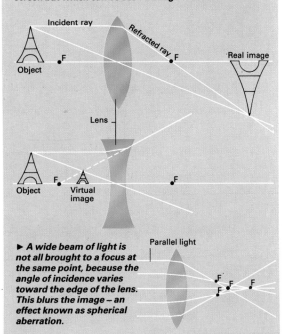

▼ **A converging (convex) lens produces an enlarged but inverted image of an object placed between the focus and a point at twice the focal length. This is a "real" image, which means that it would appear on a screen placed at its location, but cannot be seen by eye. A diverging (concave) lens, on the other hand, always produces an upright, diminished "virtual" image, which cannot be formed on a screen but which can be seen through the lens.**

Incident ray

Refracted ray

F

F

Real image

Object

Lens

Object

F

Virtual image

F

▶ **A wide beam of light is not all brought to a focus at the same point, because the angle of incidence varies toward the edge of the lens. This blurs the image – an effect known as spherical aberration.**

Parallel light

F F F F

▶ **A lighthouse uses both mirrors and lenses to produce a powerful, focused beam that can sweep across the horizon. A solid lens would need to be too big and heavy to be practical. Fresnel lenses are used instead, in which the lens surface is divided into a series of concentric circles. Relatively thin sections of glass are set in each ring at the angles that would be found in a solid lens at that point.**

◀ **This image of the focusing power of a lens was produced by superimposing a sequence of high-speed holographic photographs of light pulses, using laser pulses of 10 picoseconds each. As well as showing how light is brought to a focus by the lens, the picture shows that the light is slowed as it passes through the glass – the focused pulses are delayed relative to the original beam.**

Reflection of light

The 17th century saw the invention of the first microscopes and telescopes, and the first theories of light. By this time scientists were aware of two laws governing the behavior of light, in addition to the fact that it seems to travel in a straight line. Both these laws concern what happens to light rays when they meet a surface, for example between air and glass.

The law of reflection states that the angle of reflection equals the angle of incidence (the angle at which a light ray strikes the surface), and that the reflected and incident rays both lie in the plane that contains a line at right angles to the surface. Simple diagrams using this fundamental law show how mirrors create images. With a flat mirror the eye sees light that appears to come from behind the mirror. In fact, the light has been reflected and the image seen is not a real image, but a virtual image: no rays connect the image to the observer's eye. A convex mirror also produces a virtual image, this time reduced in size. A concave mirror can produce an image between the eye and the reflecting surface; although inverted, this is a real image, because light rays do connect the eye and the image.

Refraction and lenses

The second law concerns the "refraction" or bending of light as it crosses the boundary between two substances. This law states that the angle of refraction is in a constant relationship to the angle of incidence. The Dutch scientist Willebrord Snell (1591-1626) first enunciated this law in 1621, and it has since become known as Snell's law,

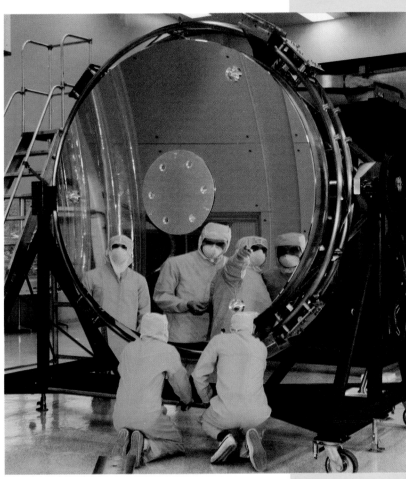

◄ **Astronomical telescopes generally use curved mirrors. The first telescopes were based on lenses but the problem of chromatic aberration (◊ page 42) led Newton to build the first reflecting telescope in 1671. Mirrors can be built with much larger diameters than lenses, and are the natural choice for large telescopes designed to collect as much light as possible. This mirror is for the Space Telescope.**

▼ **Light rays are reflected at surfaces in such a way that the angle of reflection always equals the angle of incidence. A plane mirror produces a virtual image by reflecting light so that it appears to the eye to come from behind the mirror. A convex mirror produces a virtual image, but diminished in size. The size of image produced by a concave mirror depends on the relationship between the position of the object and the focal length of the mirror – the point to which it converges light parallel to the axis. Here an inverted, reduced image is formed by a concave mirror; this image is also real and could not be seen by the eye.**

although in France it is known as Descartes' law after the French philosopher René Descartes (1596-1650), who rediscovered it some years later. The constant here depends on the nature of the substances on either side of the boundary. If the incident ray is in a vacuum, then the constant gives the "refractive index" of the refracting material. The angle of refraction is always smaller than the angle of incidence when a ray of light enters a denser medium and larger when it enters a less dense medium. The refractive index of a material is also equal to the velocity of light in a vacuum divided by the velocity of light in that material. Thus glass, with a refractive index of about 1·5, slows light down to about 200,000km/s.

Snell's law explains why a pool of water appears to be shallower than it really is, and why an object such as a spoon seems to bend as it is lowered into water. In both cases the eye sees a virtual image, of the bottom of the pool or the lower half of the spoon, and this is displaced from the position of the actual object by the bending of the light rays. Refraction also underlies the operation of lenses.

A beam of parallel light rays that pass through a convex lens converges to a point on the other side of the lens. This point is called the "focus", and the distance between the center of the lens and the focus is the "focal length". (A concave mirror converges parallel light in a similar way.) A convex lens makes a simple magnifying glass if the observer holds the lens so that the object being viewed lies between the lens and its focus. In this case, the lens forms an enlarged virtual image of the object. A concave lens diverges parallel light, so that it appears to come from the focus.

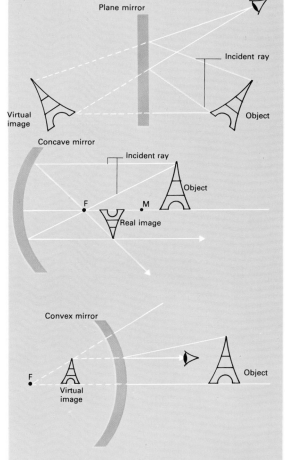

Plane mirror

Incident ray

Virtual image

Object

Concave mirror

Incident ray

Object

F M

Real image

Convex mirror

F

Virtual image

Object

Newton claimed the rainbow contained seven colors, not because they were easily distinguishable but by analogy to the notes of the musical scale

What is white?

When Newton directed a beam of sunlight through a prism he found it split into colors varying from red to purple, as in a rainbow. In analogy with the seven notes in music (A to G), he defined seven colors – red, orange, yellow, green, blue, indigo, violet – though few people find it easy to recognize seven bands of color. The colors are light of differing wavelengths, varying from 700nm for the limit of the red end of the visible spectrum, to 400nm at the violet end. The color of a non-luminous object depends on the wavelengths of light that it reflects rather than absorbs. A white object is a perfect reflector, one that reflects all light; a black object is a perfect absorber.

Sources of light, such as the Sun and a tungsten filament light, emit a broad spectrum of wavelengths, which we perceive as "white". This contrasts with a sodium lamp, for example, which emits most strongly at two closely-spaced wavelengths in the yellow region. The continuous spectra from the Sun and a tungsten light vary in intensity in a manner characteristic of a perfect emitter, or "black body" (a perfect emitter is also a perfect absorber). The intensity rises to a maximum at a wavelength that depends on the temperature of the emitter. The Sun's radiation peaks around 500nm, corresponding to a temperature of 6000K. A tungsten filament light, on the other hand, runs at a temperature of about 2000K, and its spectrum peaks at 1500nm, well into the infrared part of the electromagnetic spectrum (♦ pages 64-65). This is "heat" radiation; the visible light from the lamp comes from the higher wavelength, but lower intensity, side of the lamp's emission spectrum.

▲ Mixing colored lights is an additive process. Red, green and blue together stimulate all three types of color responsive cell in the eye, and the result is seen as white. These primary colors of light can be mixed in pairs to give secondary colors. With a paint, the process of producing a color is subtractive – the pigment absorbs light at certain wavelengths (colors); the observed color results from removing the wavelengths from white light. Thus if a pigment absorbs red, it will reflect the secondary color made from combining the other two primaries (cyan).

The color of light

Refraction reveals another property of light – its color. By the 17th century, the ability of a glass prism to produce a broad spectrum of colors from a beam of "white" sunlight was well kown. However, it was the British mathematician and physicist, Isaac Newton (1642-1727) who made the first serious study of the nature of color. He proved for the first time that color is a property of light itself, and has nothing to do with the nature of the prism or any other material. The prism "disperses" the light, refracting it according to its color. It refracts the red light the least and the violet light the most. Thus, the number quoted for the refractive index of a material depends on the color of light being used. For example, the refractive index for crown glass varies from 1·524 for red light to 1·533 for blue light.

Objects appear colored because they absorb certain wavelengths and only reflect those which go to make up the color that is seen. A colored filter absorbs all light except those wavelengths that it allows to pass through. Thus a green object viewed through a red filter appears black – all wavelengths except green are absorbed by the object, and this green is itself absorbed by the filter.

Colors appear in a different manner when white light is reflected from thin layers of material, such as patches of oil or the outer "skin" of a soap bubble. The British physicist Robert Hooke (1635-1703) studied this effect in detail and discovered that the color observed depends on the thickness of the layer. However, it was only in 1801 that the basis for a proper explanation of this effect emerged. This was the concept of the "interference" of light (♦ page 44).

▲ The British scientist Isaac Newton (1642-1727) made his fundamental discoveries about light and the nature of color – and laid the foundations of his work on gravity and motion – while at home in Woolsthorpe in 1665-6 during the Great Plague, when the university at Cambridge was closed down. He published his first scientific paper on his work with prisms in 1672, and met with great controversy, particularly from Robert Hooke (1635-1703). Only when Hooke had died did Newton publish his work Opticks in 1704. Typical of the experiments Newton describes is one that splits white light into colors with a prism, recombines them with a lens, and then splits them again to form a spectrum on a screen. In this way he showed that colors are contained within white light.

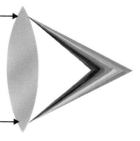

▶ Because the refractive index of a material varies with wavelength (color), a simple lens does not have a unique focus. Thus the lens forms a series of colored images of slightly different size, and the observed image appears to have a colored fringe. This effect – chromatic aberration – often occurs in inexpensive telescopes and binoculars.

▲ ▶ A spectacular demonstration of white light as a combination of colors is seen in a rainbow. Rainbows occur when sunlight from behind the observer falls on water droplets in front of him or her. Often a weaker "secondary" rainbow is seen outside the brighter "primary". In forming the primary, light from the Sun is first refracted as it enters a raindrop, then reflected from the back of the drop, and finally refracted again as it emerges, spread into the whole spectrum of colors. In forming the secondary rainbow, the light is reflected twice within the raindrop before it emerges. The additional reflection has the effect of reversing the order of the colors, so that although red appears on the outside edge of the primary bow, it is at the inner edge of the secondary rainbow.

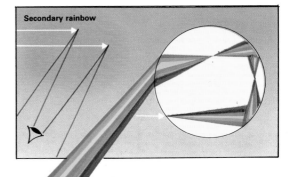

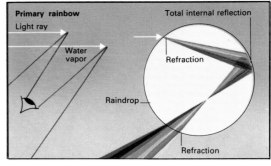

The key to understanding color – the wave theory of light – was first proposed in 1678

The wave theory of light

Thomas Young consolidated the idea that light is a wave motion, which he believed was "excited" in a "luminiferous ether (that) pervades the universe". Such ideas had been discussed by Hooke and others over a century before (▶ page 48) but Young was the first to recognize an important property not only of light but of wave motion in general. This is the "principle of superposition", which states that when two (or more) waves cross, the size of the resulting wave at each point is given by simply adding together the sizes of the individual waves at that point.

This principle is revealed in an experiment first performed by Young, in which light falls on a card with two very narrow slits that are not far apart. The two narrow beams of light emerging through the holes illuminate a screen beyond. As the beams of light originate from the same initial beam, their wave motions should be in phase (undulating in unison). According to the superposition principle, at points where the light takes paths of different lengths to reach the screen, the separate undulations can be out of phase, as the trough in one wave arrives at the same time as the peak in the other. The two waves therefore cancel each other out to give darkness at the screen. This is confirmed by the pattern of alternating bright and dark stripes that appears on the screen. The bright stripes correspond to where the difference in the paths equals an exact number of wavelengths, so that the separate beams reinforce each other. The intermediate dark regions are where the two waves cancel each other.

Such bright and dark bands occur when the two-slit experiment is performed with monochromatic light (of one spectral color). With white light, the pattern produced is a complex series of bands of different colors. This is because the reinforcing and canceling occurs at different points on the screen for different parts of the spectrum, and it shows that light of different colors has different wavelengths. The difference in the two paths necessary to reinforce (or cancel) the light must be slightly different for each color. Red light has the longest wavelength (requiring the largest path difference) while violet light has the shortest wavelength (smallest path difference).

The wave theory of light is the key to understanding color, and the concept of interference provides an explanation for the colors of thin layers. Light is reflected from both the top and the bottom of the layer, and the two reflected beams of light can interfere exactly as when they emerge from two slits. The wave theory of light also explains the phenomenon of "diffraction", first noted in the 17th century by the Italian Jesuit scientist, Francesco Grimaldi (1618-1663). Grimaldi observed that the shadows cast by narrow beams of sunlight in darkened rooms do not have precisely sharp edges but have colored fringes and these fringes spread beyond the expected edge of the shadow.

This effect is seen more clearly when monochromatic light passes through a single narrow slit to fall on a screen beyond. The light forms a pattern centered on a bright line, with a series of bright "fringes" gradually fading away to either side. Moreover, the central bright line is wider than the slit itself, the width of the line being inversely proportional to the width of the slit. The light apparently spreads as it emerges from the slit, as if the slit itself behaves like a row of little sources of light, each emitting a circular "ripple" of light. The fringes are caused by interference between the light waves from these "sources".

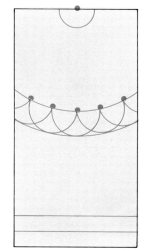

◄ The Dutch physicist Christian Huygens laid down the first foundations of a wave theory of light in 1678. He imagined that a point of light emits a spherical "wavefront", and that each point on this wavefront can be regarded as a new source of waves, and so on. The envelope of all the new "wavelets" gives the shape of the new wavefront, showing how the light spreads from the source. At large distances from the source, the wavefronts are in effect parallel. Huygens' principle successfully explained optical phenomena such as reflection and refraction, as well as interference.

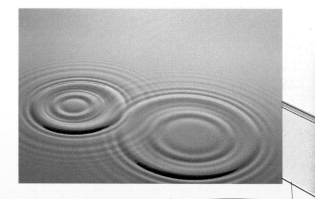

▲ Two drops produce circular patterns of spreading ripples, which create an interference pattern when they meet.

Double slits

Single slit

Ray of light

Diffracted light

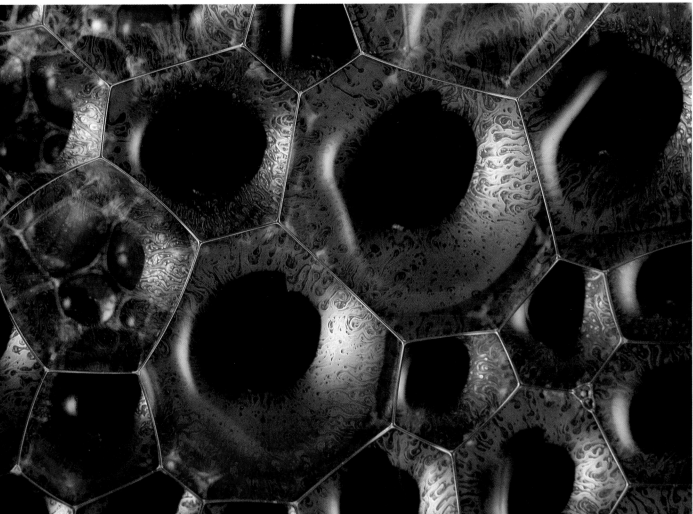

Interference patterns

▲ ▶ *Patterns in soap film form when light reflected from the top and bottom of the film interferes so that two sets of wavefronts are in step (top) or out of step.*

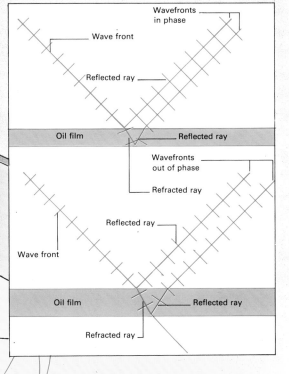

Wavefronts in phase

Wave front

Reflected ray

Oil film   Reflected ray

Wavefronts out of phase

Refracted ray

Reflected ray

Wave front

Oil film   Reflected ray

Refracted ray

Diffracted light

◀ *Light shining through a pin hole produces spherical wavefronts which create two new secondary sources of wavefronts at a screen pierced by two holes. These new wavefronts interfere to produce a pattern of bright and dark stripes – bright where the wavefronts exactly match, dark where they are out of step and cancel each other out.*

The Special Theory of Relativity

Einstein's two simple statements

If an experiment were conducted in a stationary laboratory on the Earth's surface, and then in an identical laboratory in a train moving with constant velocity along a level track, Newton's laws of motion would apply equally well in both instances. The velocity of the laboratory does not affect the results of the experiment. Indeed if the two laboratories had no windows it would be impossible to conduct any mechanical experiment to distinguish between them. This is known as the principle of relativity. All motion is relative – even the laboratory which is at rest relative to the Earth, is moving rapidly relative to the Sun. Yet Newton believed that there exists some point in space which is at "absolute" rest.

In the 19th century, the British physicist James Clerk Maxwell (1831-1879) demonstrated how electricity and magnetism are related, and showed that light is an electromagnetic wave (◊ page 65). Scientists assumed that waves could not travel through empty space so the existence throughout space of a transparent medium, called the ether, was suggested. This could conceivably be in the state of absolute rest suggested by Newton. Researchers conducted a number of experiments to find the velocity of the Earth through the ether. In 1887, two American scientists Albert Michelson (1852-1931) and Edward Morley (1838-1923) used an optical device in such an attempt, but, despite the accuracy of their measurements, they were unable to obtain a result. Among the explanations for this was a suggestion that any object moving through the ether suffers a "contraction" in the direction of motion.

In 1905, Einstein proposed his special theory of relativity, based upon two simple statements that, when taken together, lead to some extraordinary conclusions. First, he stated that no physical experiment, including optical ones, can distinguish between the two identical laboratories mentioned above; consequently there is no point in trying to detect the absolute motion of the Earth as Michelson and Morley had attempted. Second, he stated that all observers must obtain the same value when they measure the speed of light through space, even though the source of light may be approaching them or receding from them. Although supported by Michelson's and Morley's experiment, this contradicts our usual experience – a stone thrown forwards from a moving automobile moves faster relative to the ground than one thrown from a stationary vehicle.

Some of the consequences of these statements can be seen by considering two spacecraft, A and B, with B moving past A at half the speed of light. Some remarkable conclusions ensue.

A sees B's spacecraft shortened in the direction of travel. In fact a meter-rule on B's craft appears to be only 87cm long. If A were able to measure the mass of B as he passes in his craft he would discover that it had increased. His 75kg colleague would appear to have a mass of 86kg.

Einstein showed that the relative mass increase of an object was a measure of the energy imparted to it. He argued that if the additional mass is a measure of energy then surely all of the mass should have an equivalent energy value. This equivalence between mass and energy is expressed in the relationship, energy equals mass times the velocity of light squared ($E=mc^2$). A would also observe that B's clock runs slower than his. None of these effects will be noticed by B, however. In his view nothing has altered on his craft, but A (receding at high speed) seems subject to all these changes.

► The special theory of relativity rests on two basic postulates. The first is that the laws of physics are the same in all inertial (non-accelerating) frames of reference. Thus passengers on a high-speed train can pour coffee and walk about with no awareness of their rapid motion unless they look out of the window.

► Einstein's special theory of relativity set out to agree with the experimental finding that the velocity of light does not depend on the motion of the observer. In so doing, the theory made several remarkable predictions. Two of these predictions concern changes in the observed mass and length of an object as its speed increases towards the velocity of light. Lengths parallel to the direction of motion (but not those at right angles to it) decrease, or contract, while masses increase. These changes in length and mass make it impossible for anything to travel faster than the velocity of light.

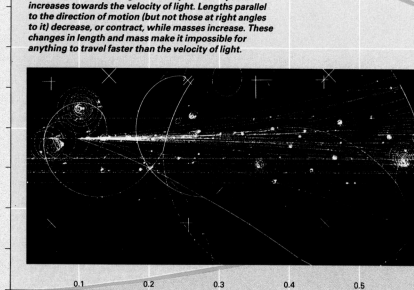

Length

Mass

◄ An object's mass is a form of energy. In this picture of tracks in a bubble chamber, the collision of a single high-energy proton with a stationary one creates a spray of new particles, as kinetic energy converts to mass energy.

| 0.1 | 0.2 | 0.3 | 0.4 | 0.5 | 0.6 | 0.7 | 0.8 | 0.9 | 1.0 |

Proportion of speed of light

Speed of light

Albert Einstein

Albert Einstein was born in Ulm in Germany in 1879. He attended a Catholic grammar school in Munich, despite being a Jew, but showed no great promise and later took himself to Switzerland to continue his studies. There he missed most of his lectures, and relied on the notes taken by a friend. When he graduated in 1901 Einstein became a Swiss citizen and obtained his first job as a junior official in the patent office in Berne.

In 1905 Einstein had a remarkable year, publishing no fewer than five important papers. In one he explained the photoelectric effect, thus laying the foundations for quantum theory (♦ page 196); in another he provided the first satisfactory explanation for Brownian motion (♦ page 31); but perhaps most famous is the paper describing the special theory of relativity. This dealt only with unaccelerated motion, and in 1915 Einstein published his general theory of relativity which covered the more general case of accelerated motion. This theory contained a new description of gravity which replaced and extended Newton's theory (♦ pages 216-217).

Einstein was appointed to professorships in Zürich and later in Berlin. In 1922 he was awarded the Nobel Prize for his work on quantum physics. He emigrated to the United States when the Nazi Party came to power in Germany in 1933 and it was from the Institute of Advanced Studies at Princeton, New Jersey, that he wrote to President Roosevelt urging the construction of the atomic bomb. However, after World War II, Einstein argued tirelessly for a world agreement to end the threat of nuclear war. Unfortunately his influence was reduced by his habit of signing almost every petition that was put in front of him. He died on 18 April, 1955, in Princeton.

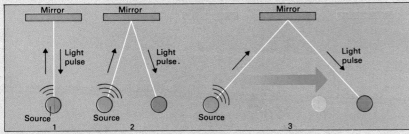

▲ The second postulate of the special theory is that the speed of light is independent of the velocity of the source. One of the consequences of this is the phenomenon of time dilation. Suppose a moving clock sends out a light pulse while coincident with a stationary clock, and the light is reflected back from a mirror (1). From the viewpoint of the stationary clock the light will appear to have traveled farther to the moving clock, so that a stationary observer will say that the time interval for the moving clock has increased (2). The faster the clock is moving, the greater the dilation will seem (3).

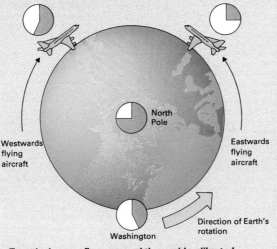

▲ Two clocks were flown around the world, calibrated against a clock in Washington. After returning, the eastward-flying clock had lost time, and the westward gained, relative to the clock in Washington. The explanation requires an imaginary clock at the North Pole, stationary relative to the rotation of the Earth. The clock in Washington is moving relative to this clock, and so loses, whereas the eastward-flying clock, moving faster than the Earth's rotation, loses more time. The westward-flying clock, moving against the Earth's rotation, seems from the Pole to be moving slower than the clock in Washington, so, though losing time relative to the Pole, it gains relative to Washington.

Flying clocks around the Earth

One of the most convincing tests of special relativity was performed in 1971, when two United States physicists, Joseph Carl Hafele and Richard Keating, flew around the world with four atomic clocks. They made two journeys – on passenger jet aircraft – one eastward and one westward, and both taking about three days. By comparing their clocks at the start and finish of the journeys with reference clocks in Washington, Hafele and Keating observed how well their clocks had kept time.

They found that the eastward-moving clocks had lost 59 billionths of a second compared with the reference clocks in Washington, whereas the westward-moving clocks had gained 273 billionths of a second. The clocks had been influenced by two effects. First, both sets had gained because they were flying at high altitudes where the gravitational force of the Earth is weaker and time runs more quickly – this is an effect predicted by Einstein's general theory of relativity. However, the eastward-moving clocks had also lost time because their velocity was greater relative to the clocks on Earth. Indeed, the eastbound clocks lost more time than they gained. The westward-moving clocks, however, gained still more relative to those on Earth, because their velocity was less than that of the clocks in Washington.

▲ **Calcite forms two images in certain directions as the crystal splits light into two rays. One ray refracts normally; the other, which here forms the displaced upper image, does not. In the line at right angles, the images are superimposed.**

▼ **Mechanical stress makes acrylic doubly refracting, so it splits light into two rays with different polarizations and velocities. This causes colored interference patterns in polarized light.**

Polarization

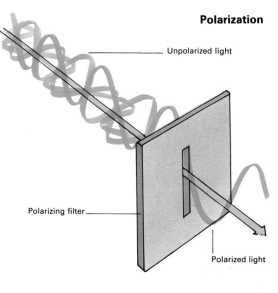

Unpolarized light

Polarizing filter

Polarized light

▲ **Ordinary – unpolarized – light consists of vibrations in all directions at right angles (transverse) to the direction of travel, but these can be resolved into two directions at right angles. A polarizing filter transmits light only in one of these directions, so that it becomes "plane" polarized. This effect is used in polarizing sunglasses, which transmit only the light plane polarized in one direction.**

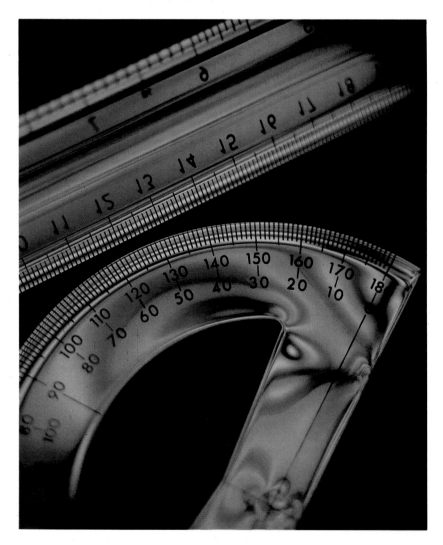

Particles, waves and polarization

In 1669, the Danish scientist Erasmus Bartholinus (1625-1698) acquired some crystals of calcite, or "Iceland spar", which had been brought to Copenhagen from Iceland. Bartholinus discovered that he could see two images through the crystals, and he concluded that the crystal split an incident ray of light in two – a process of "double refraction". One of the rays obeyed Snell's law of refraction (◊ page 40), the other did not. Another Dutch physicist, Christian Huygens (1629-1695) took these investigations further, using a second crystal to study the two beams of light emerging from the first crystal. He found that either one of the two images would become more feeble and disappear as he rotated the second crystal.

Huygens also developed a wave theory of light, first published in 1678, in which he proposed that light propagated in the form of spherical waves. Moreover, he suggested that each point on an advancing spherical "wavefront" is the source of a new spherical wave (◊ page 44). According to Huygens, this was how the waves propagated, through a "very subtle and elastic medium" that he presumed to fill all space. Huygens' theory proved successful in explaining reflection and refraction. Indeed it was an inspiration to Young more than a century later, and even now provides the simplest way to understand diffraction patterns. However, Huygens could not explain properly all the effects observed with the calcite crystals.

At the time of Huygens, however, the concept of light as a wave motion lost out to the alternative idea of light as beams of particles, which dominated thinking about optical effects for the next 100 years or so. Newton is often regarded as stressing the corpuscular, or particle, nature of light, but he seems to have preferred a bizarre mixture of corpuscular and wave theories, in which light is emitted as particles which then set up vibrations in the "ether" (equivalent to Huygen's "elastic medium") pervading all space. But even Newton was unable to explain the behavior of calcite. He postulated that rays of light have "sides", and in a sense foreshadowed what is now referred to as the "polarization" of light.

Young reintroduced the wave theories in 1801, but his ideas were at first criticized; it fell to the French physicist Augustin Fresnel (1788-1827) to show that all the optical phenomena known at the time could be understood in terms of light waves, including the double refraction of calcite crystals.

In 1808, Etienne Malus (1775-1812), an officer in the French army, was experimenting with some Iceland spar when he noticed that if he put the crystal in a beam of light reflected from a window, then the two images would vary in intensity just as Huygens had first observed with two crystals. Malus had discovered that light acquires "sides" when it is reflected. Nine years later Young pointed out to Fresnel's colleague, Dominique Arago (1786-1853), that the "polarization" of light upon reflection could be explained if it was assumed that light vibrates transversely to its direction of motion. However, the physical meaning of these transverse vibrations was not explained until Maxwell's theory of electromagnetic waves (◊ page 61).

Magnetism

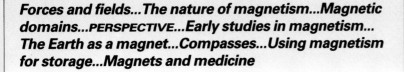

6

*Forces and fields...The nature of magnetism...Magnetic domains...*PERSPECTIVE*...Early studies in magnetism... The Earth as a magnet...Compasses...Using magnetism for storage...Magnets and medicine*

▲ *Magnets used in scrapyards are electromagnets, in which an electric current induces a magnetic field in iron. The induced magnetism does not last long once the current is switched off, and the magnet can release its load.*

A magnet is a piece of iron or other material that can attract or repel other similar materials placed nearby. If a bar magnet is suspended away from all other magnetic materials so that it can rotate in a horizontal plane, it will always tend to align itself the same way, which by convention is called the north-south direction. The end that points north is called the north (or north-seeking) pole, and the other end is the south (or south-seeking) pole. All magnets have two poles; a single magnetic pole has never been found despite speculations that such objects may exist (♦ page 204).

The magnetic field

The region round a magnet over which it exerts a force is called the "magnetic field". This field can be represented as a series of lines in three dimensions that form a pattern linking the north and south poles of the magnet. Conventionally the lines represent the way in which a (hypothetical) free north pole would move, its direction shown by an arrow. The field is strongest at the poles, and decreases with increasing distance from them. Like the electrostatic field (♦ page 53), it decreases in inverse proportion to the square of the distance. In the pattern of field lines, the lines are closest where the field is strongest.

When a piece of unmagnetized magnetic material, such as iron, is brought near to a magnet it becomes temporarily magnetized – that is, north and south poles form – and it is attracted towards the magnet. This effect is called "induced magnetism". The strength of the induced magnetism and how long it lasts depend on the properties of the material. Iron is a "soft" magnetic material: it can be magnetized and demagnetized easily, and forms a strong induced magnet, but its induced magnetism does not last for long. Steel is a "hard" magnetic material: it is more difficult to magnetize strongly but retains its magnetism for longer. Steel is therefore used for making permanent magnets, whereas iron is used for electromagnets, where the magnetism is switched on and off by an electric current (♦ page 61).

▲ *Iron filings reveal the field on the end of a magnet.*

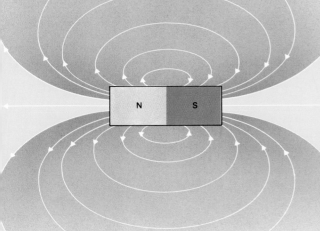

◄ *A magnetic field can be represented by lines linking north and south poles in a three-dimensional pattern. The arrows give the direction of the force on an imaginary north pole.*

▶ *Similar magnetic poles repel each other (right), giving a "neutral" point between them where there is no net force. Dissimilar poles (far right), on the other hand, are attracted to each other, with the field lines linking them together.*

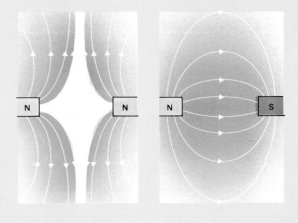

The reason why the Earth has a magnetic field has never been fully demonstrated

The Earth as a magnet

The observation that magnets align themselves along a north-south axis implies that the Earth itself must possess a magnetic field. Indeed, the Earth behaves as if it has a bar magnet at its center. Magnets and compass needles align themselves with the Earth's lines of magnetic force. These lines are not exactly along geographic meridians, so the north and south magnetic poles do not coincide with the geographic ones. Their positions are slowly changing all the time, so the declination (the angle between magnetic and geographic meridians) is variable.

If a compass is free to rotate in a vertical plane as well as a horizontal one, it dips towards the surface of the Earth at varying angles depending on latitude. At the north and south magnetic poles a dip needle points vertically while at the magnetic equator it lies horizontally.

Physicists have deduced from a study of earthquake waves that the Earth has a liquid core; they are therefore confident that there is no real bar magnet under the Earth's surface. One possible explanation for the Earth's magnetic behavior is that the liquid core has circulating electric currents inside it, which make it behave like a gigantic dynamo (♦ page 63).

The Earth's magnetic field extends some 80,000km into space and this causes many strange effects such as electric winds and aurorae, when the field traps charged particles from the Sun and outer space.

The Earth's magnetic field periodically reverses itself (that is north and south poles change places). This has been discovered in studies of magnetized rocks of various ages, some of which have a polarity opposite to the present Earth's field. Because the magnetization was produced when the rocks were formed it is possible to draw up a timechart of polarity changes. This, together with information relating to polar position, intensity of field, and angle of inclination, can then be used for dating rocks and determining their origin.

▼ **The Earth has a magnetic field similar to that of an imaginary bar magnet located at its center, with the magnet's south pole in the Northern Hemisphere. Magnetized needles free to rotate about a pivot align themselves with the Earth's magnetic field, which is horizontal at the magnetic equator and vertical at the magnetic poles. The magnetic and geographic poles are usually at slightly different locations.**

▼ **The value of compasses has been known since at least the 12th century. The main component is a magnetized needle pivoted about its center and free to rotate in the horizontal plane. In the absence of other magnetic fields, it will align with the horizontal component of the Earth's field. Instruments known as dip circles, in which a needle rotates in a vertical plane, indicate the vertical component of the field.**

▲ **The Earth's magnetic environment becomes visible in spectacular fashion in the displays of Aurora Borealis (in the Northern Hemisphere) and Aurora Australis (Southern Hemisphere). Aurorae occur at altitudes of 100-300km when electrons spiral down the Earth's magnetic field toward the poles. The electrons excite molecules, which emit light at different colors as they return to the ground state (♦ page 196).**

Earth's magnetic field

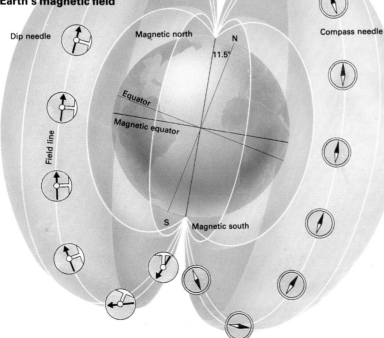

Dip needle
Magnetic north
N
11.5°
Compass needle
Equator
Magnetic equator
Field line
S
Magnetic south

Early studies in magnetism
The attractive and repulsive properties of naturally magnetic rock, lodestone or magnetite (Fe$_3$O$_4$), were probably known in prehistoric times, but it is believed that the Chinese discovered its directional properties. As early as 2500 BC a piece of magnetite was supposedly used as a primitive compass by a Chinese emperor to guide his troops through fog.

The Greeks were aware of the attractive properties of lodestone, and their literature contains fables referring to mountains that could draw nails out of ships or make it difficult for a shepherd (with iron tacks in his shoes) to move. By tradition the Greek philosopher Thales (624-546 BC) was the first to observe and study this effect.

It is not certain how or when Europeans discovered the directional properties of lodestone. By the 12th century there are references to the use by navigators of a compass consisting of a needle-shaped magnet floating on a reed or splinter of wood in a bowl of water. The more familiar form of compass, with a needle resting on a pivot over a circular scale, is due to the French scholar Petrus Peregrinus (c.1240-unknown).

Peregrinus also determined north and south poles and found that like poles repel and unlike attract. He also broke magnets in half, and found that each half had two poles. However, he believed that the geographical poles were in the heavens and did not know that the Earth itself is magnetized.

The English physicist William Gilbert (1544-1603) showed that a compass needle dips towards the Earth, and suggested that the Earth itself is a great magnet. His researches into the properties of lodestone, and the differences between magnetism and static electricity (◆ page 55) were extensive, but poorly followed up until the 19th century.

Magnetism is intimately related to electrical effects. In 1820 the Danish physicist Hans Christian Oersted (1777-1851) discovered that an electric current flowing through a wire sets up a magnetic field around the wire (◆ page 61). If the wire is twisted to form a coil, then the shape of the magnetic field outside the coil is similar to that of a bar magnet. The strength of the field can be increased by raising the size of the current or the number of turns of wire in the coil. The field strength can also be increased by placing a core of soft magnetic material inside the coil. Such a coil is called an electromagnet.

How does a material become magnetized?

In his electromagnetic theory (◆ page 65), the British mathematician and physicist James Clerk Maxwell (1831-1879) proved that moving electric charges – that is, electric currents – are the source of magnetic fields. The electrons and protons that constitute atoms are moving electric charges and are therefore themselves sources of magnetic fields. Different materials exhibit different magnetic properties and respond in different ways to external fields. Such differences can be traced to the electronic configuration and arrangements of the atoms. In magnetic materials the atoms form groups called domains, each of which contains many molecules. In a domain, the "atomic magnets" line up so that the assembly has a net magnetization. In general, neighboring domains are magnetized in opposite directions. However, the application of an external magnetic field encourages the domains aligned to this field to grow in size at the expense of their neighbors. In the absence of the external field, the agitation of the atoms and the forces at the poles disturb the alignment of the domains.

Only a few materials (iron, nickel, cobalt and some alloys and rare-earth elements ◆ page 96) show any permanent magnetism. These are called ferromagnetic materials and become magnetized in the direction of even a weak external field. "Paramagnetic" materials become weakly magnetized in the same direction as a very strong external field. "Diamagnetic" materials become weakly magnetized in the opposite direction to a very strong external field. Indeed, all materials exhibit diamagnetic properties, but usually these are obscured by paramagnetic or ferromagnetic behavior.

Magnetic domains

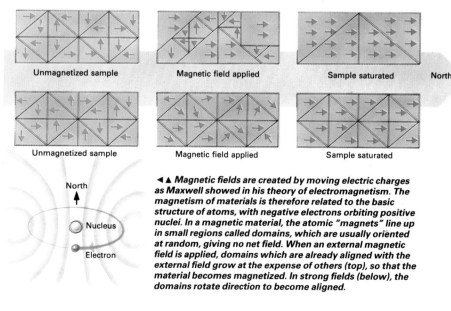

Unmagnetized sample   Magnetic field applied   Sample saturated   North

Unmagnetized sample   Magnetic field applied   Sample saturated

North
Nucleus
Electron

◀▲ *Magnetic fields are created by moving electric charges as Maxwell showed in his theory of electromagnetism. The magnetism of materials is therefore related to the basic structure of atoms, with negative electrons orbiting positive nuclei. In a magnetic material, the atomic "magnets" line up in small regions called domains, which are usually oriented at random, giving no net field. When an external magnetic field is applied, domains which are already aligned with the external field grow at the expense of others (top), so that the material becomes magnetized. In strong fields (below), the domains rotate direction to become aligned.*

Magnetism for storing information

An oscillating electrical signal can be used to create a varying magnetic field between the poles of a narrow-gap electromagnet. In this fact lies the principle of storing sound, images and computer data on magnetic media. The storage medium usually consists of finely-divided particles of iron or chromium oxide mounted on a tough plastic substrate in the shape of a tape or disk. As the medium passes the small gap (or tape head) the varying magnetic field induces varying magnetization in the metal oxide particles. The system works in reverse during playback.

Audio signals can be stored by driving a tape past a stationary head at a speed of about 5cm/s, although better quality recordings are produced at higher tape speeds. Video signals have higher frequency components and therefore require a much higher scanning speed. This is achieved by moving the tape slowly past a fast moving tape head, and results in the picture signal being stored in diagonal stripes across the tape. Computer data is in digital form with the signal either "on" (high magnetization) or "off". Such signals are pure and contain no "noise" due to the demagnetization of the tape with time. Such techniques are now also being used in the digital recording of sound and video.

▼ *In this experimental medical use of magnetism, a tumor cell in human bone marrow is surrounded by magnetic "microspheres" to allow doctors to draw the diseased cell from the body quickly, efficiently and without the disturbance to the body that irradiation or chemotherapy might cause.*

▶ *Tiny particles of iron or chromium oxide store the information contained in sound waves as varying patterns of magnetization on the tapes used in audio cassettes. Magnetic effects also play a part in the transduction of sound and electricity in microphones, speakers and headphones.*

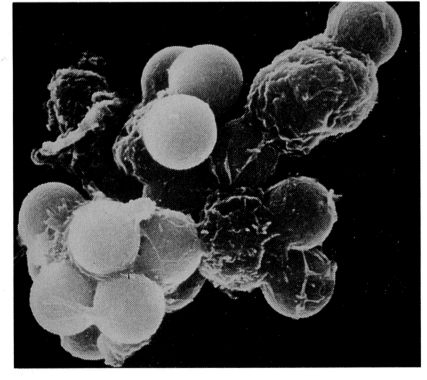

Magnetism and the human body

In recent years doctors have found increasing applications of magnetism as an aid to treatment of disease; the fact that relatively strong magnetic fields can penetrate the body without damaging living tissue has proved a great advantage in this field. Nuclear magnetic resonance (◆ page 201) takes advantage of this fact to permit a cross-sectional or three-dimensional image to be created of the internal organs.

In the mid-1980s a new technique was developed in Britain to assist doctors in the treatment of cancers, particularly tumors in bone marrow. This technique takes advantage of so-called monoclonal antibodies, cells originating from the human immune system but processed by genetic engineering techniques so that they are "targetted" to attach themselves to specific cells in the human body. Monoclonal antibodies, designed to locate the tumor cells, are coated in tiny magnetic microspheres and inserted into the diseased bone marrow. As a result, the tumor cells effectively become responsive to a magnetic field, and can be drawn out of the bone marrow by application of a magnet. This technique, known as magnetic depletion, has proved highly effective in clinical trials, and may become an important form of treatment in the future.

Electricity

7

Electric charge, a fundamental of all matter...Attractive and repulsive forces...Field lines...Potential difference... Current and resistance...Capacitance...PERSPECTIVE... Static electricity...How a battery works... Semiconductors...Electrolysis

The world is built from atoms created from a small number of types of elementary particle. The complexity of the world results from the ability of these particles to affect the motion of other particles through fields of force that surround them. It is these interactions that ultimately produce the familiar structures of the everyday world. Beyond the scale of the atomic nucleus, there are only two forces that cause these interactions: the electromagnetic force and the gravitational force (◆ page 215). The gravitational force is very weak on the scale of particles: the electromagnetic force, on the other hand, determines the structure of atoms, of solids, liquids and gases, and of the complex molecules in living matter.

Most elementary particles are surrounded by an "electric field". Such particles are said to carry an electric "charge". The field is proportional to the strength of the charge, but falls off with the square of the distance from the charge. This "inverse square law" is named for the French engineer Charles Augustin Coulomb (1736-1806), who tested its validity in the 1780s.

The electric field due to a charged particle has a direction as well as a strength. The force between identical particles is repulsive, and it acts along the line joining the two particles. The field of each particle exerts a force on the other to push them apart.

There are also, however, attractive forces that bind particles together. The electric force can attract as well as repel because there are two kinds of electric charge. These are labeled "positive" and "negative". Whereas two charges of the same kind repel, two opposite charges attract. In an atom a positively-charged nucleus is surrounded by negatively-charged electrons, which are attracted to the nucleus but repelled by each other (◆ page 72). Overall the atom is electrically neutral. The fact that an atom contains electric charge is important in forming molecules from a number of atoms (◆ page 77).

▲ *Electric charge exists in two varieties – positive and negative – which behave in opposite ways. Here, particles with identical mass but opposite charge move under the influence of a magnetic field in a bubble chamber (◆ page 206). The particles are electrons (green tracks), each carrying one unit of negative charge, and positrons (red tracks), which are antielectrons and carry positive charge.*

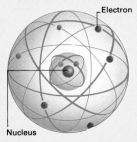

▲ *The origin of electricity lies in the structure of matter. Matter is made up of atoms. Each is neutral overall, with negatively charged electrons orbiting a positive nucleus.*

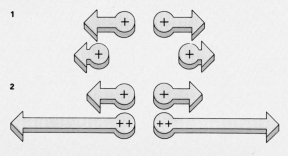

▲▶ *In the 1780s, Coulomb, using a torsion balance (right), established the basic law of force between two charged particles – the force decreases in proportion to the square of the distance between the charges (1), and it is directly proportional to the product of the charges (2). Double the distance between the charges and the force falls by one quarter; double the charges and it increases by a factor four.*

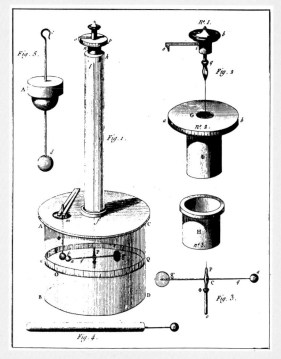

The word electricity derives from the Greek name for amber – elektron

The British scientist Michael Faraday (1791-1867) proposed a method of visualizing electric fields as lines in space. The lines point in the direction of the field, and are closer spaced where the field is stronger. Such imaginary field lines wind continuously through space, beginning only on a positive charge and ending only on a negative charge. The number of lines starting from, or ending on an electric charge, is proportional to the strength of the charge. These are properties unique to a field that obeys an inverse square law. The "electric potential difference" gives a measure of the amount of electric field between two points. For a constant field, the potential difference grows as the separation between the points increases, and it is proportional to the strength of the field. Electric potential difference is familiarly known as "voltage difference" or "voltage". Its unit is the volt (V), named for the the Italian physicist, Alessandro Volta (1745-1827), who invented the electric battery in 1800.

Voltage is a measure of work done (◀ page 22), in this case the work done in moving a positive charge of one unit against the electric field between two points. This is like work done in walking from the bottom of a hill to the top, although in that case the work is done against the Earth's gravitational field. If the electric field between the points varies with position, or if the path is not a straight line, it is still possible to compute the voltage by dividing the path into tiny pieces and adding together the voltages across each piece. Such calculations show that the potential difference between two charges does not depend on the chosen path.

◀ ▼ An electric field can be represented by lines in space. The lines are closer together where the field is stronger, while arrows indicate the direction of the field, which runs in towards negative charges (1) and out from positive charges (2). Like charges repel (3), whereas unlike charges are attracted towards each other (4). Note how the field lines always run from positive to negative charge. In a uniform field, as between the plates of a capacitor (◆ page 58), the field lines are parallel (5).

▲ Experiments into static electricity were conducted around 1700 by the British physicist Stephen Gray (1666-1736). The "charity boy" (orphan) was suspended off the floor and charged electrostatically. Pieces of paper were attracted by his positive charge (depletion of electrons).

Static electricity

Electric charge may be deposited on an object only by transferring charge to it from some other piece of material. Only a small proportion of charges need to be transferred to create observable effects. Suppose charge is driven from one plate to another 1cm away until a potential difference of 100V has built up across the plates. Only one hundredth of a millionth of a millionth of the free electrons in the metal will have been transferred.

Friction can separate small quantities of electric charge. By rubbing two materials together electrons may be transferred from the atoms of one material to those of the other, if the second material has a greater affinity for electrons than the first. The ancient Greeks knew that a piece of amber rubbed with fur would attract small pieces of hair. The word electricity derives from the Greek name for amber – elektron.

Spectacular results of the transfer of electric charge occur in lightning. Here, an electrically-charged cloud suddenly discharges, generating a current which passes through the air, raising it to very high temperatures. This causes both the flash of lightning and a shock wave of pressure changes in the air, producing the sound of thunder. The energy used in charging up the cloud initially is drawn from the atmosphere.

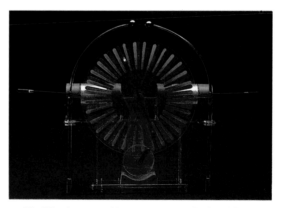

▲ The Wimshurst machine generates static electricity as two insulating disks rotate in opposite directions. Charges of opposite sign are collected by metal combs at each side. When sufficient charge has built up, the electric field between the metal spheres at the top causes a spark to fly.

◄ Lightning brightens the sky over Tucson, Arizona.

An electric battery resembles a water cistern, producing a voltage rather than a head of pressure

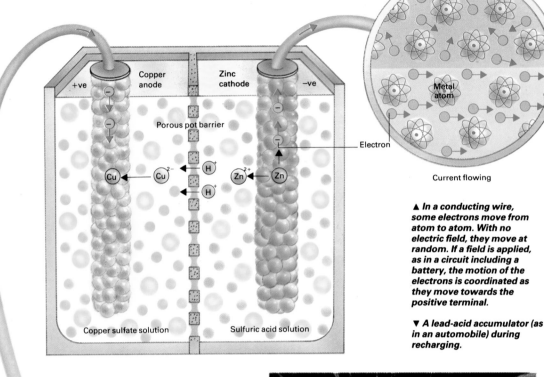

▶ The so-called Daniell cell or battery contains sulfuric acid, the molecules of which divide into hydrogen and sulfate ions when dissolved (♦ page 80). If the anode and cathode are connected via a circuit, a current flows as the atoms in the zinc cathode tend to give up electrons, thus becoming positive ions and joining with the sulfate ions. The hydrogen ions are attracted to the anode, where they attract electrons from the copper to form hydrogen molecules. The result is a net flow of electrons from cathode to anode. The flow continues until the cathode has been eaten away completely, or a film of non-conducting hydrogen bubbles on the anode causes internal resistance to build up.

No current

Metal atom

Electron

Current flowing

▲ In a conducting wire, some electrons move from atom to atom. With no electric field, they move at random. If a field is applied, as in a circuit including a battery, the motion of the electrons is coordinated as they move towards the positive terminal.

▼ A lead-acid accumulator (as in an automobile) during recharging.

Current and resistance

Electric potential difference is like a pressure difference in liquids. A battery behaves like a cistern, producing a voltage rather than a head of pressure. The voltage produced by a dynamo or electricity generator (♦ page 62), on the other hand, resembles the pressure produced by a pump. If a battery is connected up so that it completes an electric circuit – in a bicycle lamp, for instance – the battery supplies a voltage (an electric potential difference) between the two ends of the circuit to which it is attached, or sets up an electric field in the rest of the circuit. This field makes a current flow and lights the lamp as resistance to the flow in the bulb causes the filament to glow.

Electric currents usually arise from the flow of mobile electrons. To generate a current, both a "voltage source", to provide an electric field to direct the electrons, and a "conductor", in which the electrons can move freely, are required. In metals, electrons bound loosely on the periphery of atoms wander easily from atom to atom (♦ page 78). A small electric field will organize their normally chaotic wanderings into a steady flow in one direction.

The basic unit of electric charge is called the coulomb (C), and it is equal to the total charge of 6.25×10^{18} electrons. The unit of current is named the ampère, or amp, for the French professor of mathematics, André Ampère (1775-1836), who did important work on the magnetic effects of electric currents. One ampère corresponds to one coulomb of charge flowing each second across a cross sectional slice through a wire.

Electrons flowing through a conductor, such as a piece of copper wire, are not unhindered. They can collide with the atoms of the metal and with impurity atoms. In these collisions, the electrons lose some of their energy, transferring it to the chaotic motion of the atoms. This raises the temperature of the metal, since higher temperatures are linked to faster atomic movements (♦ page 32). In ordinary filament lights, a coil of very fine wire is kept deliberately at a high temperature within a vacuum by an electric current, and the wire glows bright yellow. In an electric fire, a coil of thicker wire offers less hindrance to the current; the wire is thereby maintained at a lower temperature, and it glows red.

How a battery works

The basic operation of a battery is to store chemical energy and convert it to electrical energy as required. It utilizes the natural random movement in a liquid of free ions – atoms with either too few or too many electrons. The familiar kind of battery, used for example in a torch, is a "dry" battery, its liquid components being in the form of a paste. It consists typically of a carbon rod – the positive terminal – surrounded by a muslin bag containing a mixture of manganese dioxide and powdered carbon; around this, and in contact with the zinc cannister, is a layer of ammonium chloride paste.

When the terminals of the battery are connected via a circuit, electrons flow from the zinc container to the carbon rod: an electric current flows from the positive (carbon) terminal to the negative (zinc).

The electrons are liberated when negative chloride ions (with one additional electron each) in the paste make contact with the zinc, and form zinc chloride. The positive ammonium ions (each with an electron missing) move from the paste through the bag to the carbon terminal, where they pick up electrons and form hydrogen and ammonia. The manganese oxide prevents an insulating layer of hydrogen bubbles from building up around the carbon. It reacts with the hydrogen to produce manganese trioxide and water. The net effect is that electrons flow from the zinc to the carbon, while the ammonium chloride is decomposed.

This type of dry cell is known as a primary cell: the chemical changes cannot be reversed. In a secondary cell, the changes are reversible. One familiar secondary cell is the lead-acid accumulator used as a car battery. This consists of six cells connected in series. Nickel-cadmium "rechargeable" batteries are dry secondary cells.

Resistor

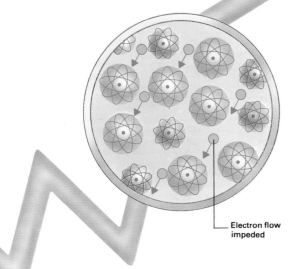

Electron flow impeded

◄ *An ordinary light bulb (here shown broken but still working) glows as the electric current encounters high resistance in the fine tungsten filament, and heats to more than 2,000°C.*

▲ *Resistance occurs when the flow of electrons is impeded by collisions with the metal atoms or with impurity atoms. The electrons lose energy to the atoms.*

A fluorescent lamp

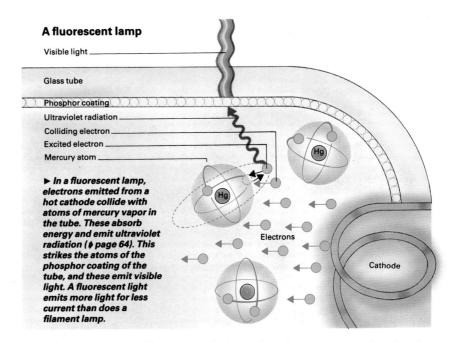

Visible light

Glass tube

Phosphor coating

Ultraviolet radiation

Colliding electron

Excited electron

Mercury atom

Hg

Hg

Electrons

Cathode

▶ *In a fluorescent lamp, electrons emitted from a hot cathode collide with atoms of mercury vapor in the tube. These absorb energy and emit ultraviolet radiation (♦ page 64). This strikes the atoms of the phosphor coating of the tube, and these emit visible light. A fluorescent light emits more light for less current than does a filament lamp.*

The ability of an object to conduct an electric current is related to its "electrical resistance": the higher the resistance, the more difficult it is for a current to flow. The resistance depends on the nature of the material in two fundamental ways. The more electrons that are free to drift in a given volume, the lower the resistance; the more obstacles in the same volume, the higher the resistance. Electrical conductors have low resistance whereas good electrical insulators have very high resistance.

Electrical resistance is a form of friction (♦ page 18) and causes the moving charges in a current to lose energy irreversibly. The energy lost appears as heat in the conductor carrying the current, and for the current to be maintained, the energy loss must be made good. A battery not only sets up an electric field so that current can flow through a lamp; it also replenishes the energy radiated as heat and light.

The unit of electrical resistance is called the ohm, after the German physicist Georg Simon Ohm (1787-1854). Ohm performed a series of experiments to study the flow of electricity, measuring the voltages and currents for wires of different lengths and thicknesses. The result was his celebrated law, according to which the resistance of a piece of wire, say, is given by the potential difference required to sustain a current, divided by that current. Ohm's law reflects some basic features of electrical resistance in wires and similar conductors. For example, the current through a wire is the amount of charge that crosses a cross-section through the wire each second. The larger the area of the wire, the greater the current that can be carried, and the lower the resistance. Moreover, longer wires have higher resistance than shorter ones. Indeed, wires carrying electric current are like pipes carrying water. Large currents of water can gush through short, fat pipes, but only small currents can travel along long, narrow pipes.

Ohm's law applies only to materials in which the resistance does not depend significantly on the current or the voltage applied. However, even in such materials the resistance depends on temperature, because the atoms jiggle more at higher temperatures, and this increases the number of obstacles that the electrons encounter. Semiconductor devices and conducting gases behave very differently. This is because the number of mobile electrons depends on the voltage supplied.

A light-emitting diode and a photoelectric cell are essentially similar devices that work in opposite directions

Capacitors – storing electricity

A simple capacitor consists of two parallel metal plates. It can be charged up by connecting it to a voltage source, such as a battery. The potential difference draws electrons along the connecting wires from one plate and supplies them to the other, making one plate positive and the other negative, until the potential difference between the plates balances that of the source. Then the flow of charge stops, provided the voltage applied to the capacitor is constant. A capacitor will not pass a steady current (♦ page 68). The source supplies energy in charging the capacitor, but this energy can be recovered.

If the two plates are now connected via a wire, with no voltage source, the potential difference across the plates drives a current through the wire and the charges on the plates fall to zero as electrons flow from the negatively charged plate to the positive. The energy that the capacitor stored is dissipated as heat in the wire connecting the plates. Thus, the capacitor acts like a temporary voltage source.

The field in a capacitor
If the two parallel metal plates in a capacitor receive equal amounts of opposite charge the charge spreads across the plates to give a constant electric field, proportional to the charges on each plate divided by the area of each plate. The electric potential of the plate carrying positive charge relative to the negative plate – in other words, the voltage across the capacitor – is given by the field multiplied by the distance between the plates. Capacitance is defined as the amount of charge a capacitor can hold divided by the potential difference that the charge produces. If a capacitor holds a charge of one coulomb on each plate, and has a potential difference of one volt between the plates, then its capacitance is one farad (F), the unit of capacitance, which is named in honor of Faraday. In practice, capacitances as high as one farad or so are rare.

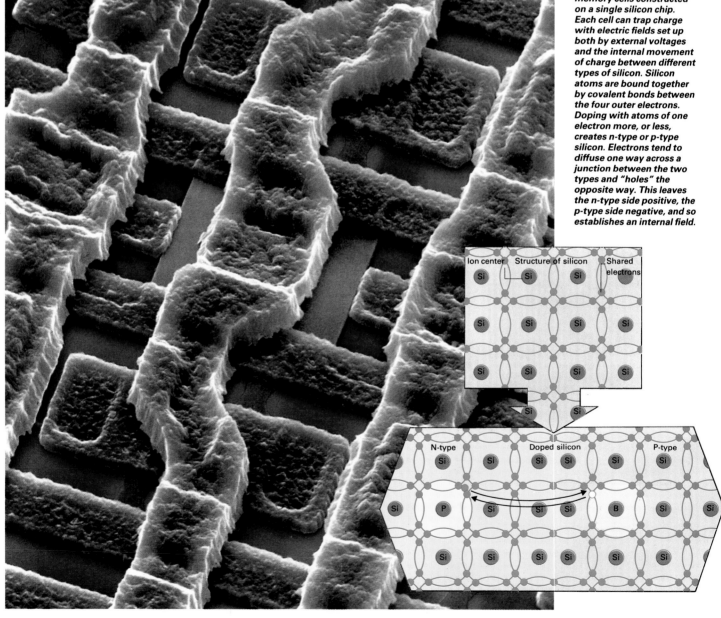

◄▼ *Semiconductor memories are thousands of memory cells constructed on a single silicon chip. Each cell can trap charge with electric fields set up both by external voltages and the internal movement of charge between different types of silicon. Silicon atoms are bound together by covalent bonds between the four outer electrons. Doping with atoms of one electron more, or less, creates n-type or p-type silicon. Electrons tend to diffuse one way across a junction between the two types and "holes" the opposite way. This leaves the n-type side positive, the p-type side negative, and so establishes an internal field.*

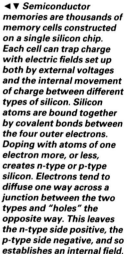

Using capacitors

Capacitors are used for a variety of purposes both to store electrical charge and energy and to control the response of electrical circuits. For example, a camera with an electronic flash unit contains a capacitor to store the energy for the flash. Moreover, capacitances as big as one farad are nowadays used in place of batteries to back up semiconductor computer memory during power failures. And because capacitors can store charge they can be used as tiny "dynamic" memory elements in microchips.

Capacitors also have important applications in conjunction with other circuit elements. In particular, one of their key functions is to control varying electric currents (♦ page 68). They are also used, for instance, in "filters" to remove unwanted signals in sensitive pieces of equipment such as television sets and audio amplifiers.

Semiconductors and silicon chips

In a metal the outermost one or two electrons in each atom are free to move through the material under the influence of an electrical force. In an insulator none of the electrons are free to move in this way. There is also a third class of materials, which are insulating at low temperatures, but which become conducting as the temperature rises. At higher temperatures, the electrons are more energetic, and some become free to drift readily through the material, just as in a metal. Such materials are "semiconductors". Understanding conduction in semiconductors and indeed in conductors, has been possible only by applying quantum mechanics to the solid state.

One way to provide a semiconductor with mobile electrons is to add ,or "dope", small quantities of another element – an "impurity" – in which the atoms contain more outer electrons. It is also possible to add atoms that contain fewer outer electrons than the basic semiconductor. In this case, the impurity atoms capture electrons from the semiconductor, leaving "holes" among the outer electrons in the atoms of the semiconductor. These holes simulate positive charges and can thus drift freely under the influence of an electric field, just as the negatively-charged electrons do.

Electronic engineers can accurately control the electrical properties of a semiconductor by doping it with impurities that donate extra electrons, or which mop up electrons and thereby provide holes. Further control is possible through the appropriate application of voltages. Thus in a transistor, the current flowing through the semiconductor can be altered or even switched off by suitably applying a voltage that is different from the one maintaining the flow of current. This effect allows transistors to operate as switches and as amplifiers.

Silicon (Si) is an important semiconductor, which can be grown into large single crystals – as big as 10cm across. A thin slice of such a crystal is the basis of the silicon chip. Impurity atoms can be diffused into the surface of the silicon wafer to make a very large number of tiny circuit elements such as transistors or capacitors, on the same chip. With appropriate connections, vast amounts of information can be stored and processed.

Both the voltage and the current fall to zero as the charge on the plates decreases, rapidly at first and then more slowly. The time taken for the capacitor to charge up and to discharge is inversely proportional to the resistance and to the value of the device's capacitance.

The capacitance can be increased by filling the space between the plates of a capacitor with a "dielectric". This is a material that does not conduct much electricity, but can be electrically polarized – the molecules within it can be pulled apart a little by an electric field. Inside a charged capacitor, the negative charges in a dielectric are pulled towards the positively-charged metal plate, and the positive to the negatively-charged plate. Thus, a layer of negative charge forms near the positive plate and a layer of positive charge near the negative plate. Both the electric field between the plates is reduced and the potential difference between them. The plates' charge is the same, however, so the capacitance – charge divided by potential difference – increases. Commercial capacitors contain dielectrics to give higher capacitances.

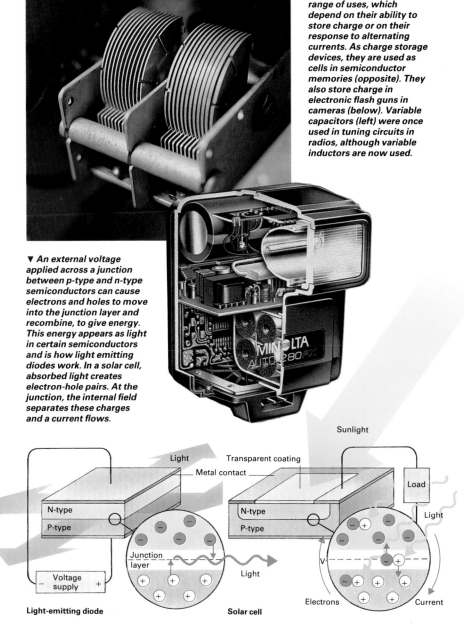

◄▼ *Capacitors have a range of uses, which depend on their ability to store charge or on their response to alternating currents. As charge storage devices, they are used as cells in semiconductor memories (opposite). They also store charge in electronic flash guns in cameras (below). Variable capacitors (left) were once used in tuning circuits in radios, although variable inductors are now used.*

▼ *An external voltage applied across a junction between p-type and n-type semiconductors can cause electrons and holes to move into the junction layer and recombine, to give energy. This energy appears as light in certain semiconductors and is how light emitting diodes work. In a solar cell, absorbed light creates electron-hole pairs. At the junction, the internal field separates these charges and a current flows.*

Light-emitting diode

Solar cell

Electrolysis

When an electric current is passed through a solution of ionic solid such as a metal salt, the different ions in the solution migrate towards either the positive electrode (called the anode) or towards the negative electrode (called the cathode). Negatively-charged ions move towards the anode and positively-charged ions move towards the cathode. When ions reach the electrodes they either take up electrons, if they have an overall positive charge, or give up electrons, if they have an overall negative charge, to form atoms or molecules. This phenomenon is known as electrolysis. It is used in industrial chemistry to form compounds such as sodium hydroxide (♦ page 128), and in electroplating to build up a coating of layer upon layer of metal atoms.

Under the influence of the electric current, ions in the solution move towards the anode and cathode. The positively charged metal ions travel to the cathode and are deposited on the object to be plated. The anode degrades to replenish the electrolyte with metal ions.

Faraday and electrolysis

The knowledge of electrolysis stems from the experiments performed by the British physicist and chemist Michael Faraday (1791-1867) in the first half of the 19th century.

Faraday determined the quantitative relationship between the amount of electricity passed through an electrolytic cell and the chemical change it produces. He showed that the weight of a substance converted by electrolysis is proportional to the quantity of electricity used. One mole of a substance that gains or loses one electron during electrolysis, for example, is deposited by 96,485 coulombs, or ampère seconds, of electricity. In general, 96,485 times n coulombs of electricity are needed to liberate one mole of substance – where n is the number of electrons lost or gained by each atom or molecule during the reaction. The quantity 96,485 coulombs is known as a Faraday but in practical electroplating the more important unit is the ampère-hour, a unit equivalent to 3,600 coulombs. One Faraday equals 26·8 ampère-hours.

Coating metals

Many metal objects and tools can be coated with a thin layer of another metal to make them more resistant to wear and attack by corrosion. These coatings can be applied by dipping in a bath of molten harder or more resistant metal, or by using an electrolytic technique known as electroplating. Electroplating can give an extremely thin and uniform coating to a host of metals.

Objects to be electroplated may be suspended from the cathode of an electrolytic cell in a bath or vat containing the solution of a metal salt – the electrolyte. An anode, often made from the metal to be electroplated, is also suspended in the vat and a low DC electric current is passed through.

All sorts of metal objects are electroplated, from nuts and bolts through to tableware and parts of automobiles. The metal fenders of automobiles are usually made of steel, and then electroplated with nickel and chromium. The nickel and chromium not only help to make the steel more resistant to corrosion, but they also give a shiny finish which is used as decoration. Silver-plated tableware is common. The silver coating protects the base metal of the tableware from attack.

Electron

▼ Many metals can be plated using simple plating baths. Often the process is continuous and an integral part of the manufacture of the metal object or tool. Strips of steel are electroplated with tin to provide the tin plate from which "tin" cans for food are made; some steel drills are given extremely thin plates of nickel and chromium to make them harder. Metal alloys can be electroplated if the metal salt solutions are mixed in the electroplating bath. Plated brass made in this way is indistinguishable from cast brass.

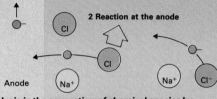

▶ Electrolysis is the separation of chemical species by passing a current through them, following a process that resembles that of an electric cell or battery. If a current is passed through a solution of common salt (NaCl) in water (H_2O), the negative Cl ion (♦ page 80) is attracted to the anode, where it gives up its excess electron and is given off as chlorine gas. Similarly, the positive Na ion is attracted to the cathode, where it picks up an electron and bonds with the water molecule to form sodium hydroxide (NaOH), while hydrogen gas is given off.

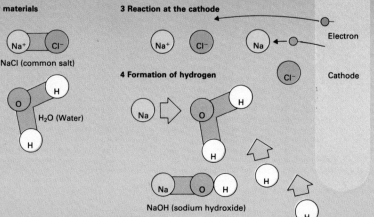

Anode

2 Reaction at the anode

1 Raw materials

NaCl (common salt)

H_2O (Water)

3 Reaction at the cathode

Electron

Cathode

4 Formation of hydrogen

NaOH (sodium hydroxide)

Electromagnetism

8

Magnets, currents and fields...The electromagnetic spectrum...Alternating current...Transformers... PERSPECTIVE...Michael Faraday...Motors and dynamos... Light and the electromagnetic spectrum...The transmission of power and signals...Inductors and capacitors

Electric charge in motion – electric current – is the origin of magnetic fields (◀ page 49). Indeed, magnetic fields are not truly distinct from electric fields; rather, both are aspects of a general "electromagnetic field". In 1820, the Danish physicist Hans Christian Oersted (1777-1851) discovered that a current flowing in a wire deflected a compass needle nearby. Before the end of the year, Ampère had extended Oersted's work and concluded that all magnetism is due to tiny electric currents. The form of the magnetic field generated by a tiny piece of wire carrying an electric current is complicated. The direction of the field is at right angles to the wire, and at right angles to the line joining the wire to the point in space at which the field is measured. The strength of the field is proportional to the current and to the length of the wire, but it also depends on distance according to the inverse-square law, and on the angles between the wire and the joining line. The strength of the field is greatest at right angles to the wire. To calculate the field due to the current in a specific wire, such as a loop or a coil, means adding all the contributions from tiny portions of the wire, but in some instances the answer has a relatively simple form. For example, the field due to a solenoid – a coil with many turns – resembles that of a bar magnet with a north pole at one end and a south pole at the other (◀ page 49). Indeed, the magnetic fields produced by permanent iron magnets are due to electric currents, not from coils of wire but from electrons spinning on a microscopic scale.

▲ *Faraday is known as one of the most outstanding experimenters of all time. He is remembered for his work on electromagnetism, static electricity and chemistry.*

Michael Faraday

Michael Faraday's life story is the classic tale of a great scientist's rise from humble origins. The son of a blacksmith, born in London in 1791, he became a bookbinder's apprentice at the age of 14. However, in 1813 he was taken on as assistant to the chemist Humphry Davy (1778-1829) at the Royal Institution in London. By 1825, Faraday had become Director of the Laboratory there.

Faraday was a prolific experimenter, and his name is attached to many phenomena both in physics and chemistry. Though no mathematician, his ideas on electric and magnetic forces gave birth to the theory of electromagnetic fields. Following the work on magnetism by the English scientist William Gilbert (1540-1603) (◀ page 51), Faraday conceived the idea of electric as well as magnetic "lines" of force, to depict the fields around electrified and magnetized objects. He was one of the first scientists to recognize the reality of these fields, and his work was later of great value to the Scottish physicist James Clerk Maxwell (1831-1879) in his synthesis of electricity and magnetism (◆ page 64).

Faraday's work ranged from the study of capacitance, to the chemical effects of electric currents in the process of electrolysis, to the discovery of benzene. His most important discovery is arguably that of electromagnetic induction. In 1831, he found that if a loop of wire moves through a magnetic field, then an electric current is induced in the wire. This is the principle that underlies the production of electricity.

Electricity and magnetism

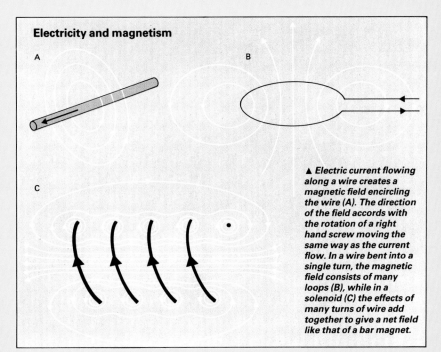

A

B

C

▲ **Electric current flowing along a wire creates a magnetic field encircling the wire (A). The direction of the field accords with the rotation of a right hand screw moving the same way as the current flow. In a wire bent into a single turn, the magnetic field consists of many loops (B), while in a solenoid (C) the effects of many turns of wire add together to give a net field like that of a bar magnet.**

The effect of a magnetic field on moving electrons is common to the workings of both generators and motors

The interaction of electrical and magnetic fields

The action of a magnetic field on an electrically-charged particle is wholly different from that of an electric field. The force exerted on the charged particle is proportional to its charge, to the velocity of the particle, and to the part of the field that is at right angles to the particle's motion. The force acts at right angles to both the magnetic field and the motion of the particle. Such forces can move matter in bulk, by acting on the electrons flowing in a current. This force underlies the electric motor, in which a coil of wire carrying a current is influenced by the field between the poles of a magnet.

The reverse effect is put to good use in the electric generator (◗ page 66), where mechanical energy is turned to electrical. The electric fields described earlier are generated by electric charges, and the magnetic fields by charge in motion – in other words by electric current. Electric fields are also generated by magnetic fields that change with time. The electric fields produced in this way are at right angles to the magnetic fields. In a similar way, magnetic fields are not only generated by electric currents, but also by electric fields that change with time. The generated magnetic fields are at right angles to the changing electric fields, and again the magnetic field lines form closed loops about the changing electric field.

▲ *A reconstruction of one of the earliest devices to study the connection between electrical and magnetic fields, built by the physicist H. Pixii in 1821.*

▶ *A magnetic-levitation (maglev) train is an application of the electric induction motor; producing a linear rather than a rotational motion.*

▶ *Two rules, based on the directions of the thumb and first two fingers pointing at right angles, describe the directions of current flow, magnetic field and motion in motors and generators. The left-hand rule pertains to motors, the right-hand rule to generators. The forefinger points to the field, the thumb to the motion, and the middle finger to the current.*

Field

Force

Current

Field

Force

Current

Motion

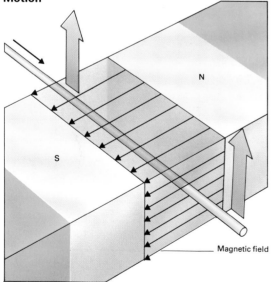

N

S

Magnetic field

▲ *When a wire is moved in a magnetic field, a current flows if the wire is part of a circuit, because the electrons in the wire experience a force. This is the principle that underlies the operation of generators.*

▶ *Many children's toys incorporate simple direct-current electric motors operating a low voltage, whether from batteries or adapted from the mains supply by means of a transformer (◗ page 67).*

Dynamos and motors

Generators (dynamos) and motors are related in the way they work. A dynamo converts energy of motion to electrical energy. An electric motor converts electricity to motion. The common factor between the two pieces of equipment is the effect of a magnetic field on moving electrons.

The electrons in a wire moving through a magnetic field feel a force as they travel through the field. The force sets the electrons in motion along the wire. This is how a dynamo works. The force on the electrons creates an electrical pressure like the water pressure generated by a pump.

In electric motors, a current is set up through a wire held in a magnetic field. The moving electrons then feel a force due to the magnetic field, which makes the wire itself move. If the wire is in the form of a loop, the forces acting on the two sides of the loop make it spin in the magnetic field, and can keep it spinning as long as the current continues.

A simple motor requires an alternating current to work – as the coil turns between the poles, so the current is reversed to ensure continued rotation. For direct current, a commutator reverses the current through the coil of each half rotation.

Direction of rotation
Flow of current through coil
Commutator

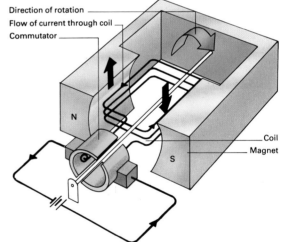

Coil
Magnet

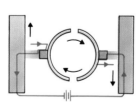

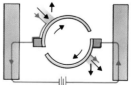

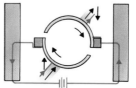

◄▲ In a simple direct current motor, current passes from a battery through one or more loops of wire held between the poles of a magnet. The magnetic field exerts a force on the electrons carrying the current and this makes the coil begin to rotate – as the force is in opposite directions on opposite sides of the coil. If the current remained in the same direction, the forces on the coil would make it rotate in the opposite direction once it was past the vertical position between the poles. To keep the coil rotating, the current must reverse direction between half-cycles. This is achieved by an arrangement of a commutator – a split ring with carbon brushes on either side as contacts.

The Electromagnetic Spectrum

Electromagnetic radiation

The fact that magnetic fields feed on changing electric fields and vice versa means that electromagnetic "waves" can propagate through space. The electric and magnetic fields in these waves are at right angles to each other and to the direction in which the wave is traveling. Such electromagnetic waves are familiar as light, radio-waves and X-rays. The only difference between these phenomena is the wavelength of the waves. The source of all such waves is the movement of electrons. Stellar radiation, generated by electrons moving randomly around in the hot gases at the stellar surface, can travel many light years across empty space. In a broadcasting station, however, the correlated motion of electrons in the trans-mitting antenna produces the radio waves.

► Electromagnetic waves consist of mutually perpendicular electric and magnetic fields that oscillate at the same frequency and propagate in the same direction at a velocity of about 300,000 km/s – the speed of light – in a vacuum. They are transverse, unlike the longitudinal sound waves (◊ page 26): the electric and magnetic fields oscillate at right angles to the direction of propagation. The waves arise as a result of the relationship between electric and magnetic fields, changes in one giving rise to changes in the other.

▼ Electromagnetic waves have a huge range of possible wavelengths or frequencies, giving rise to an electromagnetic spectrum that covers more than 19 decades in wavelength and frequency. The spectrum stretches from gamma rays at the shortest wavelengths through X-rays and visible light to microwaves and radiowaves. Shown below is the Orion Nebula seen with radiation in different regions of the spectrum, from the X-ray image on the left, through the optical and infrared pictures, to a radiowave image (right).

▲ ► In 1800, the British astronomer William Herschel (1738-1822), best known for discovering the planet Uranus, made another important discovery. He was studying the heating effect of the Sun's radiation, using a thermometer with a blackened bulb, when he found that beyond the red end of the visible spectrum, the thermometer continued to absorb heat and show an increased temperature. This was the first evidence for infrared radiation – "beyond" the red – with wavelengths longer than visible light. Infrared radiation is emitted by all hot objects, from stars to electric fires, and infrared imaging provides a way of "seeing" at night.

Wavelength (m)	10^{-16}	10^{-15}	10^{-14}	10^{-13}	10^{-12}	10^{-11}	10^{-10}	10^{-9}	10^{-8}
	Gamma rays						X-rays		Ultra-violet radiation
Frequency (Hz)	10^{24}	10^{23}	10^{22}	10^{21}	10^{20}	10^{19}	10^{18}	10^{17}	

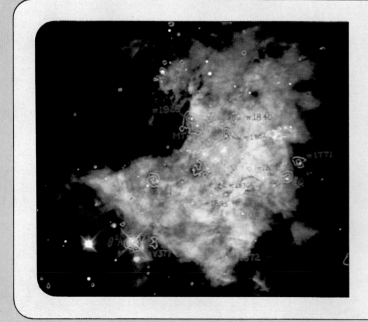

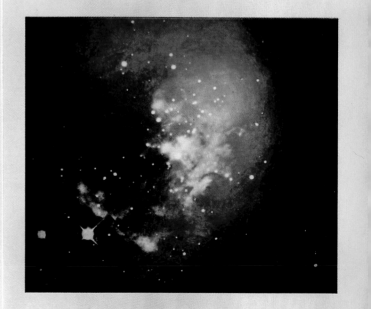

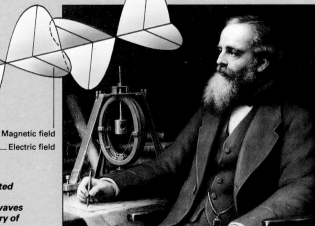

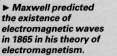

Direction of wave

Magnetic field
Electric field

▶ Maxwell predicted
the existence of
electromagnetic waves
in 1865 in his theory of
electromagnetism.

a b

A   B

The discovery of electromagnetism

In 1865, the Scottish physicist James Clerk Maxwell (1831-1879) brought together all the known laws of electricity and magnetism and reduced them to four basic equations. He thus revealed a symmetry between electrical and magnetic phenomena, and also introduced a new ingredient. It was already known that a changing magnetic field generates an electric field and that an electric current in a wire creates a magnetic field. Maxwell's extra claim was that a changing electric field can also produce a magnetic field. Including this effect meant that the equations can be seen to describe a kind of wave.

Wave motion is familiar as the ripples on water, but in the case of Maxwell's waves nothing is moving materially. The equations describe the changing wavelike pattern of fields across space.

An electrical or magnetic "disturbance" can set off such a pattern of changing fields. Maxwell found that the speed of propagation must be close to the velocity of light, and so he proposed that the two speeds are equal: in other words, light is a form of electromagnetic wave. In taking this step, Maxwell acknowledged the influence of the English scientist, Michael Faraday (1791-1867), who 20 years earlier had considered light as an electrical phenomenon.

Maxwell's idea of light as an electromagnetic wave was not fully accepted until 1889. The German physicist Heinrich Hertz (1857-1894) demonstrated that electric currents could generate "signals" – radio waves – that move at the speed of light. These waves exhibited the phenomena of reflection, refraction, diffraction and interference, that the wave theory of light had already explained.

▲ Maxwell's theory of electromagnetic waves was proved experimentally in the late 1880s by Heinrich Hertz. Hertz connected the terminals of an induction coil (lower left) so as to charge up a capacitor. Sparks eventually passed between the two small balls, discharging the capacitor, but in an oscillatory fashion due to the presence of the inductor in the same circuit. This oscillating current produced electromagnetic waves which could induce a spark between small spheres in an almost closed loop of wire some distance away.

| 10^{-6} | 10^{-5} | 10^{-4} | 10^{-3} | 10^{-2} | 10^{-1} | 10^{0} | 10^{1} | 10^{2} | 10^{3} |

| Visible light | Thermal radiation | | | Microwaves | | | | Radiowaves | |

| 10^{15} | 10^{14} | 10^{13} | 10^{12} | 10^{11} | 10^{10} | 10^{9} | 10^{8} | 10^{7} | 10^{6} |

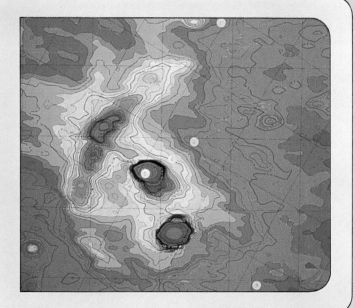

A message moves down a telephone wire at the speed of light, although the electrons in the wire travel at only 0.1 millimeters per second

Alternating current and transformers

The electricity supplied to most homes and factories is in the form of an "alternating voltage", which changes regularly between a positive value and an equal but opposite negative value with respect to a zero defined by the Earth's natural potential (referred to as "earth" or "ground"). Thus, whereas in direct current the electrons pass in a single direction down the wire, in alternating current they move back and forth as the potential changes. This alternating voltage is generated at power stations by rotating generators which extract energy from high-pressure water (hydroelectric generators) or from high-pressure steam. In the latter case the steam is heated by burning coal or oil, or with heat derived from nuclear fission (page 228).

The main reason for supplying electricity with an alternating voltage is related to the problem of transmitting electrical power with minimum losses. It generally pays to transmit at much higher voltages than are practical to generate or to use, so some device is needed to "step up" the voltage prior to transmission, and to "step down" the voltage at the receiving end. The device known as the "transformer" proves to be very efficient at changing voltages from one level to another, but it works only with alternating current.

In a typical transformer, two coils of wire surround the opposite sides of a square loop of iron. An electric current flowing in one coil generates a magnetic field, which is proportional in strength to the current and to the number of turns in a coil. The iron amplifies and traps this field which then threads the second coil. If the trapped magnetic field is changing – which it is if the current in the first coil is alternating – it generates a circulating electric field, which sets up a voltage within the second coil. This voltage is proportional to the rate at which the magnetic field is changing, and to the number of turns in the second coil. If the second coil has more turns than the first, the voltage is increased; if it has fewer turns, the voltage is decreased.

The AC generator

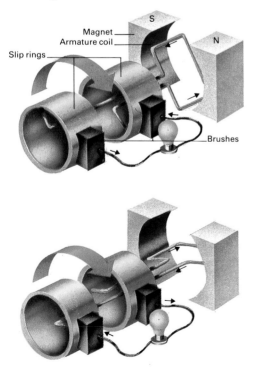

◄ *An alternating current generator resembles an AC motor: as a coil is turned by mechanical means between the poles of a magnet, a current is generated in the coil. As the coil turns between the poles the current is reversed – if the rotation of the coil is kept constant, a steadily alternating current results.*

► *A simple transformer consists of two coils wound on linked iron cores. An alternating voltage across one coil (the primary) sets up an alternating current in the coil and hence induces a changing magnetic field in the iron core. This varying field induces an alternating current in the other coil (the secondary). The size of the voltage in the secondary coil depends on the number of turns in the two coils. With the same number, the voltages are equal; with more coils in the secondary, the induced current is increased (a step-up transformer), while with fewer turns the induced current is decreased.*

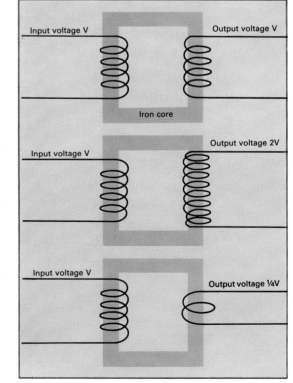

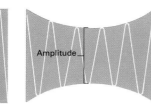

◄ *The Croatian-born physicist Nikola Tesla (1875-1943) emigrated to the United States in 1884, where he worked with T.A. Edison. In 1888 he developed the AC induction motor, and improved the techniques of power transmission.*

Transmission of signals and power

One of the most profound ways in which electromagnetism affects daily life is in communications. Electrical effects can pass information between two distant points by transmitting signals that can be interpreted as simple messages, as sounds, or even as color images together with sound, as in a television signal.

Signals can be transmitted by "modulating" electromagnetic waves (that is, by modifying them slightly) or by changing the level of current flowing along a wire. The earliest practical communication system of this kind was the telegraph, developed in 1844 by the United States inventor, Samuel Morse (1791-1872). In "Morse code", a message is translated into a series of "ons" and "offs", where there are two kinds of "on", one lasting for a longer time (a "dash") than the other (a "dot"). A compass needle at the end of a length of wire is sufficient to reveal if the current has been switched on or off at the other end.

Nowadays, radio and television signals, are transmitted via electromagnetic waves although ultimately they travel along wires. The signals in the electromagnetic waves induce currents in a TV or radio aerial which then propagate signals through a cable to the receiver. Local telephone calls are transmitted along wires, but international calls may involve a stage in which the signal is carried by electromagnetic waves to a relay satellite.

Electric currents can be used to carry electrical power, rather than information. Large quantities of power are distributed to homes and factories in towns and cities along wires – power cables – sometimes across distances of hundreds of kilometers. These cables have electrical resistance, which means that some energy is lost in transmitting the power. To minimize these losses, very high potential differences (voltages) are used. The power lost in overcoming resistance is proportional to the resistance multiplied by the square of the current. The transmitted power is equal to the voltage generated multiplied by the current supplied. Thus to maximize the transmitted power, while minimizing the power lost, it pays to keep the current low and the voltage as high as is practical – at 400,000V, for example. However, such high voltages are not suitable in most industrial or domestic applications. When the power lines reach their destination, the voltage is reduced to an appropriate level, with little power loss, by using transformers.

Both the signals transmitted along telephone lines and the energy transmitted along power lines travel at the speed of light, not at the speed at which the electrons drift through the wires (in the region of 0·1mm/s). The wires and the drifting electrons they contain act as guides for electromagnetic fields, which carry both signals and energy at the speed of electromagnetic waves, that is, the speed of light.

Frequency modulation

Amplitude modulation

Wavelength

Amplitude

◄ *A signal is carried on an electromagnetic wave by modifying the frequency of a wave with constant amplitude (FM), or by modifying amplitude with constant frequency (AM).*

▼ *Television signals are carried from transmitter to receiver as FM waves.*

Inductors

An electric current flowing through a coil of wire generates a magnetic field that threads the coil. If the current changes with time, the magnetic field also changes, and therefore generates an electric field. However, the force that the electric field exerts on the electrons is such that it opposes their motion and impedes the changing current.

If such a coil – known as an inductor – is connected to a constant voltage source, then the current through the coil will rise slowly to a value equal to the applied voltage divided by the resistance of the circuit. Unlike a capacitor, an inductor does not impede a steady current, although in practise the wire will have some resistance. However, the potential difference generated in the inductor by the changing current is proportional to the rate at which the current changes. Thus, again unlike a capacitor, an inductor will not pass alternating currents with very high frequencies. This contrasting behavior of inductors and capacitors means that the two devices can be used together as a "filter". Hifi equipment and other devices that seek to filter a clear signal out of its background "noise" make use of this effect.

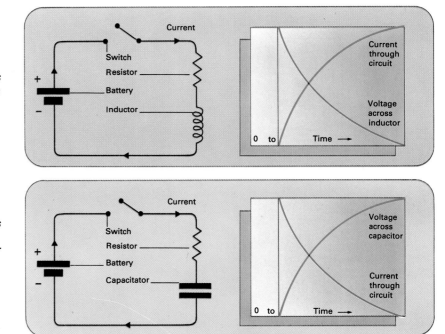

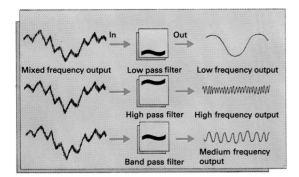

◄ **The complementary responses of capacitors and inductors can be combined together in filter circuits which operate on alternating current. The values of capacitance and inductance can be chosen so that the circuit passes current only at low frequency, high frequency, or in a range of frequencies. Conversely, the filter can be designed to block unwanted signals of particular frequencies.**

▲ **Inductors and capacitors respond in opposite ways when a voltage source in the circuit is switched on. In an inductor (top) the current rises as the voltage across the coil falls. With a high frequency AC supply, the inductor "blocks" the current. In a capacitor, the current falls as the voltage rises to its full value, but at high frequencies the capacitor "passes" current.**

▼ **The electromagnets that guide beams of subatomic particles around accelerators require DC voltage, which can come from the grid, but only after rectification, converting AC to DC. This process still leaves "ripples" on the DC voltage, which must be reduced to less than 0·01 percent of the total voltage for the magnets to produce fields stable enough to keep the particles on track. Filters are therefore crucial.**

Atoms and Elements

9

The elements...Theories of the atom...Structure of the atom...Nuclei and electrons...Atomic weight and number...Electron shells and orbitals...The Periodic Table...PERSPECTIVE...Dalton's atomic theory... Calculating atomic mass...Predicting unknown elements...Features of the Periodic Table

Every substance in the universe, from hydrogen to the complex proteins that are essential to life, is made of either a pure element, or a number of different elements bonded together to form molecules (♦ page 77). The distinction between molecules and elements is not an obvious one, but the identification of elemental materials, notably some metals, was achieved by the alchemists. In the 18th century the French chemist Antoine Lavoisier (1743-1794) identified the element oxygen as being responsible for combustion in air, and more elements were discovered throughout the 19th century.

The study of matter on the molecular or elemental level was greatly assisted by the revival of the atomic theory by the English chemist John Dalton (1766-1844). According to this theory, all matter is made up of atoms. The atom is one of the basic structural units of matter, and is the smallest particle of an element that can take part in chemical combination to form a compound. This concept was originally proposed by the Greek philosopher Democritos (*c.* 400 BC), who argued that there existed minute particles of which all matter was made. These particles, he suggested, were indivisible and indestructible. The word atom is derived from the Greek word meaning "that which cannot be divided".

Today it is known that, although atoms are the smallest fragments into which an element may be divided without losing its particular properties, they are not themselves the fundamental units of matter. Rather, they have a structure and are made of more fundamental particles – protons, neutrons and electrons. At the center is a very small, dense nucleus which contains the relatively heavy protons and neutrons (♦ pages 187-194). These have almost equal masses, but protons possess a positive charge, whereas neutrons have no charge at all. Overall, however, atoms are electrically neutral, since the positive charge of the protons is balanced by the negative charge of the electrons. These exist at a relatively large distance from the nucleus, and the space within which they can move is, in atomic terms, vast. We know from mathematical calculation of probability that electrons exist in volumes around the nucleus; these volumes are known as atomic orbitals (♦ page 72).

Thinking of the atom

In 1802 the English chemist John Dalton (1766-1844) presented his theory of atoms to the Literary and Philosophical Society of Manchester. Although the ancient Greeks had thought of atoms, Dalton was the first to provide evidence for them, based on the relative combining weights of the elements.

Dalton's atomic theory was published in 1805. In it he said that all matter is made up of particles which are indestructible and indivisible. The atoms of an element are identical in weight and chemical properties; but atoms of one element are different from those of all other elements.

Dalton argued that when elements combine to form compounds, they do so in simple ratios. Scientists had realized that compounds had fixed compositions and that certain weights of the elements combined to form compounds; they had already drawn up a series of "combining weights". But Dalton was the first to see that elements combine in simple ratios. He also saw that when compounds decompose atoms can be released of the elements that they comprise, and that these can be used to take part in further reactions, unaffected by compound formation and decomposition.

Dalton also published atomic weights of the elements which were, unfortunately, often wildly inaccurate. But his ideas were used to calculate formulae for chemical compounds. Slowly, though, evidence accumulated that atoms were not the indestructible spheres that he had thought.

The elements of water

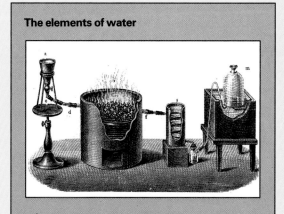

▲ Antoine Lavoisier used this apparatus to show in 1783 that water is made up of hydrogen and oxygen. The English chemist Joseph Priestley (1733-1804) had prepared oxygen in 1774 but did not recognize its role in combustion; Lavoisier named it oxygen and showed that combustion consisted of oxidation (♦ page 92).

► Dalton saw atoms as solid objects like billiard balls. When Thomson discovered the electron, he imagined electrons as "plums" in a "pudding" of positively charged matter. Rutherford discovered the nucleus with electrons in orbit (♦ page 187); Bohr showed these orbits are arranged in distinct "shells" (♦ page 73).

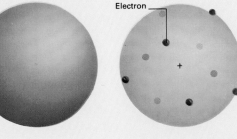

Dalton: "billiard ball" 1803

Thomson: "plum pudding" 1901

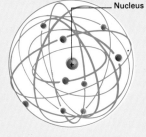

Rutherford: "electron cloud" 1911

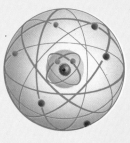

Bohr: "shell" 1913

Until the 19th century only a handful of chemical elements had been clearly isolated

Transmutation of the elements

Elements can change into other elements, or transmute. Medieval and oriental alchemists were continually trying to find a way to transmute base metals, such as lead, into silver and gold. The transmutation of some metals takes place naturally, but by performing nuclear reactions in laboratory conditions the elements can also be transmuted artificially. Radioactive isotopes decay in various ways, emitting particles and energy from their nuclei. Three natural radioactive decay series are known, starting from isotopes of uranium and thorium. One series starts from the isotope of uranium which has 238 protons and neutrons in its nucleus (written as ^{238}U or U-238); the second from the uranium isotope with 235 protons and neutrons (^{235}U); and the third from the thorium isotope ^{232}Th. All the natural decay series end with a stable, or non-radioactive, isotope of lead.

Each of these natural series occurs through the sequential decay of radioactive isotopes with the emission of alpha particles or beta-minus particles (♦ page 190).

▲ With their primitive apparatus, seen in this 18th-century painting, alchemists could perform only the most rudimentary chemical experiments. They searched for the elixir of life, and the "philosopher's" stone, to turn base metals into gold. Despite a few claims to the contrary, this conversion of one element to another was a hopeless task until Lord Rutherford detected nuclear decay (♦ page 188).

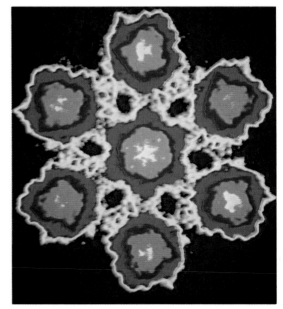

► Modern techniques permit the photographing of individual atoms which are only 30 billionths of a centimeter across, like this group of seven uranium atoms. The picture was taken using a scanning tunneling microscope (STM), which was developed in the 1980s.

The mass spectrometer

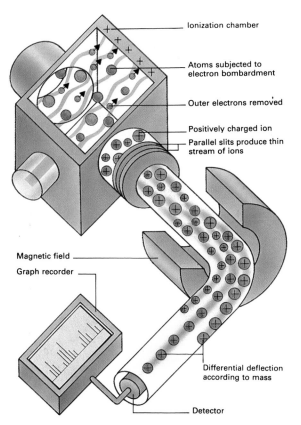

Ionization chamber

Atoms subjected to electron bombardment

Outer electrons removed

Positively charged ion

Parallel slits produce thin stream of ions

Magnetic field

Graph recorder

Differential deflection according to mass

Detector

▲ ► *The mass spectrometer can count and weigh atoms and molecules. Within its vacuum chamber, a substance is ionized by a strong electric field. This knocks off some of its atoms, and some of their electrons, giving the fragments a positive charge. As these charged ions move out of the chamber and through a magnetic field they are deflected, to an extent determined by their mass and charge. A detector records the mass spectrum. Even though most molecules are fragmented, once the pieces are identified they can be reassembled to discover the original molecule.*

Atomic masses

Atoms are very light, so scientists use a shorthand notation to represent their mass. This shorthand gives a scale of the masses of atoms of the elements starting with hydrogen as the lightest, but using the common isotope of carbon, ^{12}C, as a reference. Even using this notation, none of the atomic masses calculated are whole numbers. One might expect the atomic mass of carbon to be 12 on the scale, as this is the reference mass. But in fact carbon does not exist as a single isotope, and the atomic mass calculated for it represents an average of the masses of the two isotopes for carbon found naturally – ^{12}C and ^{13}C. Some 98·89 percent of carbon is found as ^{12}C and 1.11 percent as ^{13}C: if a weighted average of these percentages is taken, the figure of 12·011 is calculated. Other atomic masses are calculated in a similar manner, but relative to the mass of carbon. Not all elements occur naturally as mixtures of isotopes. The atomic masses of these elements are not weighted averages, but are simply calculated with reference to the mass of ^{12}C.

The atomic nucleus

Atoms range in size from 0·1 to 0·5 nanometers (nm). Nuclei are some 100,000 times smaller, and have a diameter of roughly 10^{-6}nm. Most of the mass of the atom is concentrated in the nucleus, and is made up of the masses of protons and neutrons. The masses of protons and neutrons have been measured to be about $1·674 \times 10^{-27}$ kilograms. An electron weighs about 9.110×10^{-31} kilograms and is 1,836 times less heavy than a proton.

The atoms of an element are identical, but different from those of all other elements. Hydrogen atoms are the simplest and lightest: they comprise a nucleus of a single proton, and one electron. The heavier elements have a more complex nucleus, with increasing numbers of protons. The number of protons in the atom characterizes it, and is known as the atomic number. Hydrogen, with one proton, has the atomic number 1, and uranium has 92. Elements with even more protons are known, but they can usually only be produced artificially by nuclear reactions.

Atoms and isotopes

The mass of an atom is another frequently quoted property. This is the sum of the number of protons and neutrons in the nucleus. The number of neutrons in the atoms of a particular element can vary, whereas that of protons cannot. Chlorine atoms, for example, always have 17 protons, but there are two distinct forms of chlorine atom known to exist naturally. In one, the nucleus contains 18 neutrons; in the other, 20.

These two forms of chlorine are examples of isotopes. Some isotopes are unstable and degrade to form isotopes of other elements, and emit radiation as they do so (♦ page 191). The isotopes of some elements in particular are especially stable, and this is found when the number of neutrons in the nucleus is in the series 2, 8, 20, 28, 50, 82, 126... . The numbers in this series are known as magic numbers. A magic number of either neutrons or protons makes the nucleus of an isotope stable. In rare cases, isotopes with a magic number of both neutrons and protons are found; the lead isotope with 82 protons and 126 neutrons is one example, and is particularly stable.

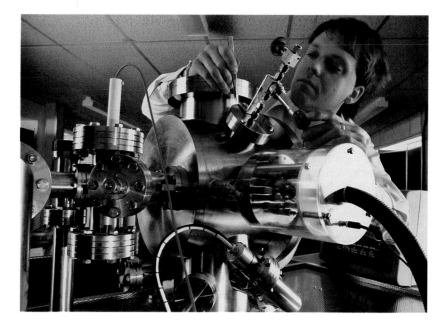

The arrangement of electrons around atoms other than the simplest is too complex even for large computers to describe

Electrons and their orbitals

Electrons are fundamental particles with negative charge, which were once thought to orbit the nucleus much like the planets orbit the Sun in our Solar System. The Danish physicist Niels Bohr (1885-1962) was the first to suggest this, basing his ideas on the emission spectrum of hydrogen. Bohr postulated that electrons moved in clearly defined orbits around the nucleus, and that in each of these orbits they had a certain energy (◗ page 196). Electrons absorb or emit energy by jumping between orbits. This theory was able to account for the spectrum of hydrogen, but it could not be used to explain the emission spectra of the heavier elements.

However, later work took into account his ideas and led to the current thinking on the behavior of electrons in atoms. Using a wave equation for particles devised by Erwin Schrödinger (◗ page 198), volumes of space can be mapped out within which electrons with a particular energy can be said to exist, even though their exact location cannot be stated. These volumes are known as *atomic orbitals*. For hydrogen with its single electron, a number of these volumes can be derived. Each type of volume, or orbital, is named according to lines observed in the hydrogen emission spectrum – sharp (s), principal (p), diffuse (d) and fundamental (f).

The simplest atomic orbital is spherically symmetrical around the nucleus and is known as the s orbital. Electrons with more than one energy can exist in larger spherically symmetrical orbitals or in other orbitals which have different shapes. The so-called p orbitals have twin lobes. Electrons can exist in either of these lobes, but the probability of finding an electron near to the nucleus tends to zero. There are three p orbitals for hydrogen atoms and these are mutually perpendicular – lying along the x, y and z cartesian coordinates. Five d orbital volumes have been calculated. These volumes again are strongly directional, and consist of lobes around the nucleus. There are seven f orbitals with shapes which are very hard to represent graphically.

The "buildup" principle

Although the Schrödinger wave equation for the sole electron in a hydrogen atom can be solved fairly easily, it is less simple for atoms with more than a few electrons, even using the largest computers.

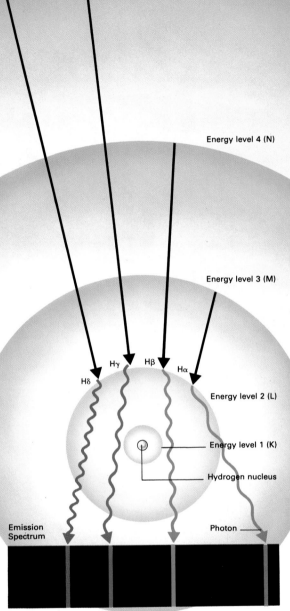

Energy level 5 (O)

Energy level 4 (N)

Energy level 3 (M)

Hγ   Hβ   Hα

Hδ

Energy level 2 (L)

Energy level 1 (K)

Hydrogen nucleus

Emission Spectrum

Photon

▲ Around the hydrogen atom there are several energy levels. The sole electron of hydrogen resides in the lowest energy level (K), nearest the nucleus, but it can be excited to the outer levels. On returning to a lower level it emits a photon of light. When it returns to the L level it emits visible light where a series of four lines are found. In the ultraviolet region, lines correspond to the electron returning to its original K shell.

◀ Joseph J. Thomson (1856-1940) discovered the electron. Using a cathode ray tube he was able to show that all elements contained these particles.

▶ Although electrons are particles they also behave like waves. Thus two pictures of electrons in atoms have emerged. The particle model has electrons orbiting the nucleus, like planets round a sun, and these are shown for some of the lighter elements (right). The first orbit can hold two electrons, the second eight, and so on. The wave picture gives a very different image (upper right). The wave form of an electron depends upon a series of quantum numbers. The shapes are called orbitals of which there are four types: s spherical (1); p dumb-bell (2); d four-leaf clover (3) or hour-glass and ring (4); and f very complex.

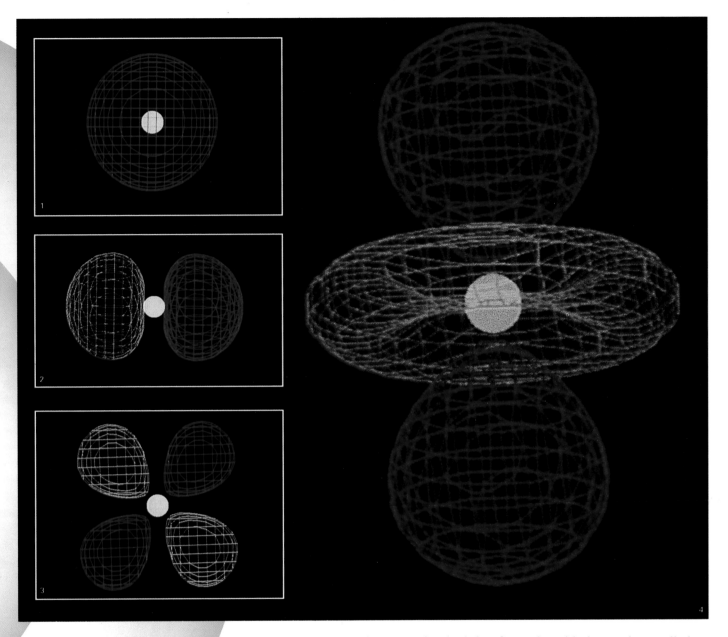

The "buildup" of electrons

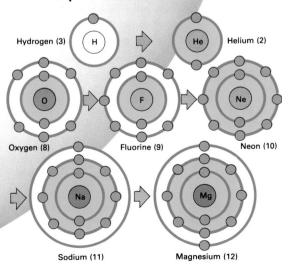

Hydrogen (3)   H   ⇒   He   Helium (2)

Oxygen (8)   Fluorine (9)   Neon (10)

Sodium (11)   Magnesium (12)

Even so, the general principle of atomic orbitals can be applied to atoms other than hydrogen. Orbitals similar to those of hydrogen are assumed to exist for all other elements, and models of these elements' atoms are built up by adding one electron at a time. This principle, developed by the Austrian-born physicist Wolfgang Pauli (1900-1958), is known as the Aufbau, or buildup, principle. Pauli proposed that no two electrons could exist in an atomic orbital with the same energy. However, they can exist side by side if they have opposite spin (♦ page 201). As electrons are added to atomic orbitals they pair up, and their spins are mathematically canceled out. Thus each orbital can contain only two electrons. If electrons have two equal possibilities, they move into an orbital of their own. Thus the p, d and f orbitals of electrons other than hydrogen in their normal state fill individually before any electron pairing takes place. This buildup of electrons in atomic orbitals allows scientists to explain many of the chemical and physical properties of the elements, and lies at the root of the modern understanding of chemistry (♦ page 74).

The Periodic Table

The Periodic Table

If the elements are arranged in order of increasing atomic number, surprising correlations are found between their chemical and physical properties. Elements with similar properties are found at definite intervals of atomic number, or periods, and it is possible to draw up a table, known as the Periodic Table, which shows these similarities.

In a Periodic Table the elements are separated into horizontal periods and vertical groups. Elements in each group are found to be very like one another. Across the periods of the table, the properties of the elements are seen to change steadily. The table shows the distinction between metallic and nonmetallic elements and, when first introduced, it allowed chemists to predict the existence of elements which were then unknown. The Periodic Table includes all the naturally occurring elements but, by extrapolating from the known elements, it can be used to predict the properties of as yet unknown superheavy elements.

The origins of the Table

The Periodic Table originated from the desire of chemists in the 19th century to present the observed similarity in chemical and physical properties of the elements in a systematic way. By the middle of that century almost 70 of the 92 naturally occurring elements were known. Substances such as iron (Fe), tin (Sn), lead (Pb) and gold (Au) had been known as elements for some time, but it was very hard to prove that an element was actually an element. It was easier to show that an element was not an element than that it was. Only with the advent of spectroscopy and modern laboratory analytical techniques has conclusive proof been available.

However, the Russian chemist Dimitri Mendeleev (1834-1907) first illustrated the periodic nature of the elements by placing them in order of increasing atomic weight. In his periodic law, he said that the properties of the elements vary in a systematic way, and are related to the atomic weights. Mendeleev arranged all the elements he knew into a table of eight vertical groups and 12 horizontal periods. Often he left gaps when the atomic weight of an element did not fit his scheme. These were later filled in by newly discovered elements. For example, he predicted that an element which he called ekasilicon should exist, placed in the table below silicon (Si) and above tin (Sn). This was later discovered, and is now known as germanium (Ge); it occupies the place in the table that Mendeleev predicted, and has physical properties similar to both silicon and tin. Mendeleev's work was very accurate, and led to the search for new elements and for improved values of atomic weight and density of known elements. Several new elements have been discovered, and predictions made of new transuranium elements.

Nevertheless, his ideas were not entirely correct, as an element's properties depend on its atomic number – the number of protons in the nucleus – rather than the atomic weight. The periodic nature of the configuration of the outermost electrons of atoms explains observed similarities in chemical and physical properties.

◄ **The s block consists of group 1, the alkali metals, and group 2, the alkaline earth metals. These elements are known for their chemical reactivity and are only extracted as metals by processes that require high energy. They have one or two outer electrons which they can relinquish to form positive ions such as Na^+ or Ca^{2+}, and this is how they are found in nature, as salts.**

▼ **The d block elements are all metals, in which the d shell, underlying an outer s shell, is being filled with electrons. In their compounds these elements exhibit several oxidation states, depending on the number of their electrons transferred to other atoms. Thus manganese ranges from two [Mn(II)] as in Mn^{2+} salts, up to seven [Mn(VII)] as in potassium permanganate, $KMnO_4$.**

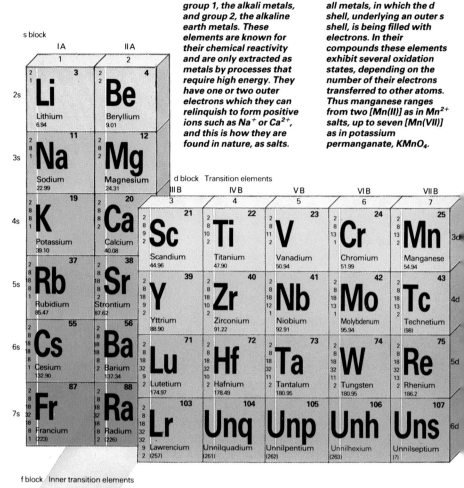

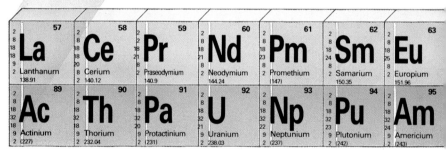

▶ **The Periodic Table reflects the electron make-up of the atom. Electrons occupy shells around the nucleus. The inner shell can hold two electrons, the next eight, and so on, up to 32 for the outer shells. Within these shells are subshells of 2, 6, 10 or 14 electrons with the symbols s, p, d and f – terms originally from lines in the atomic spectrum. The Aufbau principle sees the table built up by a step-by-step addition of electrons to atoms, from the nucleus outwards. When a shell is full it results in an atom of great chemical inertness – the rare gases.**

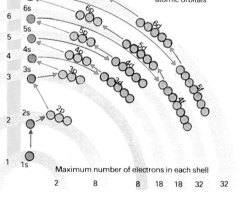

The Aufbau principle

Order of filling of atomic orbitals

Energy level

Maximum number of electrons in each shell

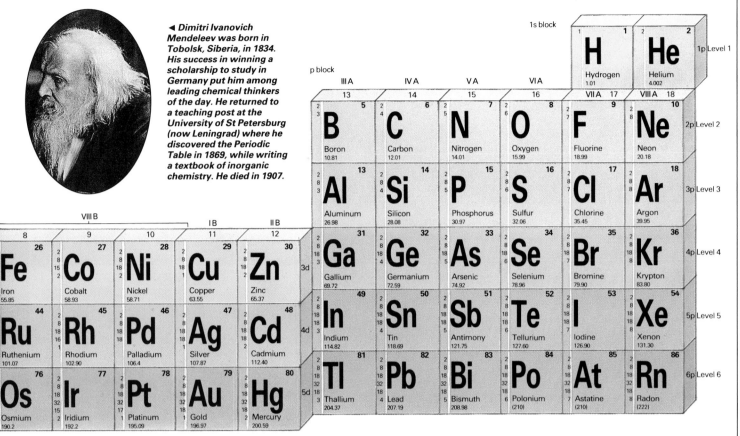

◄ Dimitri Ivanovich Mendeleev was born in Tobolsk, Siberia, in 1834. His success in winning a scholarship to study in Germany put him among leading chemical thinkers of the day. He returned to a teaching post at the University of St Petersburg (now Leningrad) where he discovered the Periodic Table in 1869, while writing a textbook of inorganic chemistry. He died in 1907.

KEY

Electronic configuration (principal levels) — Group number
Atomic number
— Orbital designation
— Chemical symbol
— Name of element
— Relative atomic mass

Period
Group

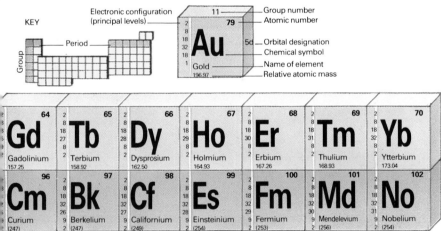

◄ The f block metals consist of the lanthanides (top row), and the actinides (bottom row). The lanthanides are also called the rare earths, although some are quite abundant (cerium is more common than lead). They are chemically very similar, a fact that led to much confusion among 19th-century chemists of the last century. The actinides are all radioactive elements, but thorium and uranium have sufficiently long half-lives that deposits of these elements are still found. The elements beyond uranium are all artificial.

▲ The p block is divided diagonally into metals and nonmetals. The nonmetals bond to one another by electron-sharing to form discrete molecules – the basic components of matter. Their chemical behavior is determined by the completeness of their shells. The elements of group 18 have a filled shell and are most stable, while those of group 17, the halogens, are extremely reactive. The 1s block, hydrogen and helium, are at this end of the table because both are nonmetals and helium is an inert gas.

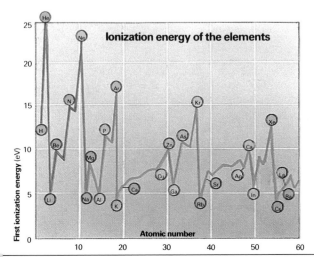

◄ Elements arranged in a line according to their atomic number show repetition of certain properties. This is illustrated by their ionization energies – the energy needed to pluck the outermost electron away from an atom. Some elements are very loath to give up their electrons, especially if they have a full shell. Thus in the graph the pinnacles are all capped by the rare gases. The s block elements of group 1 offer the least resistance to losing their sole outer electrons, and so these are at the low points.

Using the Table to predict new elements

The Periodic Table has always been useful in predicting the existence of as yet unknown elements. Fourteen artificial, or transuranium, elements have been discovered. Elements with the atomic numbers 93-103 are the remaining elements in the actinide series; elements 104-118 begin a new transactinide series: in an expanded form of the Table these may be placed under the row beginning with hafnium and ending with radon.

The chemical properties of these transactinide elements should be similar to those in the groups of the representative elements in which they have or might be found. Around element 121 it has been predicted that a "superactinide" series might start. In this series, which would be similar to the actinides, the g orbital would fill with electrons.

See also
Molecules and Matter 29-38
The Structure of Matter 77-90
Chemical Reactions 91-8
Studying the Nucleus 187-94
The Quantum World 195-204

The structure of the Table

The modern Periodic Table represents three distinct types of elements, called representative elements, transition elements and inner transition elements. The representative elements are perhaps the most familiar. They include both metals, such as tin (Sn) and lead (Pb), and nonmetals such as oxygen (O), sulfur (S) and chlorine (Cl). The transition metals are a series of metallic elements, including iron (Fe), nickel (Ni), copper (Cu), zinc (Zn) and gold (Au), with properties that are very similar to one another, but which definitely change as the atomic number rises. The inner transition metals, sometimes called rare earths or lanthanides and actinides, have very similar properties and have often been difficult to separate and identify.

The groups of the Periodic Table may be numbered 1 to 18, or the representative elements arranged into eight vertical groups numbered I to VII and O. The transition metal groups can also be numbered I to VIII. The elements are arranged into seven horizontal periods.

The group 18 elements – helium (He), neon (Ne), argon (Ar), krypton (Kr), xenon (Xe) and radon (Rn) – lie to the right of the table. Collectively they have been called the noble or inert gases, because of their lack of reactivity; now they are more often called the rare gases. These elements are unreactive because their atoms have a complete outer shell of electrons, filled with six electrons. This is an extremely stable structure and it is very hard to remove electrons to allow chemical bonding and the formation of compounds.

To the left of the table lies a group known as the alkali metals – lithium (Li), sodium (Na), potassium (K), rubidium (Rb), cesium (Cs) and francium (Fr). All of these are very reactive. The atoms of each contain one s-orbital electron in the outermost electron shell; this electron is easily lost, giving rise to a positively charged atomic particle or ion with the structure of an inert gas. This can take part in chemical bond formation. Group 17 comprises fluorine (F), chlorine (Cl), bromine (Br), iodine (I) and astatine (At). Their atoms have almost full outer p orbitals containing five electrons. They can easily gain an electron to form a negatively charged ion, again with the structure of an inert gas. Such ions can also readily take part in bond formation.

This gain and loss of electrons to form ions is shown across the periods of the table. Elements are grouped together because of their physical and chemical similarities, and their ability to form ions with the same positive or negative charge. The ability of an atom to lose or gain an electron, its "ionization energy", increases along the periods and decreases down the groups, but it is only one of many properties that change "periodically".

Most elements are metals, but groups 13 to 16 contain elements neither metallic nor nonmetallic in character – the semimetals. In each of these groups the metallic nature of the elements increases down the group. In group 13, for example, boron (B) is a semimetal whereas aluminum (Al), gallium (Ga) and other elements in the group are all metals. The other semimetallic elements are silicon (Si) and germanium (Ge), arsenic (As) and antimony (Sb), and tellurium (Te).

Metals and nonmetals

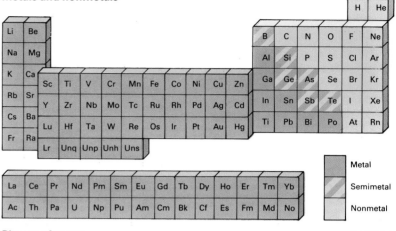

Phases of matter

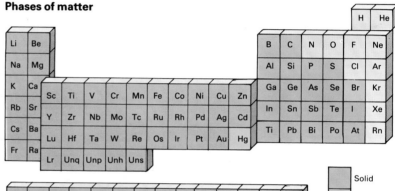

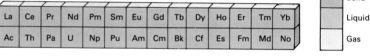

Acidity

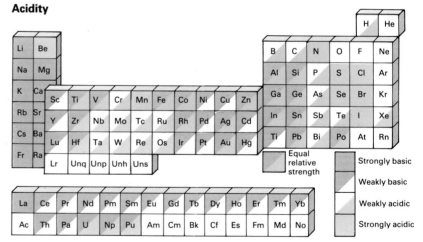

▲▲▲ *Recognizing a metal is fairly easy. It is hard, dense, shiny and gives out a characteristic "ping" when struck. Chemists, however, define a metal by its ability to conduct electricity. Conductivity stems from the free moving electrons that are a feature of the chemical bonding in metals. A few elements are semi-conductors, and these straddle the boundary between the metals and non-metals.*

▲▲ *The three states of matter, solid, liquid and gas, are very dependent upon temperature and pressure. At 25 Celsius and atmospheric pressure, the conditions that humans find most agreeable, only two elements in the middle table are depicted as liquids – bromine and mercury. At 30 Celsius cesium and gallium would also have been shown as liquids. These melt at 28·5 and 29·8 Celsius respectively.*

▲ *Almost all the elements form oxides, and how these behave in water is a guide to the chemistry of the element. If the oxide reacts with water to give an alkaline solution then it is described as a basic oxide; if it reacts to form an acid solution it is described as an acidic oxide. Metals give basic oxides, non-metals give acidic ones. Amphoteric oxides, such as aluminum oxide, show both kinds of behavior.*

The Structure of Matter

10

Compounds and mixtures...Types of chemical bonding...Metals and metallic bonding...Salts and ionic compounds...Covalent bonds...Carbon at the heart of large molecules...Diamonds and graphite...Organic compounds...Hydrogen bonding...Substances in solution...PERSPECTIVE...Ways of depicting molecules...Spectroscopy...Classes of organic compounds

The countless different substances that exist in the Universe consist of collections of fewer than 100 naturally-occurring elements. The structure and properties of each of these substances – whether they are hard or soft, colored or transparent, liquid or solid under normal conditions – reflect the way in which the atoms that make them up are combined. The ways in which elements can combine to form compounds is therefore crucial to any study of chemistry.

As atoms of different elements approach one another their outermost electrons can interact. When this happens a *chemical reaction* is said to occur (◆ page 99), and electrons can be shared between two or more atoms, or exchanged. This sharing or exchange creates a strong link between the atoms known as a *chemical bond*, and leads to the formation of a *molecule*. A molecule may comprise a single element, as in the case of the hydrogen molecule H_2, or it may involve numbers of different elements, in which case it forms a chemical compound. Several types of chemical bond are known, which between them are responsible for the millions of known chemical compounds.

Scientists in the 19th century knew that atoms combine in simple ratios to form compounds. This combining capacity is known as valence. The valence of an atom indicates the number of electrons the atom has available for bonding in its outermost electron shell (◆ page 73). Oxygen has a combining capacity, or valency, of two in H_2O and in most other compounds; chlorine has a combining capacity of one. Sulfur has a valency of four in sulfur dioxide (SO_2) and six in sulfur trioxide (SO_3). Knowledge of the valence of an atom in different compounds and configurations allows the chemist to predict the shape of the resulting molecules. Calculations of the behavior of valence electrons thus provide the backbone of theories of chemical bonding.

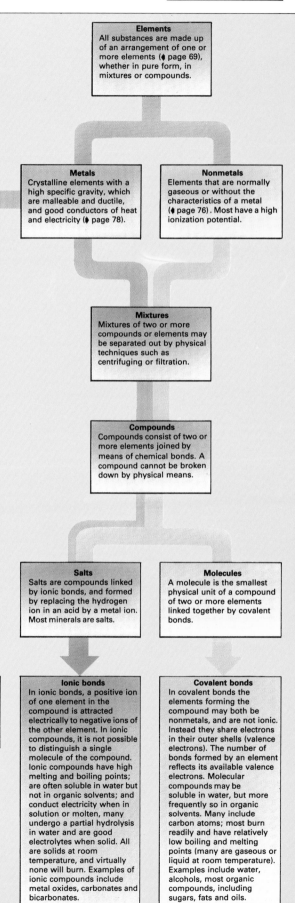

Elements
All substances are made up of an arrangement of one or more elements (◆ page 69), whether in pure form, in mixtures or compounds.

Metals
Crystalline elements with a high specific gravity, which are malleable and ductile, and good conductors of heat and electricity (◆ page 78).

Nonmetals
Elements that are normally gaseous or without the characteristics of a metal (◆ page 76). Most have a high ionization potential.

Mixtures
Mixtures of two or more compounds or elements may be separated out by physical techniques such as centrifuging or filtration.

Compounds
Compounds consist of two or more elements joined by means of chemical bonds. A compound cannot be broken down by physical means.

Salts
Salts are compounds linked by ionic bonds, and formed by replacing the hydrogen ion in an acid by a metal ion. Most minerals are salts.

Molecules
A molecule is the smallest physical unit of a compound of two or more elements linked together by covalent bonds.

Metallic bonds
The atoms in a metal are closely packed in a regular formation, with their outer electrons moving freely between the atoms (◆ page 78).

Ionic bonds
In ionic bonds, a positive ion of one element in the compound is attracted electrically to negative ions of the other element. In ionic compounds, it is not possible to distinguish a single molecule of the compound. Ionic compounds have high melting and boiling points; are often soluble in water but not in organic solvents; and conduct electricity when in solution or molten, many undergo a partial hydrolysis in water and are good electrolytes when solid. All are solids at room temperature, and virtually none will burn. Examples of ionic compounds include metal oxides, carbonates and bicarbonates.

Covalent bonds
In covalent bonds the elements forming the compound may both be nonmetals, and are not ionic. Instead they share electrons in their outer shells (valence electrons). The number of bonds formed by an element reflects its available valence electrons. Molecular compounds may be soluble in water, but more frequently so in organic solvents. Many include carbon atoms; most burn readily and have relatively low boiling and melting points (many are gaseous or liquid at room temperature). Examples include water, alcohols, most organic compounds, including sugars, fats and oils.

◄► *All the materials in the Universe are made up from fewer than one hundred elements, just as a child creates unlimited designs with a few building blocks. Substances are either elements, compounds or mixtures. A compound can only be broken down via a chemical reaction, but a mixture can be separated by physical means.*

Close packing of metal ions

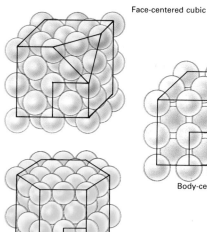

Face-centered cubic

Body-centered cubic

Hexagonal close packed

▲ ▶ *Using a scanning electron microscope, the surface of a piece of pure aluminum metal shows surface irregularities and hints of an underlying order (above). Using a field ion microscope to examine a single crystal of iridium it is possible to locate the actual atoms (right). These are the black dots. The green dots are probably atoms of gas which have been absorbed by the metal.*

◀ *Pure metals differ from other substances in that they consist of a close-packed array of a single type of atom. There are three ways in which perfect spheres of the same size can be stacked, and these have their counterparts in metal crystals. Aluminum and iridium prefer to stack in the face centered mode.*

Metallic bonding

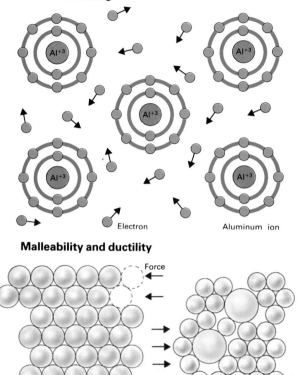

Electron   Aluminum ion

Malleability and ductility

Force

Force

Metal alloy

◄ The bonds between metal atoms are very different from other forms of chemical bonding – each metal atom contributes its outer electrons to a common pool. This "sea of electrons" explains a key property of metals – their ability to conduct electricity. It is this sea which allows the passage of a current of electrons through the metal. Metallic bonding is said to be nondirected, in other words an atom is not specifically bonded to a neighboring atom. This means that deforming a piece of metal does not lead to fracturing of the type experienced with molecular and ionic substances. Thus metals can be bent, hammered or drawn into wires without breaking. The inclusion of atoms of another metal, or even nonmetal, can alter the physical properties of a metal dramatically, simply by causing changes to the packing arrangement of atoms. Indeed the effect of alloying a metal can be to change its luster, density and resistance to corrosion: thus just a little carbon turns iron to steel.

▲ ▼ The Statue of Liberty, 46m tall and weighing 280 tonnes, is made from 300 copper plates, which have the blue-green patina that copper acquires when exposed to the weather. This is basic copper sulfate, $CuSO_4.3Cu(OH)_2$. However, while the century-old lady was outwardly protected by her patina, inside she was being affected by another chemical change – galvanic corrosion. This occurred where the copper plates came into contact with the iron framework. Originally, shellac and asbestos were used to insulate these metals, but over the years this broke away, rain water seeped in and the iron rusted rapidly due to the galvanic action. For her hundred-year facelift the statue was reequipped with stainless steel ribs and Teflon insulators.

Metallic bonding and the structure of metals

Many substances have a crystalline structure, with a regular, ordered arrangement of atoms. Among this class of substance are metals. Metals are generally characterized by their density, strength, thermal and electrical conductivity, and the fact that they are malleable (can be beaten to form new shapes) and ductile (drawn out into wire). Although metals are made up of a single element, their properties are related to the ways in which their atoms combine.

The geometrical structure of metals can be described by considering the ways in which spheres of the same size can be closely packed together. There are three common structures – body-centered cubic, hexagonal close-packed and face-centered cubic or cubic close-packing. Aluminum (Al) and gold (Au) atoms adopt face-centered cubic packing, sodium (Na) and vanadium (V) body-centered cubic packing, and magnesium (Mg) and cadmium (Cd) the hexagonal close-packed structure.

It has been suggested that the nuclei and nonvalence electrons of metal atoms are arranged in crystal structures held together by a "sea" of valence electrons not associated with any one atom. This helps to explain the high electrical and thermal conductivity of metals (◀ page 56) and also why metals are malleable and ductile. In another theory, metals are looked upon as giant molecules in which all the atomic orbitals of a particular type interact to form delocalized orbitals that extend throughout the structure. The energy of these delocalized orbitals is thought to spread into bands which overlap. This would explain the good conductivity of the metallic elements. This "band" theory of metals is often used to explain the electrical properties of the semiconducting elements such as silicon and germanium (◆ page 104), elements with both metallic and nonmetallic characteristics.

The strength of the bonds in ionic substances such as salt makes their boiling and melting points very high

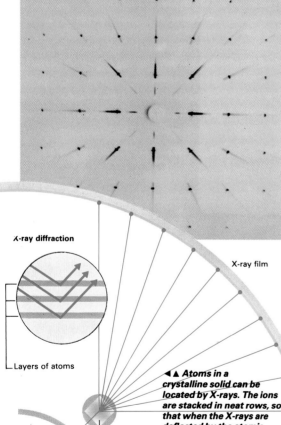

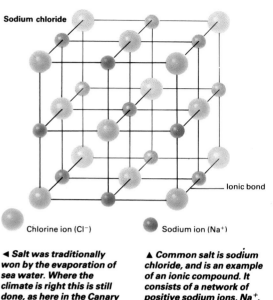

Sodium chloride

Chlorine ion (Cl⁻)   Sodium ion (Na⁺)   Ionic bond

◄ *Salt was traditionally won by the evaporation of sea water. Where the climate is right this is still done, as here in the Canary Islands. Such "sea salt" is relatively expensive. It is identical with ordinary salt from mines, which was also produced by solar evaporation millions of years ago.*

▲ *Common salt is sodium chloride, and is an example of an ionic compound. It consists of a network of positive sodium ions, Na⁺, and negative chloride ions, Cl⁻, in which the ions are located at the alternate corners of stacked cubes. This structure maximizes the electrostatic attraction between these ions.*

X-ray diffraction

X-ray film

Layers of atoms

Crystal

X-ray source   Lead shield

◄▲ *Atoms in a crystalline solid can be located by X-rays. The ions are stacked in neat rows, so that when the X-rays are deflected by the atomic nuclei they emerge from the crystal at the same angle. Analysis of the emergent X-ray pattern (above) reveals the structure of the crystal.*

Ionic compounds

Some atoms readily lose or gain electrons to form positively charged ions (cations) or negatively charged ions (anions). Bonds formed between these ionic species may be completely electrostatic in character. Such atoms swap electrons and then form ions held together by an ionic bond. The bond between sodium (Na) and chlorine (Cl) in sodium chloride (NaCl) or common salt is an example of an ionic bond. Sodium gives up one of its outer electrons to form the sodium cation (Na⁺) and chlorine gains this electron to form the chloride anion (Cl⁻). Electrical attraction between the two then bonds them.

Ionic solids consist of infinite arrays of positively and negatively charged ions. The size of the ions making up the ionic solids determines the geometry of the structure to a large extent. The most stable arrangements occur when the ions are in contact with one another and arranged symmetrically.

The structure of ionic solids is determined by the coordination number of the positively and negatively charged ions. This is the number of ions surrounding, in three dimensions, the central ion. In sodium chloride, the coordination number of both sodium and chloride ions is six: each ion is arranged octahedrally around the other. Other structures are also found in ionic solids. In cesium chloride (CsCl) crystals, the coordination number of both the anion and cation is eight, and a cubic arrangement is seen of ions around each other. Zinc sulfide (ZnS) crystallizes in two forms – the zinc blende and the wurtzite structure – but in both cases the anion and cation have a coordination number of four. Titanium dioxide (TiO_2) and one of the oxides of lead (PbO_2) crystallize with the rutile structure, in which the coordination number of the cation is six and that of the anion is three. Anions are arranged octahedrally around cations, whereas cations are arranged trigonally around anions. The electrostatic forces holding together ionic solids and, to a lesser extent ionic liquids, are very strong. Their melting and boiling points are consequently high.

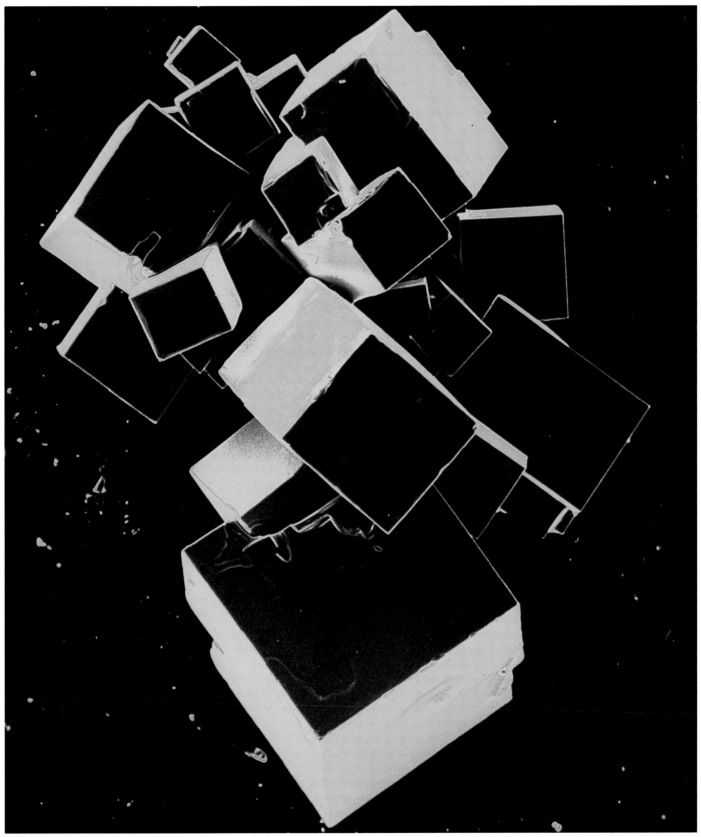

Ionic bonding

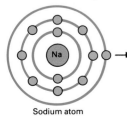

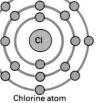

Sodium atom

Chlorine atom

Sodium ion

Chlorine ion

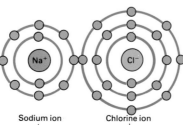

Sodium chloride molecule (NaCl)

◄ *Sodium is a very reactive metal and chlorine is a poisonous and corrosive gas. Take the outermost electron from a sodium atom and make it positively charged, and give it to a chlorine atom, which then becomes negative; and the two form an ionic bond. The result is sodium chloride, very unreactive and safe.*

▲ *Pure sodium chloride crystals conform to the underlying arrangement of the sodium and chloride ions. This micrograph reveals the cubic nature of salt, and also shows that the stacking of the ions can sometimes go wrong, resulting in dislocations in the lattice. Impurities often cause such imperfections.*

Computer models have proved very effective at describing the orbitals of covalent bonds

Theories of covalent bonding

Two theories have been used to predict the characteristics of covalent molecules – the valence bond theory and the linear combination of atomic orbitals (LCAO) theory. The LCAO theory suggests that after a covalent bond is formed electrons can exist in clearly defined regions of space known as molecular orbitals. Molecular orbitals of a molecule are similar to, yet different from, the atomic orbitals derived for atoms. They are concerned only with the outermost valence electrons of the atoms involved in the chemical bond. The LCAO theory has been applied to many covalent bonds and has become an especially important and powerful tool in recent years as computers have developed sufficiently to allow very complex mathematical calculations to be made of the behavior of electrons in molecules.

The symmetry of atomic orbitals is important in chemical bond formation. Bonds are formed when electrons have similar energies and when symmetry allows maximum overlap between the atomic orbitals. Often, as is the case with the carbon atom in methane, atomic orbitals are thought to mix, or become hybridized, to satisfy the symmetrical requirements of bond formation.

Molecular orbitals derived using the LCAO theory provide a useful pictorial method of considering the formation of covalent bonds. A combination of two s atomic orbitals gives rise to two sigma molecular orbitals. One of these is called a bonding orbital, the other, which is of higher energy, is called an antibonding orbital. Both orbitals can hold two electrons. In a simple sigma electron pair bond, the electrons forming the bond will go into the sigma bonding orbital. Sigma bonds and antibonds can also be formed when p atomic orbitals combine if overlap and symmetry are correct, and when a p orbital from one atom combines with an s atomic orbital from another.

A pi molecular orbital is formed from the lateral overlap of p atomic orbitals from two atoms. Again, a combination of two p atomic orbitals gives a bonding and an antibonding pi molecular orbital. The pi bond is not as strong as the sigma type, and molecules with pi bonds are usually more reactive than those without them.

In benzene (C_6H_6), mixing or hybridizing the carbon s and p atomic orbitals allows carbon-carbon sigma bonding around the benzene ring, and carbon-hydrogen sigma bonding between each carbon and hydrogen atom. Carbon p atomic orbitals overlap laterally to give a continuous pi orbital above and below the plane of the ring of carbon atoms. Electrons in this orbital are not associated with any particular pair of carbon atoms, and are said to be delocalized.

Covalent bonding

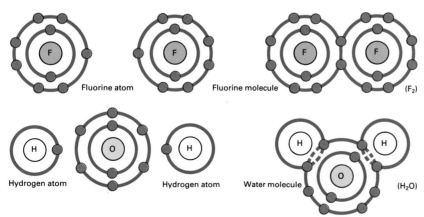

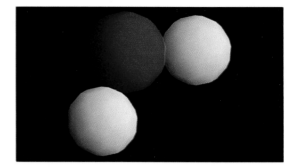

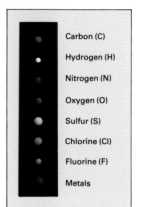

●	Carbon (C)
●	Hydrogen (H)
●	Nitrogen (N)
●	Oxygen (O)
●	Sulfur (S)
●	Chlorine (Cl)
●	Fluorine (F)
●	Metals

▲ ◄ *One of the most visual ways to represent molecules such as water (H_2O) is to imagine atoms as spheres, in what are called space-filling models. This book uses a system according to the color code (left). Another way is to show the bonds between atoms as rods, colored in the same manner.*

► *This representation of the benzene molecule shows the pi orbitals above and below the ring, in which the electrons are not tied to any one carbon atom.*

Chemical diagrams

The formula of a chemical compound shows how many atoms it contains, but gives little information about its structure or architecture. Formulae of ionic solids show only the simplest ratio of atoms within the structure, but do not tell how many ions surround each ion in the solid network. Similarly the formulae of covalently bound molecules give no indication of molecular geometry. Several types of molecular diagrams can be drawn up to represent the architecture of the molecules.

The simplest type shows the atoms joined by straight lines. Water (H_2O) and hydrogen sulfide (H_2S) are represented in this way as linear molecules, or, more accurately, the oxygen and sulfur atoms can be displaced to indicate the bond angle between the two H-O and H-S bonds, of 105° and 92° respectively. Multiple bonds are shown using two or three lines as necessary.

This type of representation can also be used to give some idea of the arrangement of the molecule in three dimensions. In two dimensions, methane (CH_4) and ammonia (NH_3) can be drawn, but it is known that in methane the hydrogens surround the carbon atom tetrahedrally, and that the ammonia molecule has trigonal pyramidal geometry. To represent this three-dimensional form, wedge-shaped bonds are used to indicate a bond pointing towards the viewer, and dotted bonds to show a bond pointing down into the page.

◄ *The formation of molecules is all to do with the "pairing" of electrons and the filling of electron shells around the atoms. In the case of water (left) oxygen is two electrons short of a filled shell and each hydrogen is one electron short of a filled shell. By coming together to form water both the oxygen and the hydrogens can achieve complete shells – as is the case with fluorine in F_2.*

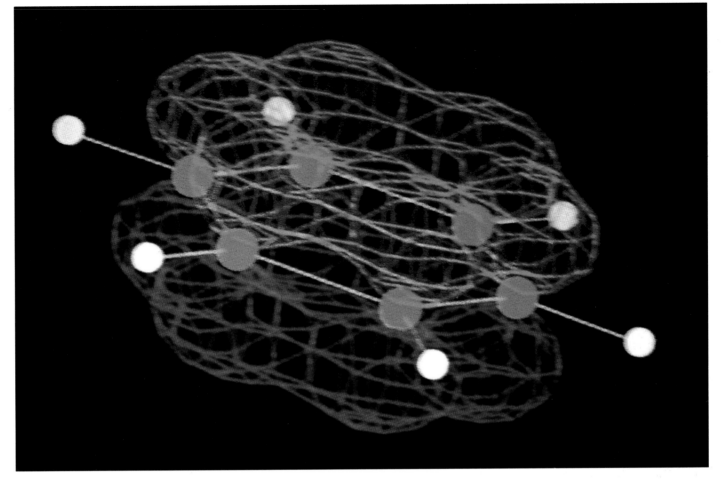

Other ways of representing molecules include the "ball-and-stick" model, which provides a useful three-dimensional structure of the molecule; and a space-filling model, which gives a clearer concept of the overall shape, but the individual bonds may be less easy to see.

G.N. Lewis (1875-1946) developed electron dot formulae for chemical compounds to help in the understanding of covalent bonding. Dots are used to show the number of valence electrons of an atom, and the shared electron concept of covalent bonding is easily represented using these "Lewis structures". In some cases single Lewis structures can be drawn for both molecules and ions, but in some cases one diagram is insufficient.

If two or more Lewis structures can be drawn for a compound they are called resonance structures and the compound is considered to be a resonance hybrid of these structures.

Two resonance structures can be drawn for the benzene molecule, showing alternate double bonds between carbon atoms around the six-membered ring. Measurement of the carbon-carbon bond lengths in benzene show that they are all the same. Double bonds are shorter than single bonds, but measurements show that the carbon-carbon bond lengths in benzene are all the same. The true benzene molecules are therefore a resonance hybrid of these structures. It is often represented using a special full or dashed line notation.

▶ Each way of representing a molecule has its advantages. The electron-dot way is still a useful teaching aid. Or the molecule can be drawn flat, which is easy to draw on paper, but misses the subtlety of its structure. This can be revealed by a ball-and-stick figure. Finally, perhaps the best picture of a molecule is the space-filling diagram – how we might expect a molecule to look if we really could photograph it!

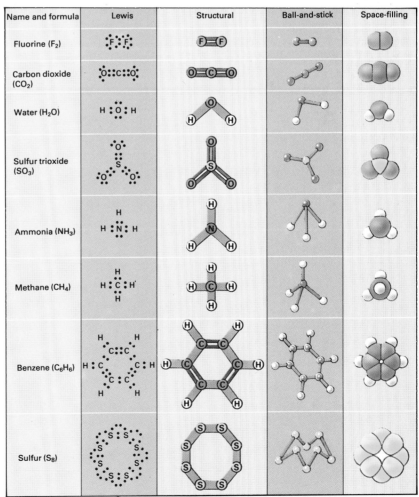

Name and formula	Lewis	Structural	Ball-and-stick	Space-filling
Fluorine (F_2)				
Carbon dioxide (CO_2)				
Water (H_2O)				
Sulfur trioxide (SO_3)				
Ammonia (NH_3)				
Methane (CH_4)				
Benzene (C_6H_6)				
Sulfur (S_8)				

Ethane and derivatives

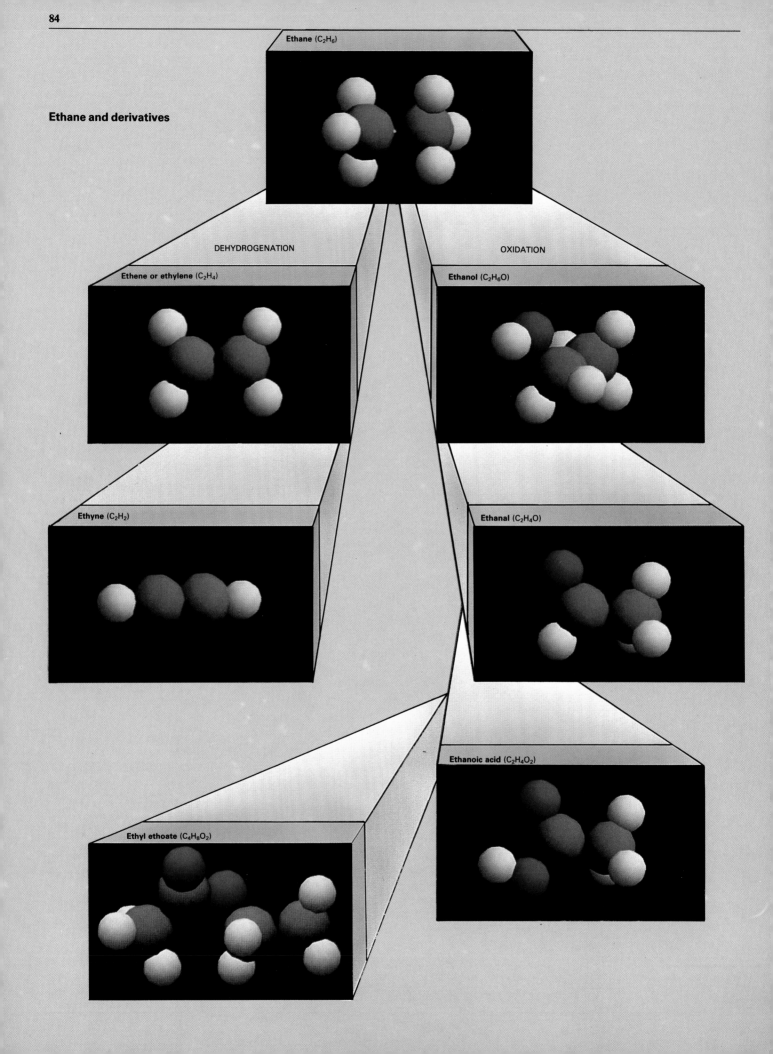

Ethane (C₂H₆)

DEHYDROGENATION

OXIDATION

Ethene or ethylene (C₂H₄)

Ethanol (C₂H₆O)

Ethyne (C₂H₂)

Ethanal (C₂H₄O)

Ethanoic acid (C₂H₄O₂)

Ethyl ethoate (C₄H₈O₂)

◄ *Although millions of organic compounds are known, most fall into a few well-defined classes. These are identified by the functional groups which the molecule possesses. These are groups of atoms in a molecule that confer upon the compound properties, recognizable as typical of their class. Often one class is related to another in quite a simple way. Thus ethane (C_2H_6) is an alkane, related to the alkene ethene (C_2H_4), which has a double bond, and the alkyne ethyne (C_2H_2) with a triple bond. This sequence can be seen as the removal of hydrogen atoms. Another series of compounds can be related to ethane by oxidation. This leads to molecules that are alcohols, aldehydes and carboxylic acids. From carboxylic acids another series of compounds arise, called esters, known for their fruity smell and flavor.*

Representing the structure of molecules

Studies of the electronic structure of atoms, and predictions of how different atoms bond together, help in identifying the architecture of molecules. So too do studies of chemical reactions. In addition, scientists can use spectroscopy to view the makeup of individual molecules and crystals.

Molecular spectroscopy is used to investigate the structure of many different molecules. By subjecting a sample to radiation of varying wavelengths, vibrations, rotations and electron jumps can be induced in the molecules. These movements are often characteristic of individual atoms or groups of atoms bonded in a certain way. Thus measurement of the absorption of radiation gives an insight into the structure of the molecules.

If samples are excited with visible and ultraviolet radiation, electronic transitions take place in atoms. Radiation in the near infrared and infrared parts of the spectrum sets up vibrations in the atoms of molecules just as if the bonds were made of elastic. Absorption of far infrared or microwave radiation makes the molecules rotate.

Measurement of infrared radiation absorption in different portions of the molecule can be made. For example, the radiation may induce oxygen-hydrogen and carbon-hydrogen bonds to stretch and bend. A trace of wavelengths absorbed indicates the bonds present in the molecule.

Different types of spectrometry give invaluable information about the structure of molecules, including ideas of the distance between the atoms, and the strength of individual bonds. Infrared spectrometry has become useful in investigating slight chemical changes in large molecules.

In the 1950s, physicists began to measure the magnetic properties of atomic nuclei. They found that the effects they were observing depended not only on the type of atom but also on its chemical environment (the types of atoms that surrounded it). Since that time the technique of nuclear magnetic resonance (NMR) spectroscopy has been developed from this observation. With the help of computers NMR can be used to investigate the arrangement of atoms in large, often complex, molecules such as insulin and some proteins.

The most powerful technique, though, for investigating the structure of compounds is X-ray diffraction (◆ page 80). Over 4,000 new structures each year are identified by X-ray diffraction.

EMISSION SPECTROSCOPY

ABSORPTION SPECTROSCOPY

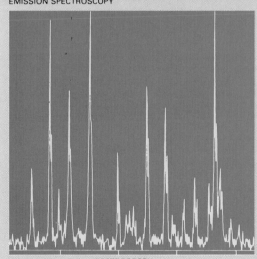

X-RAY DIFFRACTION SPECTROSCOPY

NMR SPECTROSCOPY

▲ ◄ *Father and son William Henry and William Lawrence Bragg (above) were the first chemists to probe the structure of matter, using X-rays to work out the positions of atoms in crystals. Today there are many ways of analyzing and identifying compounds. Most rely on studying the effect of light. Absorption spectroscopy identifies elements by the wavelengths of the light which they absorb. In emission spectroscopy it is the light given out by an excited atom which tells what element it is, and how much there is of it. Even X-ray diffraction is used for identification purposes since the pattern of the scattered rays is typical of a particular substance. Another useful technique for studying molecules is nuclear magnetic resonance (NMR) spectroscopy, which relies on radiowaves and strong magnetic fields.*

It took chemists more than 70 years to produce the first synthetic diamond

▶ Carbon as the pure element is found in nature in two very different forms – diamond and graphite. Diamond is the hardest natural material, yet graphite is so soft that it is used as a lubricant. The way their atoms are packed reveals why this is so. Diamond (below) has every carbon atom bonded to four other atoms in a rigid three-dimensional lattice. Graphite (right) has each carbon atom bonded to only three others, and this produces planes of atoms. Since there is no bonding between the planes, they can slip over one another.

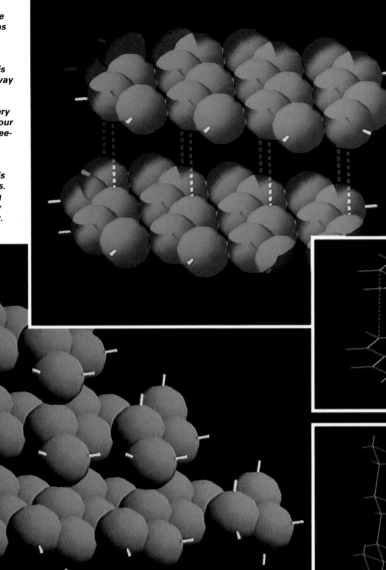

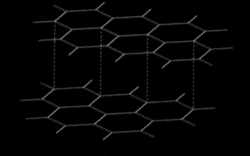

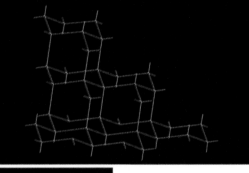

▶ This scanning electron microscope enlargement of the diamond surface of a dentist's drill shows why it is so effective at cutting and abrading. Diamond is used industrially on a large scale for cutting tools. Of the two forms of carbon (left), graphite is the more stable thermodynamically, and it requires great heat and pressure to convert it to diamond. Natural diamonds are found in old volcanic pipes. Artificial diamonds can be produced by simulating these conditions, but they are very small and suitable only for industrial use.

▲ The chemical bonds in diamond and graphite show two kinds of covalent bonding. In diamond (above), the four electrons pair off with four other electrons, one each from four surrounding atoms. The result is very strong localized bonds. Graphite (top) on the other hand uses only three of its valence electrons for bonds to other carbons, and donates the fourth electron to a pool of delocalized electrons which lie above and below the planes of atoms. Because of these delocalized electrons graphite, a nonmetal, can conduct electricity.

Making diamonds
Diamond and graphite are the most common allotropes of carbon. Graphite is made up of two-dimensional molecules – six-membered rings of carbon atoms – held together by weak intermolecular forces. Diamond is a much more tightly packed form of the element, in which each carbon atom is surrounded tetrahedrally by four others. Graphite has a soft, slippery feel, due to its sheetlike structure; the tight packing of diamond, on the other hand, makes it one of the hardest materials known.

Diamonds have been found in mineral deposits in India, the Soviet Union, Brazil and South Africa. The hardness of diamond makes it an important industrial material, used as an abrasive in saws, drills and grinding wheels. Single crystals of diamond are used as drill crowns and in tools used to shape other, less hard abrasive materials.

Very high temperatures and pressures were needed to make the diamonds naturally. This fact was recognized when the first attempts were made to construct the diamonds artificially. Diamond is most stable at very high pressures, but by investigating the very different solid, liquid and gaseous phases of carbon (◀ page 29), scientists calculated that diamonds could be made at very high temperatures and pressures. From 1880 many attempts were made to make diamonds from graphite, but none was successful until 1955.

Diamond can be made directly from graphite at pressures of about 12.5GPa and temperatures over 3,000K. These can only be achieved in the laboratory using special techniques involving shockwaves from high-pressure gases and explosives. For production on a commercial scale, metal catalysts are used to reduce the pressure to about 6GPa, and the temperature to 1,700-1,800K.

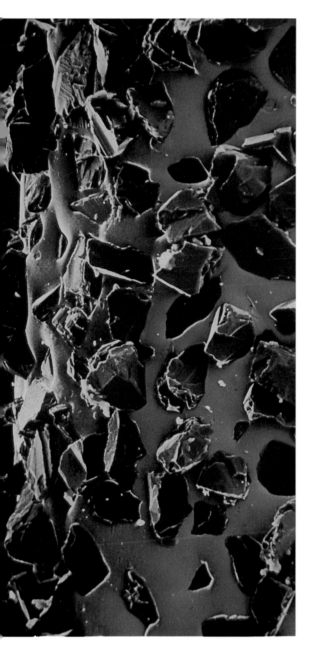

Molecules containing carbon

Carbon (C) is able to link itself better than any other atom to form straight and branched chains and rings. The vast majority of molecules that exist naturally contain carbon.

The chemistry of carbon-containing molecules is so large a subject that it is split into two areas, organic chemistry and biochemistry. Today the study of the chemistry of living organisms is known as biochemistry, whereas the field of organic chemistry includes the study of the myriad compounds of carbon, including those with and without biological activity. All life on Earth is carbon-based, just because of the ability of carbon to bond with itself in so many ways. Living organisms are a complicated collection of carbon-based molecules, which include sugars, carbohydrates, enzymes and other proteins.

The size and electronic structure of the carbon atom are responsible for its bonding characteristics. There are four outer, or valence, electrons surrounding the nucleus. It can bond easily with hydrogen, and with atoms to its right in the Periodic Table (◀ page 74), and chemists have succeeded in bonding carbon with many other elements, deriving a class of substances known as organometallic compounds, which have a carbon-metal bond. Often a carbon atom can form covalent bonds with four other atoms – with hydrogen in methane (CH_4), for example. It can also become involved in multiple bonds with other carbon atoms, using electrons in pi molecular orbitals to form double or triple bonds which make the molecule reactive in various ways. In benzene and many other ring compounds of carbon, delocalized as well as localized carbon-carbon bonds can be formed. Carbon compounds with multiple bonds of the type found in benzene are known as aromatic compounds. Aromatic rings are involved in much organic chemistry. Aliphatic compounds, on the other hand, have their carbon atoms arranged in straight lines or branched chains.

Nitrogen (N) and boron (B), elements in the second period of the Periodic Table, are similar in size to carbon. Silicon, like carbon, has four valence electrons. These three elements might be expected to bond in a similar manner to carbon and to form a similar number of compounds, but they do not. Each may form carbonlike compounds, but none has carbon's combination of valence and inner electrons. Their carbonlike compounds are often short-lived and unstable.

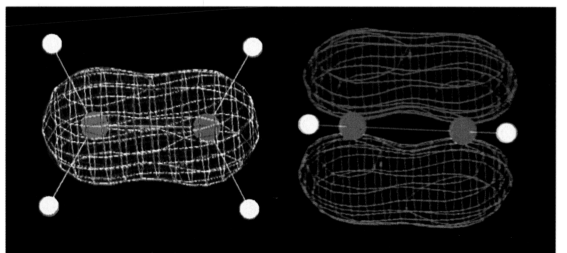

► When a molecule has more electron-pairs than bonds between its atoms, it can use the extra electrons to give the molecule stability by forming double, and even triple, bonds. Ethylene, C_2H_4, is the simplest example, with five bonds and six pairs of electrons. The sixth pair forms a double bond between the two carbons. The electron density of the primary bond between these atoms is shown. The extra bond is called a pi bond, and its electron density spreads above and below the plane of the molecule.

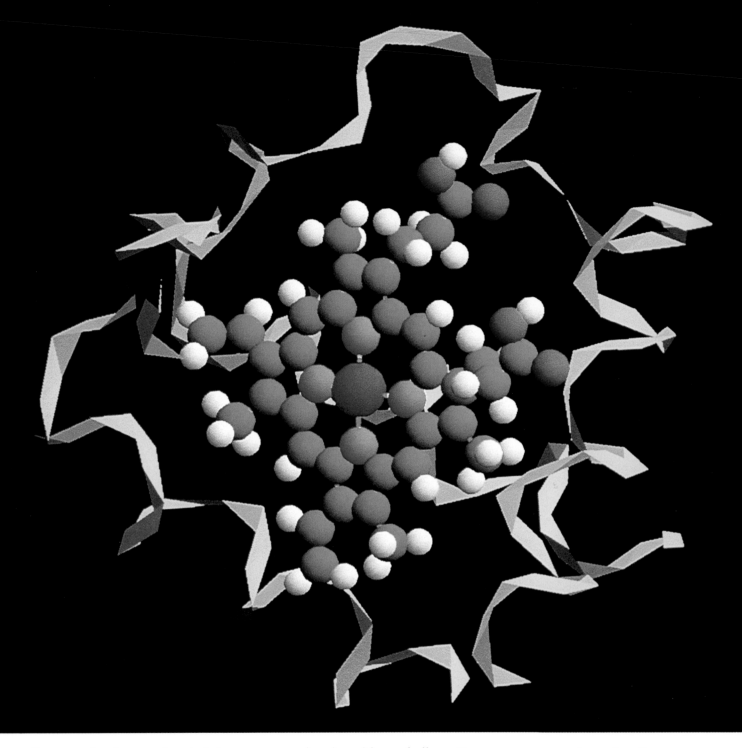

Some molecules are bonded in a manner that is neither wholly covalent nor wholly ionic. The electrons are not shared between the atoms, as in a covalent bond, but two electrons can be donated from one atom to another to form a coordinate bond. The atom or group of atoms donating the electrons acquires a positive charge, and the atom accepting electrons acquires a negative charge. Coordinate bonds are found in so-called complex molecules, especially those formed by the transition metal elements (◀ page 75).

Complex compounds do not seem to adhere to the simple ideas of valency. A central atom in a complex with vacant bonding orbitals is able to accept a donated pair of electrons – or a lone pair of electrons – from another atom, molecule or ion forming a coordinate bond. In doing so the valence of the central atom is often exceeded.

Molecules or chemical species surrounding the central atom in a complex are called ligands. Ligands can be simple or complicated chemical species, and can bond to the central atom through one or

How many compounds are there?

In the last 50 years there has been a spectacular increase in our knowledge of chemistry and molecular architecture. In 1930, just over one million chemical compounds were known; today there are more than eight million. In the 11 years between 1969 and 1980 the number of known chemical compounds doubled. The previous doubling of this figure had taken 14 years, and the doubling before that, 24 years.

In the 1940s, 1950s and 1960s, knowledge of organic chemistry (of compounds based on chains and rings of carbon atoms) grew rapidly. New plastics were developed from petrochemicals and new drugs were found to combat disease.

Organic chemists can extract natural chemical products from plants, microorganisms and human and animal sources, but they continually try to find new ways to make these

◄ *The molecule at the heart of hemoglobin (left) is called a porphyrin. Ideal for holding a metal atom, in this case iron, its flat shape lets molecules approach easily and undergo reaction with the metal. Oxygen from the air is attracted to the iron atom (purple) at the center of hemoglobin; once attached it is then taken around the bloodstream to where it is needed. On its journey back to the lungs the hemoglobin carries the waste gas carbon dioxide, formed during oxidation, which is then breathed out. Some gases cling to the iron atom but will not let go; carbon monoxide is an example. Breathe in too much of this and you die.*

► *Animal blood is red due to the color of iron-porphyrin. Spiders' blood is blue because it uses copper-porphyrin to carry oxygen. In the chlorophyll of plants it is green magnesium-porphyrin (right), which traps sunlight to use as energy for photosynthesis and so turns carbon dioxide of the air into carbohydrate.*

► *Models of chlorophyll (above right) may look very complicated, but they are in fact made up of sub-units. At the center is magnesium, bonded to four nitrogen atoms, which are part of four pyrrole rings. These four rings joined by four carbon atoms in turn make the perfectly flat porphyrin ring system. Attached to the outside of the porphyrin are other groups of atoms. Chlorophyll-a is shown here. Chlorophyll-b also exists, in which the methyl CH_3 on the left pyrrole is oxidized to a CHO group.*

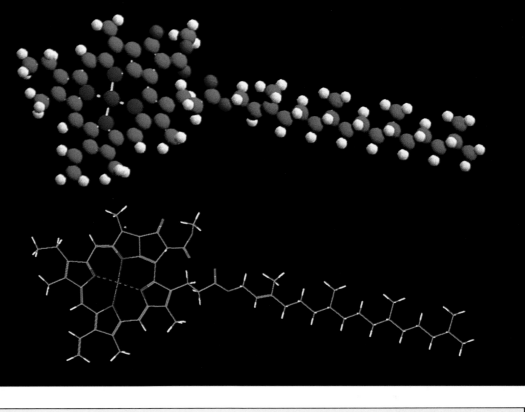

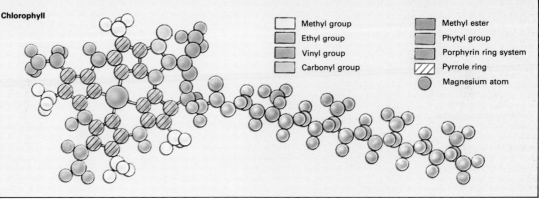

Chlorophyll

Methyl group
Ethyl group
Vinyl group
Carbonyl group
Methyl ester
Phytyl group
Porphyrin ring system
Pyrrole ring
Magnesium atom

chemicals in the laboratory. They can also tailor reaction pathways (► page 94) to give products which may have similar properties to naturally-occurring chemicals but which are not themselves found in nature.

The field of bioinorganic chemistry is growing rapidly. Many new organometallic compounds have been prepared by reactions between chemical species that were once thought to be too inert to react. New environments under which ligands can bond to metal atoms have been discovered.

Of all the new chemical compounds discovered each year, only a very small proportion appear immediately useful. However, as knowledge increases, new syntheses may lead to the preparation of new drugs, pesticides and herbicides which can help to fight disease, increase food production and make life more comfortable for those who can afford the products.

more lone pair of electrons. Ligands that bond through one lone pair of electrons are called monodentate. Some more complex molecules and ions can act as ligands and form two or more bonds with another atom. These are called multidentate ligands. Ethylene diamine $(NH_2.CH_2.CH_2.NH_2)$ bonds through lone pair electrons on its nitrogen atoms and can act as a bidentate ligand, forming two coordinate bonds with another atom. Usually, lone ligand pair electrons bond through sigma molecular orbitals to another atom, but pi coordinate bonds are found in some molecules.

Ligands that bond through more than one pair of electrons are called chelates. One important chelate is the ethylene diaminetetra-acetate (EDTA) ion. This can bond to some metals through six co-ordinate bonds. Its attraction for metals is so strong that it is used as a scavenger for calcium, magnesium, iron and copper. EDTA is used to remove trace metals from distilled water; it can also remove the metal ion from an enzyme, thus destroying its enzymatic activity.

Bonds between molecules

Some atoms have a greater attraction for electrons, or electro-negativity, than others. In a simple two-atom covalently bonded molecule, electrons in the bond will be attracted to the most electro-negative atom, giving it a partial negative charge. Conversely they will move away from the less electronegative atom, and leave it with a partial positive charge. Bonds and molecules containing such bonds are said to be polar. The greater the polarity of a bond, the more ionic in character it becomes. Most molecules are polar to some extent and so they attract one another and can also attract positively and nega-tively charged ions. This is important in that ionically bonded mol-ecules (◀ page 80) dissolve in polar substances. The polar nature of these substances also affects their melting and boiling points, and solubility in solvents.

Another type of attraction between molecules is known as a hydrogen bond. This is formed between molecules with highly polar bonds involving hydrogen. The chemical bonding in water (H_2O), though considered to be covalent, is highly polar because of the electronegativity of oxygen. In water and ice the H_2O molecules are held together by hydrogen bonds, which attract the hydrogen atoms of one molecule to oxygen atoms of other molecules. This type of bonding makes H_2O very different from hydrogen sulfide (H_2S), which is an analogous compound.

The thermal properties of water, on which the Earth's ecosystem depends, are a direct result of this hydrogen bonding, which keeps water liquid rather than gaseous at a wide range of temperatures. The fact that the hydrogen bond is weak enough to be overcome by thermal agitation at normal temperatures and pressures also means that hydrogen bonding is important in biological systems, for example in bridging and holding together protein molecules.

Dissolution of salts in water

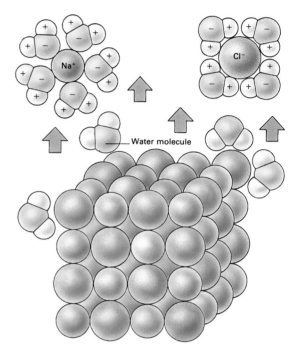

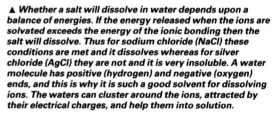

▲ *Whether a salt will dissolve in water depends upon a balance of energies. If the energy released when the ions are solvated exceeds the energy of the ionic bonding then the salt will dissolve. Thus for sodium chloride (NaCl) these conditions are met and it dissolves whereas for silver chloride (AgCl) they are not and it is very insoluble. A water molecule has positive (hydrogen) and negative (oxygen) ends, and this is why it is such a good solvent for dissolving ions. The waters can cluster around the ions, attracted by their electrical charges, and help them into solution.*

◀ *Although atoms come together to form molecules, molecules themselves are attracted to one another by weaker chemical forces, found both in liquids and in solids. Among these forces hydrogen bonding is key. It occurs when a molecule contains hydrogen atoms that are covalently bonded to nitrogen, oxygen and other electron-attracting elements. Hydrogen bound to these elements finds its electrons attracted away towards the other atom, so giving the hydrogen a small positive charge. This it then uses to attract electrons in other molecules to form hydrogen bonds. Nowhere is this more apparent than in water, which is why this tiny molecule has such unusual properties. Its relatively high melting and boiling points are due to hydrogen bonding in ice and water. Ice floats on water for this same reason – unlike other substances where the solid form is denser than the liquid. The hydrogen bonding in ice creates a very open lattice, which lowers its density.*

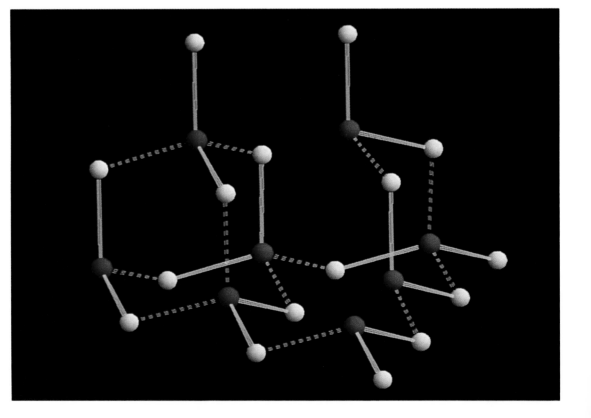

Chemical Reactions

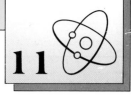

11

What is a chemical reaction?...Factors determining the rates of reactions...Activation energy...Catalysts and reaction rates...Chain reactions and complex reactions... Studying the pathways of reactions...PERSPECTIVE... Types of reaction...Natural catalysts, enzymes and zeolites...Free radicals...Studying very fast reactions

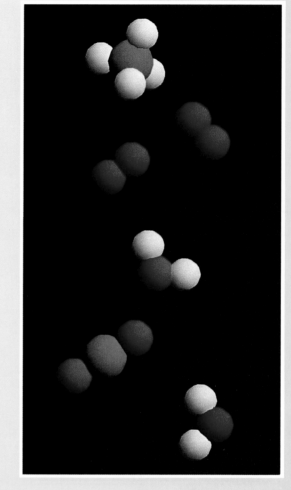

Chemical reactions involve the formation or breaking of bonds between atoms or molecules. Such bonds form when their outermost or valence shell of electrons mix and electrostatic forces are strong enough to hold them together. To reach this state they need to be very close to one another, and to gain sufficient energy to overcome the repulsion between the electron clouds that surround their nuclei.

In gases, liquids and solids the atoms are moving – they have kinetic energy (◀ page 22). Chemical bonds form if the two colliding atoms are moving fast enough for their valence electrons to mix on contact.

From studies of the kinetics of reacting gases, liquids and solids, scientists can tell how reactions occur between atoms and molecules. They can also tell what route a reaction takes and how fast it happens. If this study is linked with an investigation of the initial and final states of the reaction and a study of the structure of the reactants and products, much can be learned of how molecules react and form.

Collision theory is used to explain the likelihood and speed of a reaction. Collisions between atoms and molecules during a reaction may be effective or ineffective, and the speed or rate of the reaction is directly linked to the collision frequency, or total number of collisions occurring between the reactants in a certain period of time. Atoms may collide with sufficient energy to make or break bonds, yet fail to do so if the collision occurs at the wrong atom in a molecule, or if the arrangement of groups of atoms in the molecule blocks the collision. The architecture of the molecule, and also its orientation in space, therefore also affect the rate of reaction.

▲▼ *When methane gas burns, either productively as a source of heat or wastefully as surplus gas in the desert, it produces carbon dioxide. The equation for the reaction is: $CH_4 + 2O_2 = CO_2 + 2H_2O$. Chemical kinetics is the name given to the study of how a reaction like this actually happens. Although it occurs very rapidly it has been possible to deduce the many steps involved.*

Types of Reactions

▼ Several kinds of chemical reaction are encountered in the everyday world, such as oxidation-reduction and acid-base neutralization. In the first of these an electron exchange (e^-) takes place, the oxidizing agent gaining electrons, the reducing agent losing them. In acid-base reactions an exchange of hydrogen ions (H^+) occurs. The acid gives up its hydrogen, the base accepts it. The product of this reaction is a salt plus water.

▼ Since the atmosphere contains elemental oxygen it is described as an oxidizing atmosphere. The effects of this are seen in the ease with which iron rusts back to iron oxide, as in this wrecked tanker. The chemistry taking place is the oxidation of the metal (Fe), which loses electrons to form Fe^{2+}, and even Fe^{3+}, and the atmospheric oxygen which gains them. The process eventually gives red-brown Fe_3O_4.

▶ We can see the effect of an acid-base reaction when an Alka-Seltzer tablet is dropped into water. The acid is citric acid, the base is sodium bicarbonate, and this releases carbon dioxide as it is neutralized. The measure of acidity is the pH scale which ranges from 0, strong acid, to 14, strong base. The concentration of acid H^+ and base OH^- in neutral water is 10^{-7} moles per liter, expressed as 7 on the scale.

Oxidation of iron

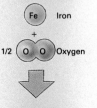

Fe — Iron
+
1/2 O O — Oxygen

⟶ Fe O — Rust

Fe — Electron transfer
O

⟶ Fe^{2+} O^{2-}

Acid neutralization

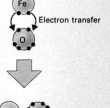

H^+ — Acid
+
O H — Base

⟶ O H H — Water

The pH scale

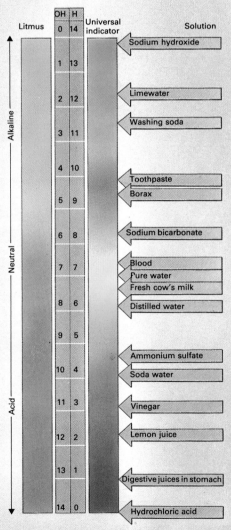

Litmus	OH	H	Universal indicator	Solution
	0	14		Sodium hydroxide
	1	13		
Alkaline	2	12		Limewater
	3	11		Washing soda
	4	10		
	5	9		Toothpaste / Borax
	6	8		Sodium bicarbonate
Neutral	7	7		Blood / Pure water / Fresh cow's milk
	8	6		Distilled water
	9	5		
	10	4		Ammonium sulfate / Soda water
Acid	11	3		Vinegar
	12	2		Lemon juice
	13	1		Digestive juices in stomach
	14	0		Hydrochloric acid

▼ Although polymerization is thought of as an industrial process in which plastics and fibers are made, it is possible to see it at work in the home. Jam or jelly making, for example, involves quite elaborate chemistry. To make jam "set" it is necessary to extract the pectin, a carbohydrate polymer, from the fruit. Since there is more pectin in unripe fruit it often helps to have some of this in the jam.

▼ Photochemistry is the study of the reaction of light and chemicals ("photo" means light). In plants photosynthesis converts low-energy carbon dioxide gas in the atmosphere into high-energy carbohydrate in the plant. The first step involves an atom of magnesium – part of a chlorophyll molecule. This captures a photon, and uses it to excite an electron. With this energy it can power the reaction.

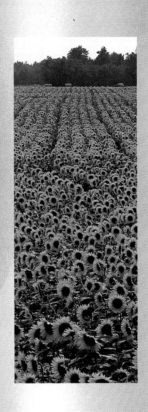

◄ Photography too depends upon the effect of light interacting with chemicals, in this case with silver halides. Having exposed the film, it is essential that no other light comes into contact with it, hence the need for a dark-room in which to develop it. However, total darkness is not necessary to prevent further exposure of the film, because red light does not react photochemically with the silver halide.

*Short-lived chemicals known as free radicals can have a debilitating effect
on the human body if not rapidly destroyed*

Free radicals

The highly reactive chemical species known as free
radicals are often formed during chemical
reactions. These play an important part in some
reactions. Their reactivity is a result of their having
at least one unpaired electron.

Free radicals are short-lived intermediates.
Mechanisms involving their formation and
combination have often been proposed, but it has
not always been easy to prove that they are actually
formed. In some instances free radicals have been
trapped as stable compounds by adding other
chemicals to the reaction system.

In one chemical method, toluene ($C_6H_5CH_3$) is
used to detect and investigate the formation of free
radicals during decomposition reactions. Toluene
vapor is mixed with the substance in question. If
free radicals are formed during any subsequent
decomposition they will react with the toluene to
form a benzyl free radical ($C_6H_5CH_2.$) by the
reaction: $R. + C_6H_5CH_3 \rightarrow RH + C_6H_5CH_2..$ In this
equation, free radicals are indicated using a dot
notation; R. represents a free radical formed during
the decomposition of the substance.

This technique only works if the bonds in the
decomposing substance are weaker than the
carbon-hydrogen bonds in toluene. The benzyl
radicals formed by this reaction are fairly stable
because of their resonance structures; they react to
form dibenzyl molecules. The presence of dibenzyl
in the reaction mixture thus proves the existence of
free radicals during the decomposition.

Free radicals are thought to play an important
part in metabolism and they have been linked with
the cause of a number of types of tissue damage in
humans. Numerous studies have been made of
their role in the action of aspirin in the body. They
are thought to play a part in the action on the body
of a number of toxic chemicals including carbon
tetrachloride (CCl_4), chloroform and DDT (◆ page
248), nitrogen-containing chemicals and alcohol.
They also play an important part in the effect of
ionizing radiation on the body (◆ page 224).

The highly reactive and potentially harmful free
radicals are usually kept at bay in the body by
radical scavengers. These are also free radicals
(including the hydroxyl radicals HO.), or
inactivating enzymes. When this protection is
broken, disease and cell destruction may result.

▲ ▶ *The speed of a reaction
can be explosively rapid or
extremely slow. The picture
above shows one that is
over in a fraction of a
second, the one on the right
(showing rust under layers
of paint) takes years. The
explosion reveals the
dramatic effect of the rapid
release of gases from
unstable molecules. The
heat released also adds to
the expansion of the gases,
and the damage. Rusting is
the slow formation of iron
oxide by the oxygen in the
air, a process aided by
water. Paint delays the
process but ultimately the
chemical reaction wins.*

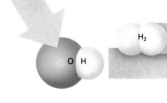

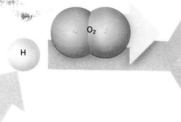

$2H_2 + O_2 \rightarrow H_2O$

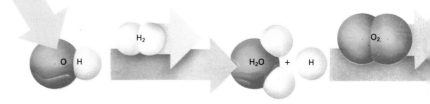

▶ *When hydrogen and oxygen combine to form water, the
chemist writes this as a simple equation (arrow above). This
reaction may be explosive because of the large amount of
energy that is released which causes the water (as steam) to
expand rapidly. A chain reaction occurs, in which colliding
molecules break into fragments or atoms with enough
energy to react with more molecules, which in turn release
yet more active fragments that react with even more
molecules, and so on. The chain reactions in the many steps
to water involve highly reactive intermediates: oxygen
atoms (O), hydrogen atoms (H) and hydroxyl (OH) radicals.*

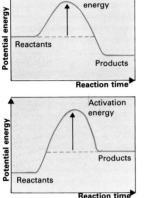

◄ Although a mixture of oxygen and hydrogen gas can be highly explosive, nothing happens until it is set off, perhaps by a spark. In so doing the kick-start of energy begins the chain reaction and releases all the pent-up energy of the system. Chemists describe the process in terms of energy levels and the energy barrier between them: the activation energy. If the activation energy barrier is high, then some external input of energy is needed to start the reaction.

Chemical equilibrium

All chemical reactions occur in two directions: just as the reactants react to form products, at the same time the products are themselves reacting, or breaking down, to form the original reactants. At a certain point the rate of reaction of the products to form reactants equals that of the reactants to form products; this state is known as chemical equilibrium.

The French chemist Henri le Châtelier (1850-1936) proposed the important principle that if an external stress is applied to a reaction system, then the point of chemical equilibrium will change to counter-act the effects of the stress and restore the initial equilibrium position. Le Châtelier's principle applies to both temperature and pressure changes. If, for example, the reaction releases heat (is exothermic), then heating favors the reverse rather than the forward reaction. The formation of products is thus impeded.

Activated complexes

Chemical species need a certain amount of energy, known as the activation energy, before they can react. This activation energy can be seen as an energy "hill", over which chemical species need to pass, the minimum amount of energy needed before any reaction occurs. Some activation energy is needed to drive all reactions, whether the whole reaction releases heat to its surroundings or takes heat from the surroundings (is endothermic).

Much of the study of chemical kinetics concerns the passage of atoms and molecules over the activation energy barrier. Before they cross this barrier they are far apart; as they reach it they are closer together and beginning to react. Once over the barrier, new bonds have been formed and others may have been broken; the reaction products have been produced.

Complex molecular species have been found to exist for very short periods of time during the passage of some reactions. These complexes exist at high energy during the reaction, close to the activation energy, and are called activated complexes or transition state complexes. The absolute rate theory of chemical reactions (also known as the transition state theory) predicts that a reaction occurs only when an activated complex breaks up into products. According to this theory, the rate of the reaction depends on three factors: the concentration of activated complexes, the rate of their breakdown or their movement over the energy barrier, and the probability that they will not break down into reactants.

◄ Temperature has a profound effect on the speed at which a chemical reaction occurs, and may even decide whether it will happen at all. Freezing slows down bodily decay by reducing the chemical reactions performed by microbes and their enzymes until they virtually cease. At the other extreme, some chemical reactions need very high temperatures, especially if they are intended to force the reverse of a change that would normally occur in the other direction, as in the blast furnace production of iron from iron oxide.

Some catalysts may be used as molecular sieves to separate mixtures of chemicals and hold reacting molecules

The design of catalysts

Minerals known as zeolites have been found to be very useful catalysts for organic chemical reactions, particularly those used to produce petrochemicals. Zeolites are hydrated aluminosilicate minerals – substances containing alumina, silica and water bonded with a metal in a complex crystalline structure. Thirty-six zeolite structures are known, but not all are natural: synthetic zeolites have been prepared for specific purposes.

Zeolites have open structures and can absorb smaller molecules. They are often used as chemical absorbents, or molecular sieves, to separate mixtures of chemicals. As catalysts they can absorb reacting molecules in the cages and channels of their structures.

Modification of their structure allows them to absorb some molecules and not others. It can also make them selective to molecules of a particular size and shape. Much effort has been spent on altering this "shape selectivity", because of the rewards it might bring. Zeolite shape selectivity can be changed by a variety of methods, either by altering the crystal structure of the zeolite or changing the size of the cages and channels in the crystal structure. Particular attention has been paid to zeolites that catalyze the conversion of methanol to gasoline and other petrochemicals.

Natural and synthetic enzymes

The most efficient and elegantly designed catalysts are natural ones known as enzymes. Natural enzymes speed a myriad of biochemical reactions, and in recent years it has been possible to synthesize artificial enzymes that can mimic natural ones. Smaller molecules can be trapped in an enzyme's structure and immobilized, ready to react with other molecules. Enzymes can also provide sites at which certain reactions can occur. Natural enzymes catalyze specific reactions at relatively mild temperatures. In effect they hold molecules ready for reaction until a reacting molecule appears to produce a particular product with its own geometry.

Enzymes are used in a variety of chemical processes, mainly in the manufacture of food and drinks. However, they are very sensitive substances and can sometimes lose their catalytic activity very rapidly when used in a reacting solution. To overcome this problem they are often immobilized on a solid support for both storage and use. Enzyme supports can be organic or inorganic molecules. Immobilized enzymes are used in the production of high fructose corn syrup, a sweetening agent used in soft drinks manufacture.

It is today possible to prepare artificial enzymes that mimic natural enzymes. Some of these have been used for the biological synthesis of amino acids and for the digestion of proteins. Copies of biological membranes and the molecules that carry substances through the membranes have also been made. The production of novel artificial enzymes is also well under way. Organic chemists are investigating complex molecules which have holes in them that can act as catalytic binding sites. These enzymes might be used to speed reactions that now need expensive energy to drive them.

Types of chemical reaction

Energy is required to initiate all chemical reactions. Spontaneous reactions, which involve the splitting of molecules into other molecules and atoms, usually need heat to start them. Such reactions can be very fast, as in the case of some explosions, or very slow, occurring over many years. Reactions between two or more atoms need energy in the form of heat, light or electricity.

Four main factors affect the rate of the reaction. These are the nature or type of the reactants, their concentration, the temperature at which the reaction takes place, and the effect if any of a catalyst on the reaction. A catalyst is any substance that speeds the rate of the reaction without itself being consumed in the process. Catalysts speed many chemical reactions, including those in living organisms as well as many industrially important chemical processes.

The nature of the reacting chemical species determines not only whether the reaction will occur, but also its speed. The decomposition of a molecule with strong bonds (◀ page 77) needs more energy than that of a molecule with weak ones. Conversely, two oppositely charged ions may combine almost immediately because of the electrical attraction between them.

Most chemical reactions are speeded by increasing the concentration of the reactants, since this raises the number of collisions between the reactants. The rate of reaction is also directly related to temperature, since, as the temperature rises, so the average kinetic energy of the atoms and molecules increases. Again, the frequency of effective collisions thereby rises.

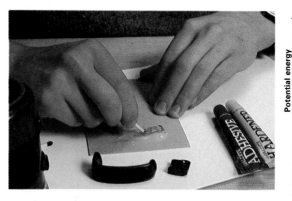

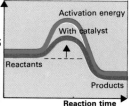

◀▲ Two energies are important in chemical reactions – the energy difference between the reactants and the products, and the activation energy, which has to be put into the system to get the reaction started. Catalysts affect the activation energy, resulting in a speeding up of the process, such as when a setting agent is added to an adhesive.

◀▶ The most amazing catalysts are the zeolites. These are important in the conversion of methanol (CH$_3$OH) to gasoline, and a plant using the zeolite process is now in operation in New Zealand. Inside the zeolite are minute channels and cavities into which the methanol molecules percolate and where they are induced to give up their oxygen atoms as water, leaving behind their carbon part which emerges as hydrocarbon chains.

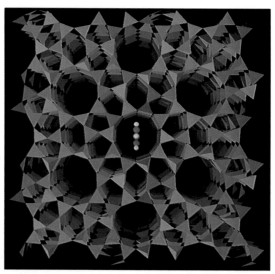

98

Investigating pathways

An apparently simple chemical reaction may in fact involve a number of steps, which can be studied mathematically. An investigation of the rate of the individual steps of a chemical reaction gives some idea of the overall mechanism, but must be coupled with other investigations before a full picture is obtained. The reaction products must always be determined. Isotopic substitution gives chemists another weapon in their armory to help determine the often complex reaction pathways (♦ page 224). Radioactive isotopes introduced into a molecule as a substitute for nonradioactive atoms can be useful in tracing the mechanism of some reactions. Isotopes can be used to tag part of a molecule and can be detected in the products of the reaction.

Isotopic labeling, together with nuclear magnetic resonance (NMR (♦ page 201) can be used to detect the making and breaking of carbon-carbon bonds, and of bonds between carbon and other groups of atoms. It can also be used to distinguish between inter- and intramolecular rearrangements, even in complex biological molecules. The molecular mechanism of photosynthesis has been determined by use of $^{14}CO_2$. Similar studies have helped in tracing the pathways of other important syntheses, such as those for some fungal toxins which are dangerous contaminants in grain and other foodstuffs.

Chemistry at the speed of light

The study of chemical processes that occur within tiny fractions of a second is fundamental to the understanding of many chemical reactions and pathways. For many years chemists have known that transient molecules or chemical species are formed in the course of some reactions, but only since the 1940s has it been possible to trace and identify some of these elusive intermediates.

These studies are often used to trace the formation and action of free radicals. They also give an idea of the chemical processes that take place during photosynthesis and even combustion.

Early work on fast chemical reactions was carried out in the 1950s. At the Max Planck Institute for Physical Chemistry in West Germany, Manfred Eigen (b.1927) developed new methods of chemical relaxation for studying very rapid reactions in solution. At about the same time, the British chemist George Porter (b.1920) pioneered the study of reactions started by short pulses of intense light.

Eigen's studies involved the development of a means of measuring the rate of very fast reactions (those occurring in less than one millionth of a second 10^{-6}). At such speeds, the mixing of reactants is important, and it is virtually impossible to investigate the kinetics of the reaction even using quite exotic mixing techniques.

In Eigen's chemical relaxation methods, the reactants are mixed and allowed to reach equilibrium with the reaction products. One of the physical parameters on which the equilibrium of the system depends is then suddenly changed, and the system is observed until it moves back to equilibrium, or relaxes. Temperature, pressure and electric field changes have been used to investigate fast reactions, as have ultrasonics.

If the shift of equilibrium of the system is small enough, the rate of the relaxation reaction equates to a unimolecular reaction. Calculations of this reaction rate offer information about the overall reaction. The rate of reactions of hydrogen ions (H^+) and hydroxyl ions (OH^-) to form water have been studied with a temperature jump method.

The pioneering work on flash photolysis by George Porter have proved of more importance today, with the development of laser technology. With this technique he was able to investigate reactions occurring in fractions of a millisecond ($10^{-3}s$). In the original method, a discharge tube was used to give a flash of high-energy light and to initiate the reaction; today lasers are used. The products of the fast reaction induced by the light flash (photolysis) – usually free radicals and atoms – are detected using a spectrometer.

In an early experiment Porter examined the reaction of chlorine (Cl_2) and oxygen (O_2). He found that a chloromonoxy radical (ClO) was formed which reacted after the flash to give chlorine and oxygen again. This suggested that an intermediate complex ClOO was formed during the reaction.

◄ George Porter using laser equipment to study very fast reactions. The introduction of lasers has allowed light pulses as short as 0·01 picoseconds to be used, and these can probe the most short-lived chemical processes. Information on free radicals, transition state species and electronically and vibrationally excited molecules can now be gleaned.

Using the Elements

12

Iron and steel...Other alloys...Making liquid oxygen... Separating the gases in the air...Silicon and germanium, the semiconducting metals...PERSPECTIVE... Metallography and archeology...Using helium, neon and argon

Materials found in nature have always been used to make life easier and safer. Over some 7,000 years, wood, bone and flint were the raw materials used for most artefacts. Since that time the elements from nature have been separated and recombined and fashioned into an almost endless list of everyday objects, tools and machines.

The separation and isolation of some of the elements on à large scale has driven the advance of technology since the Industrial Revolution. At the forefront of this progress has been the use of metals and alloys to build better and more efficient machines.

Iron is the most widely used metal. It is produced in three main types – wrought iron, cast iron and steel. Pure iron (Fe) is a white lustrous metal that is quite reactive and not very hard. It is attacked easily by air; iron hydroxide is formed and flakes away exposing more metal (◀ page 94). But when iron is made on a large scale in a blast furnace, it mixes with carbon and silica (SiO_2) to give irons of varying degrees of hardness and malleability. Iron is drawn from a blast furnace in ingots or pigs and these can be used to make the three main types of iron – wrought iron, cast iron and steel.

Wrought iron is essentially a malleable alloy, or mixture, of iron and phosphorus with small percentages of carbon, manganese, sulfur and silicon as impurities. The properties of wrought iron depend upon all these ingredients and also upon the amount of glassy fibers of slag produced from the "scum" or waste material – including silica – in the blast furnace. Cast iron contains between 1·8 and 4·5 percent carbon plus elements such as silicon, phosphorus and manganese.

The properties of cast and wrought types of iron can be changed by altering the conditions under which they are made and by reducing, or increasing, the amounts of impurities. Commercially pure iron can be prepared in either a specially-lined blast furnace, or by electrolysis (◀ page 60), but it is a very different material from pure iron. This type of iron contains only about 0·002 to 0·008 percent silicon, sulfur, phosphorus, carbon and manganese.

Iron alloys are usually produced from cast iron. Different percentages of different elements give these alloys better mechanical and physical properties when in use. The most widely used alloy of iron is steel. Steel is really an iron alloy with a low carbon content, the alloying elements being manganese, cobalt, chromium, molybdenum, nickel, tungsten, vanadium, selenium, aluminum and other metals in varying combinations for different purposes. Chromium and nickel are added to steel to improve its corrosion resistance and malleability. Very hard tungsten steels are used in high-speed cutting tools.

Stainless steels contain relatively high percentages of nickel and chromium, but other elements are often added – including bismuth, selenium, silver and lead. The addition of these elements helps make the machining of stainless steel parts easier.

▲ *The Pillar of Mehauli (near Delhi) is a column of pure iron over 10m tall. It has resisted rusting for nearly 1500 years, but this is less due to any special metallurgical knowledge of the ancient craftsmen, just a favorable climate. The heat of the day is retained by the pillar sufficiently to prevent dew condensing on it during the night.*

▼ *Electron microscope photographs of metal surfaces. On the left a piece of ferrite (iron) that has fractured as a result of being overstretched. The round particles are actually manganese sulfide and each is about a millionth of a meter in diameter. On the right is a piece of steel magnified a thousand times.*

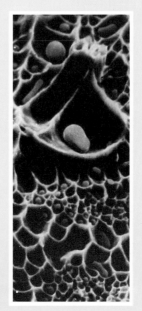

Metals, alloys and carbon

Other metals are used throughout industry, either in their pure state or as alloys. Copper can be made in commercial quantities to a purity of 99·95 percent for use in power cables, busbars and windings for electric motors and dynamos. The brasses are alloys of copper and zinc with manganese, aluminum, nickel, phosphorus and antimony sometimes added. Alloys of copper and tin, often with other elements added, are called bronzes.

Special applications require special materials and, while steel still provides the bulk of all engineering materials, other metals and alloys are used to make aircraft, spacecraft and some automobiles. Titanium is used to make aircraft but, as with steel, engineers usually find that titanium alloys are more useful than the pure metal. Aluminum is also used whenever corrosion-resistant structures are needed.

Carbon is used in industry as a pigment, additive and in the manufacture of inert coatings and electrodes. Small fibers of a polymeric form of carbon can be made, and these are very strong. These carbon fibers are used to make lightweight but strong composite materials for use in aircraft structures and many leisure goods such as tennis rackets and skis. Because carbon is inert the fibers are also being used in medical research as implants in the body, to replace tendons for example, and in parts of heart pacemakers.

Much of the carbon produced in industry is made as carbon black. This type of carbon is made today from petroleum and tar residues, in a process that can be traced back to antiquity. To make carbon black, petroleum and tar residue "fuels" are burned in an atmosphere with insufficient oxygen. A sooty deposit of carbon is formed, which settles and is collected. The types and sizes of particles formed from the carbon black process play an important part in the uses to which carbon black can be put. Most carbon black is used in the rubber industry as a reinforcement, particularly in the manufacture of automobile tires. A loose or "fluffy" form is used as a pigment, mostly for newspaper printing inks but also for paints (◆ page 170). Some special forms of carbon black are added to plastics to make them electrically conductive or to improve their ability to resist weathering and their antistatic properties. Because of its high electrical conductivity, carbon black is also used to make brushes for dynamos and electric motors and also to make electrodes.

◄ Titanium is a very light metal; its alloys are strong and have a high melting point which makes them, suitable for use in supersonic aircraft, where friction from air resistance is high. Recent experiments have been made to test its suitability for prostheses (artificial limbs); its strength, lightness and resistance to corrosion are its main advantages here.

▲ Carbon fiber has recently found many new uses, particularly for objects that need to combine lightness with strength. Many leading tennis players prefer carbon-fiber rackets (top), while carbon-fiber reinforced plastics have revolutionized light aircraft design (above); as much as 60 percent of such an aircraft by weight may be made of this material.

Chemistry and archeology

The skills of the chemist, particularly those of the metallurgist, can come to the aid of the archeologist in unraveling the achievements of early cultures.

The figure shown left, about 20cm high, was produced by the Indians of Colombia. The main material is a gold-copper alloy, which was made simply by melting gold and copper together to combine hardness with brilliance. Study of the microstructure of the metal in an artefact can reveal much information about the way in which it was made. Analysis of minute cross-sections of the metal has shown that the surface layer was treated with acidic plant juices to remove an ultrathin layer of copper. This technique, known as depletion gilding, left a bright sheen on the metal.

The object was made to be hung around the neck, and contained lime. This was used by the Indians to release the alkaloids in the coca leaves that they chewed in order to stave off hunger and induce altered states of consciousness.

In addition to metallographic analysis, spectrography, X-ray fluorescence analysis and examination of samples of the metal under the optical and electron microscopes, it has also proved possible to subject this particular piece to radiocarbon dating. The statuette was made by the lost-wax technique whereby a wax model was surrounded with clay and then baked. The wax was run off, leaving a clay cast into which the molten metal was poured. Tiny amounts of clay remain in the core, and these have been analyzed and the piece dated to between AD 600 and 1000, 500 years before the arrival of the Europeans.

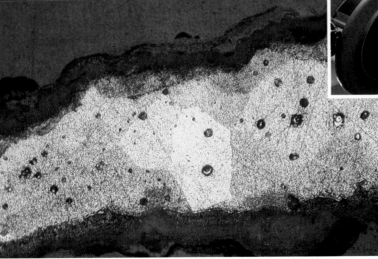

◄▲ *The gold figure was collected in Colombia, South America, and has been subjected to metallographic analysis to confirm its date and method of manufacture. The micrograph of a section through the metal shows the very thin surface layer, where the copper in the gold-copper alloy has been carefully removed to enhance its appearance.*

▲ *An archeologist submits another ancient artefact, this time a Roman coin, to metallographic analysis. By placing the coin in a mass spectrograph, the metal content of the alloy can be determined. Knowledge of the proportion of precious metal in coins is invaluable information to the historian attempting to draw a picture of the economic conditions of the past.*

Making pure oxygen

Oxygen is used in many of the process industries to make combustion more efficient. It is vital to the cost effectiveness of many chemical and metallurgical processes. Oxygen is used in the production of both iron and steel; in the chemical industry it is used to "crack" methane (natural gas; CH_4) to produce acetylene (C_2H_2) and also in the manufacture of oxygen-containing compounds.

Air contains almost 20 percent by volume of oxygen and is the raw material for almost all oxygen production. Air can be separated into its component gases if it is first liquefied and then distilled. The various gases in the air – oxygen, nitrogen and some argon – can be separated in this way because of their different boiling points. Liquid oxygen is used as a propellant in missiles and rockets and also to produce oxygen for breathing in aircraft, spacecraft and hospitals.

Two basic processes are used to liquefy gases and widely used in all oxygen-producing plants. In both processes air is first filtered to remove dust and other solid particles and then compressed. This compressed gas is further purified if necessary and then cooled, both with the help of refrigerant chemicals and by the exchange of heat with recycled gas, product and waste gas from the plant. In one process the compressed gas is allowed to expand through a throttle valve or nozzle. The volume of the gas before and after expansion is kept constant, heat is lost and the gas cools. Some of the gas is cooled sufficiently for it to liquefy, and this is either stored or passed to a distillation column to be separated into oxygen, nitrogen and argon.

In the second method, the liquefaction process is more efficient. Cooling is achieved by expanding the compressed air through an engine, usually a turbine, doing what is called "useful work". Expansion of gas through the engine is at constant entropy (◀ page 38), energy is removed from the system and the temperature of the gas drops. The engine does not produce liquefied air, but expansion through a valve and further cooling allows liquid air to be made.

When liquid air is warmed the first gas to be given off contains 93 percent nitrogen and 7 percent oxygen to leave a liquid rich in oxygen. As the temperature is raised towards the boiling point of oxygen (−183°C), more oxygen is released. The last liquid to boil contains about 45 percent oxygen. Because of this, oxygen separation plants use two fractional distillation columns, one at a relatively high pressure, to ensure the final liquid is almost pure oxygen.

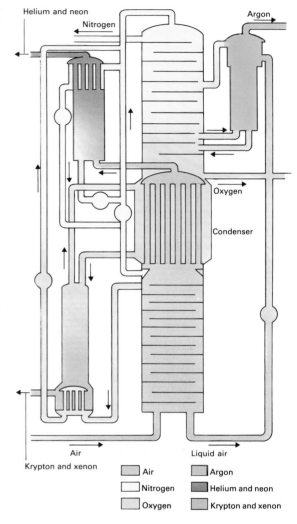

Helium and neon
Nitrogen
Argon
Oxygen
Condenser
Air
Liquid air
Krypton and xenon

	Air		Argon
	Nitrogen		Helium and neon
	Oxygen		Krypton and xenon

▲ *Six gases are extracted from the air by cooling it until it liquefies and then distilling it. Each of the component gases boils at a different temperature. Nitrogen, oxygen and the noble gas argon are the main industrial products, and are produced in very large quantities, but the rarer gases that are extracted, neon, krypton and xenon, also find commercial and scientific use.*

◀ *Provided the water leaving a sewage works contains enough dissolved oxygen it can safely be returned to the environment at the end of the treatment. The purification of water is one of the major uses of purified oxygen.*

▶ *When the gas acetylene is burned in pure oxygen, enough heat is given out to melt and weld steel even in the coldest climates, like here on the Alaska pipeline. Sometimes the inert gas argon is used to protect metals such as titanium and zirconium from being attacked by the air while they are being welded.*

Helium, argon and neon

Not all the so-called "rare" gases are particularly rare (♦ page 74). Helium is produced on a quite large scale from some natural gases. Argon is separated from air in much the same way as oxygen and nitrogen. The other noble gases, krypton, xenon and neon, are also obtained from air by liquefaction but special techniques are used to separate them.

These gases are used wherever a chemical, metallurgical or experimental process needs to be carried out in an inert atmosphere. Argon is most widely used for this purpose. It is used along with helium to provide an inert atmosphere under which some metals that are readily attacked by air can be welded. These "hard-to-weld" metals include aluminum, bronze and copper and some stainless steels; but the technique known as gas shielded arc welding, is most widely used for welding parts containing titanium and zirconium. The ability to weld these metals has increased their use in aircraft, spacecraft and in the nuclear industry.

The most common use of the inert gases is in the manufacture of "neon" lights. Neon emits the familiar orange light while krypton emits a pale violet light. Mixtures of the inert gases are used in incandescent and high-output lamps to give a range of colors and to improve their useful life.

Helium is used today on a relatively large scale to fill balloons and dirigibles. It is also used in rockets to replace liquid oxygen as it is burned and in the gas mixture breathed by deep sea divers. Helium boils at −268·93°C, or 4·21K, a temperature very close to absolute zero. Because of this, liquid helium is used to investigate the properties of matter close to absolute zero (♦ page 202).

▲ Hong Kong glows with the colors familiar to every modern city. The inert gas neon is the secret behind the illuminated signs that brighten cities at night. When an electric discharge is passed through neon it glows red. Other colors may be produced by traces of mercury vapor.

▶ Helium, being an inert gas and much lighter than air, is the ideal gas for use in balloons, whether for meteorological purposes or to assist in the study of the upper atmosphere or of cosmic rays. Such balloons may be required to reach heights of as much as 30,000m above sea level.

104

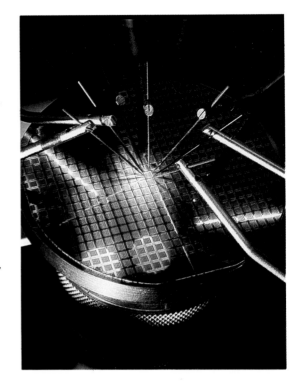

◄ *Gallium arsenide (GaAs) is now being manufactured as an alternative to silicon for microchips. This combination of elements is a semiconductor like silicon, but is electronically faster and therefore better for microprocessors. Silicon is a group 14 element, with four valency (bonding) electrons, whereas gallium is in group 13 with three electrons, and arsenic group 15 with five electrons.*

▶ ▼ *Making microchips demands extreme control over chemical components. The layers of conducting and insulating materials that are laid down on the surface of a silicon chip may be only a few atoms thick yet must perform to the highest specifications. Great care has to be taken in their manufacture (below), and each chip is checked by test probes to ensure it performs correctly (right).*

The semimetals

Some of the elements in Groups 13 to 18 of the Periodic Table (◊ page 75) behave neither as metals nor as nonmetals. They are not true conductors of electricity as are the metals, nor true insulators like the nonmetals. Because of this, they are useful in the fabrication of electronic devices.

The dramatic growth of the electronics and microelectronics industry followed the invention of the transistor in 1947 and more recently the silicon chip. These developments were dependent upon understanding and increasing production of semiconducting elements (◊ page 58). Silicon is the most widely used such element today. Single crystals of silicon are "pulled" from crucibles containing the molten element to produce wafers for the manufacture of silicon chips.

To create a silicon crystal of sufficient purity, metallurgical grade silicon, 98 percent pure, is used. This is reacted with hydrogen chloride to form a liquid, trichlorosilane, which is distilled several times to reduce impurities to a level of parts per trillion. The trichlorosilane is then heated and decomposes to form perfectly pure silicon crystals up to 125mm across.

Making integrated circuits

The silicon crystals are "doped" to produce semiconducting material with special characteristics. Doping silicon with tiny amounts of phosphorus or arsenic produces "n-type" material while doping with boron produces "p-type" material. Combinations of n- and p-type semiconductors produce so-called p-n junctions. These junctions are used to create diodes and transistors in microelectronic circuitry. Photographic and chemical techniques are used to etch wafers of silicon and other semiconducting materials to form these junctions and make ever more compact electronic devices.

Compounds or alloys of two or more semiconducting elements have unique electrical properties. One of the most exciting of these is gallium arsenide (GaAs). Because it is made from gallium, in Group 13 of the Periodic Table, and arsenic, in Group 15, gallium arsenide can be grown on a base, or substrate, of one crystal lattice layer at a time. By careful control of impurities, or dopants, and the thickness of the layers, electronic components can be tailor-made. Work on the manufacture and use of GaAs devices is burgeoning, especially because of the fast growth of optical laser communication technology.

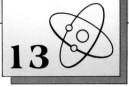

13

Natural Compounds

The first compounds on Earth... The development of the atmosphere...Water and chemistry...The molecules of life...Proteins, peptides and amino acids... Hydrocarbons...PERSPECTIVE...Stalactites and stalagmites...Left- and right-handed molecules

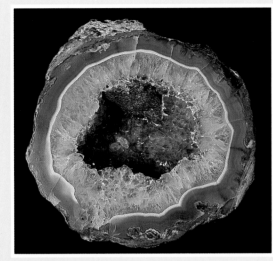

▲ *The Earth's crust is formed mainly of inorganic minerals. Most rocks are composites, but a few are pure chemical substances. Given the right conditions it is possible for dissolution and very slow crystallization to occur. This can result in beautiful crystals which are prized as gems, such as these agate and amethyst crystals which have grown inside this hollow stone, called a geode, from Mexico.*

With very few exceptions, everything that surrounds us in daily life is either a pure chemical compound or a mixture of compounds (◀ page 77). The exceptions are those chemical elements which are used in their pure form, such as gold, or which exist in mixtures, such as the gases that make up most of the atmosphere.

The majority of the chemical compounds around us are not the products of industry but of nature. Whenever we look at a spectacular piece of scenery we see the effects of the laws of physics, chemistry and biology in assembling compounds in a particular manner.

The Universe seems to have originated about 15 billion years ago. Although small atomic nuclei such as helium may have formed within the first few minutes of the Universe's existence, no chemistry occurred until it had cooled down sufficiently for electrons to stabilize around nuclei. Evidence obtained from radiochemical analyses of meteorite material indicates an age of about 5 billion years for most elements. It was another billion years later that the Earth settled down chemically, although the planet probably formed about 4·5 billion years ago.

The origins of compounds on Earth

How the planets formed is still debated. Nevertheless, it is likely that the primeval Earth was much hotter than it is now, so its chemical equilibrium was significantly different from that which is experienced today. As a result, the Earth now has a molten metallic core, composed primarily of iron, surrounded by a solid metallic layer. This in turn is surrounded by a thin crust that contains the large variety of accessible chemical compounds.

Another factor which affects the Earth's chemical composition today is biology. Biological activity, utilizing energy obtained from the Sun, maintains a chemical disequilibrium on Earth. If life were to cease the atmosphere would change dramatically over a short period (in geological terms), as oxygen combined with elements in the Earth's crust to form oxides.

The early Earth probably had an oxygen-free atmosphere. Most of the oxygen now in the atmosphere was probably combined as water in mineral hydrates. The energy of sunlight may have decomposed this water to hydrogen and oxygen and, as the lighter hydrogen escaped from the Earth's gravitational pull into space, free oxygen accumulated above the Earth's surface.

Many of the major mineral compounds formed before the oxygen entered the atmosphere, although some mineral formations are the result of biological chemistry, and are so of more recent origin. Igneous rocks, such as granites and basalts which formed from melts, are examples of compounds formed by chemical processes as the Earth cooled down.

▲ *Fingal's Cave, on the southwest coast of Staffa, an island of the Inner Hebrides off Scotland's west coast, is a cave carved out of the basalt rock. Here the dramatic effects produced as the molten magma of the early Earth gradually cooled and set in large crystalline forms can be seen. Such igneous rocks are among the oldest rocks that can be found on the Earth's surface.*

Fluoride (F⁻)
Strontium (Sr²⁺)
Boric acid (H₃BO₃)
Bromide (Br⁻)
Bicarbonate (HCO₃⁻)
Potassium (K⁺)

Calcium (Ca²⁺)

Magnesium (Mg²⁺)

Sulfate (SO₄²⁻)

Sodium (Na⁺)

Chloride (Cl⁻)

▲▶ The salinity of the sea is about 33-38 parts per thousand, and there are trace amounts of most elements, held in solution. Sodium and chlorine ions, which go to make up common salt, account for about 85 percent of the substances in solution. Most of this salt is washed into the sea from rivers, although these have an average salinity only one two-hundredth of the sea.

Dissolved salts

Water (H₂O)

Sea water

Once the Earth had cooled sufficiently for liquid water to exist on its surface, additional physical and chemical factors came into play in the formation of chemical compounds. Water dissolves many compounds, and offers conditions in which new compounds may be formed. In parts of the world, deposits of soluble compounds were left behind when climatic change evaporated inland seas. In some cases, as in the salt deposits of western Europe, subsequent geological activity buried these soluble materials under impermeable rock. This has protected them from dissolution so that they are now exploitable mineral deposits. In other parts of the world, such as Chile which is famed for its nitrate deposits, subsequent climatic conditions have never been wet enough to redissolve the exposed deposits.

Possibly the most peculiar geological phenomenon of this type is the Dead Sea. River waters bring soluble salts into this inland sea in vast quantities each year. The water then evaporates, increasing the concentration of mineral salts so that they crystalize out on the shores

◄ *All sea water is salty but not to the same extent. The saltiest of all is the land-locked lake called the Dead Sea, which as its name implies is too salty to support most animal or plant life. A satellite photograph gives some idea of the unique location of this body of water which is 295m below sea level, 80km long, with a surface area of 1049km², and is over 400m deep in places. At the bottom sodium chloride (salt) is precipitating from water that is ten times saltier than the oceans. The lake is an enormous mineral reserve and a chemical works on its shore extracts potassium, magnesium and calcium chlorides. Another works extracts bromine from the lake water. Although the lake is fed by the River Jordan, more water is lost by evaporation than flows in: this explains its saltiness.*

▲ ◄ *When seas dry up the dissolved salts are left behind. Vast deposits, several meters thick, are found in several parts of the world. Since salt is one of the basic raw materials of the chemical industry these reserves are exploited. In some countries the salt is mined, as here at Kewra in Pakistan, but where water is more plentiful the salt is obtained by pumping in fresh water and extracting it as brine.*

Chemistry in caves

Some of the world's most spectacular natural phenomena are caves filled with stalactites and stalagmites. These have been formed primarily through simple solution chemistry.

Most cave systems occur in limestone rocks, which are composed mainly of calcium carbonate ($CaCO_3$). When carbon dioxide gas dissolves in water, it forms a weak acid which dissolves calcium carbonate by converting it to calcium bicarbonate. The rate at which this reaction occurs and the extent of the solubility depend on a number of factors, including pressure and temperature.

Rocks may be dissolved internally as well as externally. Rainwater entering a rock through a fissure, caused by volcanic upheaval for example, can hollow out the rock from the inside, thus forming a cave.

Where geological changes have reduced the amount of water entering a cave system, so that most of the space is filled with air, stalactites and stalagmites may form.

If small amounts of water saturated with calcium bicarbonate percolate from the surface into cave roofs, they may enter a cave atmosphere in which the concentration of carbon dioxide gas is low. The solution then loses carbon dioxide. This reconverts some of the calcium bicarbonate to less soluble carbonate, which precipitates.

For a stalactite to form, the rate of drop formation has to be low enough for this process to take place. The carbonate precipitates on the outside of the drop and, when the drop falls, is left behind. The next drop adds more solid to the stalactite.

Provided that conditions are suitable on the floor of the cave where the drops land, further precipitation can take place there to build up a stalagmite. In some cases, the growth process continues long enough for stalactite and stalagmite to join into a single column.

Other minerals as well as calcium carbonate are susceptible to cave formation. Notable among these are dolomite $CaMg(CO_3)_2$ and anhydrite $CaSO_4$. Anhydrite expands when it is hydrated to form gypsum, and thus helps fractures to occur. Alabaster and onyx are precipitations of gypsum solutions which have reached the surface.

of the lake and on any objects, such as driftwood, found in it. This site has been exploited commercially for these chemicals for many centuries.

A major determinant of the current chemical composition of the Earth's rocks is another phenomenon that owes its development intimately to the existence of water, namely life. Living organisms produce large quantities of complex chemical substances. At the same time, through their metabolic processes such as digestion and respiration, they alter the structure of many other chemicals. This process occurs as much among the lowest forms of life as among the more developed animals and plants. Thus iron-oxidizing bacteria obtain their energy by converting naturally-occurring metallic sulfides into more highly oxidized compounds. Indeed, the ability of this type of bacteria to convert insoluble copper sulfide (CuS) into soluble copper sulfate ($CuSO_4$) is increasingly exploited in the commercial production of sulfur from lowgrade ores.

▲ *All stalactites start as thin "straws" of calcium carbonate. Subsequent growth over many hundreds of years can produce very thick stalactites, which may be impressively colored from other minerals deposited with the carbonate.*

Phospholipid membranes

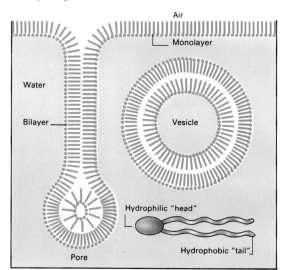

▲ Living things need the protection of membranes, and happily there is a type of molecule that will form these naturally. Such molecules have a hydrophilic (water-seeking) end, derived from an oxy-acid, and a hydrophobic (water-repelling) end, such as a hydrocarbon chain. Such molecules cluster together with all the hydrophobic ends at the water side. Phospholipids, with a phosphate head and two hydrocarbon tails, make the best membranes. Such molecules can form either a single layer (monolayer) for skin, or a double layer (bilayer) for a cell membrane.

The molecules of life

It is the production of large quantities of complex, carbon-based compounds – carbohydrates, proteins and nucleic acids – which is the most notable attribute of living organisms. Several theories have been put forward as to how these originally came about.

The small molecules from which these biological macromolecules are made can be formed from inorganic materials. The conditions required for this process may have existed on the early Earth. The key factors seem to have been an oxygen-free atmosphere, abundant supplies of energy (either as solar radiation, intense electrical storms, or the high levels of radioactivity which would have existed at the time) and a very large amount of time.

It is also possible that complex molecules formed in space before the surface temperature was sufficiently cool for them to be stable on Earth. The first molecule discovered in space was ammonia (NH_3), which was found in 1968 in a dark nebula. Dark nebulae contain a massive number of dust grains. Dust provides a very large surface area which can catalyze the formation of molecules and may also help to protect them, once formed, from decomposition by ultraviolet radiation.

Since 1968, many molecules, including methanol (CH_3OH), formaldehyde ($HCHO$) and hydrogen cyanide (HCN), have been detected in space. It is possible that the Earth may have continued to collect dust from space long after its early hot phase, and this might have brought with it many molecules which formed the building blocks for the chemicals of life.

◄ *The atoms from which we are made began their life in the nuclear reactions of stars deep in space. Evidence of simple molecules in the debris between the stars has been found by analyzing infrared radiation from galaxies.*

► *The chemicals for making cells may have been present in the Earth's early oceans. In 1953 Stanley Miller made many such complex molecules by passing an electric discharge through a mixture of gases like those of the early atmosphere.*

▼ *Amino acids are key building blocks of life, among which glycine (left) and alanine (right) are the simplest. some amino acids are found in both a left- and a right-handed form, as shown for alanine. Nature prefers the left.*

In the 19th century, the Russian chemist Alexander Butlerov (1828-1886) discovered that formaldehyde polymerizes in the presence of alkali to produce mixtures of sugars (◆ page 158). Hydrogen cyanide and ammonia, on the other hand, will readily react together to form adenine, one of the four organic bases found in the DNA molecule and so found in the genetic code.

Many such compounds can be formed under laboratory conditions believed to simulate those present on the early Earth. Different hypotheses have been tested as to which gases were present in the early atmosphere. Regardless of the mixture proposed, an array of potential biological molecules can be made to appear.

Other experiments, originally undertaken in the Soviet Union but subsequently repeated in other parts of the globe, have shown that isolated mixtures of biological macromolecules tend to form cell-like aggregates. There is a great difference between such cell-like aggregates and a reproducing organism. Nevertheless, these experiments have shown that the origin of life from inanimate matter does not violate any of the principles of physics and chemistry.

Although many biological compounds have complex chemical structures, most of them are composed of only four elements – carbon, hydrogen, oxygen and nitrogen. Compounds which have a backbone of carbon atoms are called "organic", because it was originally believed that they could be produced only by living organisms. This belief was dispelled a little over 150 years ago, when chemists began to synthesize "organic" substances in the laboratory from inorganic precursors.

Nature's left hand

Many of the molecules produced by living organisms have a particular "handedness". The atoms making up the molecule can be linked together in the same order, but produce a molecule which is spatially non-superimposable. The simplest analogy is with a pair of gloves. Mirror-image molecules are said to show chirality, and this property can be observed by seeing the way in which a solution of such molecules rotates the plane of polarized light. Where the names of molecules are prefixed with an L- or a D-, it indicates their chiral configuration.

All the amino acids used in building proteins have the L-configuration, and few D-amino acids are found in nature. On the other hand, the sugars are D-sugars, and the L-sugars occur infrequently. The different configurations of these two classes of compound are probably related. When D-glucosamine, a derivative of glucose, is converted to the amino acid alanine, it produces the L-form.

In many cases, where a substance is converted by an enzyme, only one form of the substance will convert successfully: the mirror-image is inactive. Life has a definite handedness. If, on another planet, an astronaut found life based on the opposite optical isomers, he could starve to death, as he would be unable to digest or utilize proteins made from D-amino acids. Recent research has suggested, however, that if there is life elsewhere in the Universe, and it uses similar molecules to those of life on Earth, then they will have the same handedness.

When experiments to simulate conditions on the early Earth have produced amino acids, they have all done so in the form of racemic mixtures (that is equal amounts of the L- and D-forms). For many years, scientists have thought the predominance of one form might have been due to chance. Recently, however, studies of the weak interaction in particle physics (◆ page 213) have shown that the electron and positron (positive electron) are themselves chiral particles.

What this means is that the true mirror-image of a chiral molecule is one in which all its constituent subatomic particles have opposite charge. Such true mirror images would be isoenergetic. However, where the mirror-image is composed of those particles of which our world is made, the two molecules have very slightly different energies. So one form is slightly more stable than another.

The difference in stability is very small. One mole of any substance contains about 6×10^{23} molecules; in a mole of a racemic mixture of L- and D-alanine, there would be an excess of about 1,000 molecules of the L-form, because of its greater stability. On the geological timescale, this would have been enough to swing the balance in favor of the handedness we now find on Earth. It may have been helped by the handedness of some inorganic materials, which could have acted as catalysts for the formation of complex organic compounds. There is a very slight preponderance in most rocks of quartz of a particular handedness. One question that still remains unanswered is whether this handedness existed before life, or whether it has been induced by the handedness of life.

Some 20 amino acids combine together to form all the myriads of proteins in living creatures

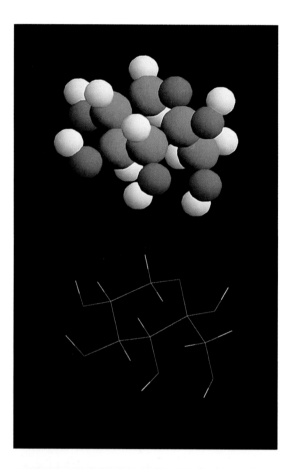

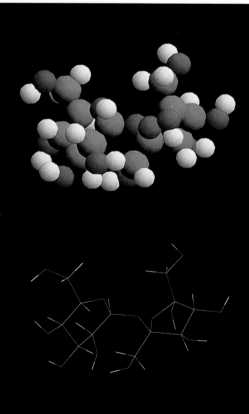

Many of the essential compounds of life are polymers: that is, giant molecules made by linking together dozens, sometimes even hundreds, of smaller units called monomers. Thus, the organic compounds known as proteins are made from about 20 different amino acids, which all have the same "core" structure, comprising a carboxyl (COOH) group at one end and an amino group (H_2N) at the other, but a different side chain. There are only about 20 common amino acids, but by combining different ones together in different proportions, an enormous range of proteins result, with widely different characteristics. Hoof, horn and hair may be seen as one extreme, while the enzymes, the catalysts of life (◀ page 96), come at the other.

In addition to the proteins, many other, smaller molecules are made of amino acids. These compounds, called peptides or oligopeptides, may contain only three or four amino acids or as many as a hundred. Many play essential roles in metabolism: the hormone insulin is one, the lack of which causes diabetes.

Proteins are made according to instructions contained in genes, which are themselves made of nucleic acid polymers. These are more complex in structure than proteins, because the monomeric unit actually comprises three smaller molecules: a sugar, a nitrogen-containing base and a phosphate. However, diversity again comes from a small number of components. Only two sugar molecules are used: ribose ($C_5H_{10}O_5$) in ribonucleic acids (RNA) and deoxyribose ($C_5H_{10}O_4$) in deoxyribonucleic acids (DNA; ◀ page 13). In DNA, only four different bases occur: adenine, guanine, cytosine and thymine. In RNA, cytosine replaces thymine, and occasionally other bases are also incorporated.

Simple sugars are also widespread in nature, as are the polymers, such as starch and cellulose, which are made from them. Cellulose is one of the most abundant compounds to occur naturally, because it is used as a structural material by plants. Although primarily a glucose polymer, naturally-occurring cellulose is not 100 percent pure. One of the purest forms, cotton, is about 96 percent pure glucose polymer.

Living systems have developed some ingenious combinations of different materials to help their survival. A polymer related to cellulose is chitin, which occurs widely in insects and crustaceans. In insects, the chitin is combined with protein to form structural components, such as the body casing of beetles. In crustaceans, the chitin provides a matrix for the calcium carbonate which is the major component of the shells.

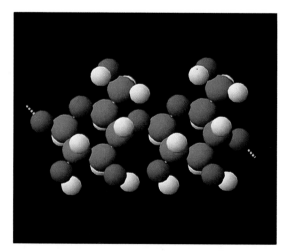

◀ **Saccharides or sugars consist of rings of four or five carbon atoms (green) and one oxygen atom (red). Attached are hydroxy (OH) groups. Sugars can be single ring substances, such as glucose (top), or two rings joined together, as in sucrose (below), better known as ordinary sugar, or hundreds of rings (starch). The figure (left) shows a tiny part of a starch chain.**

▶ **Many biological reactions are speeded up by enzymes, or natural catalysts. These are giant molecules, with cavities or active sites (inset) where the reactants are trapped.**

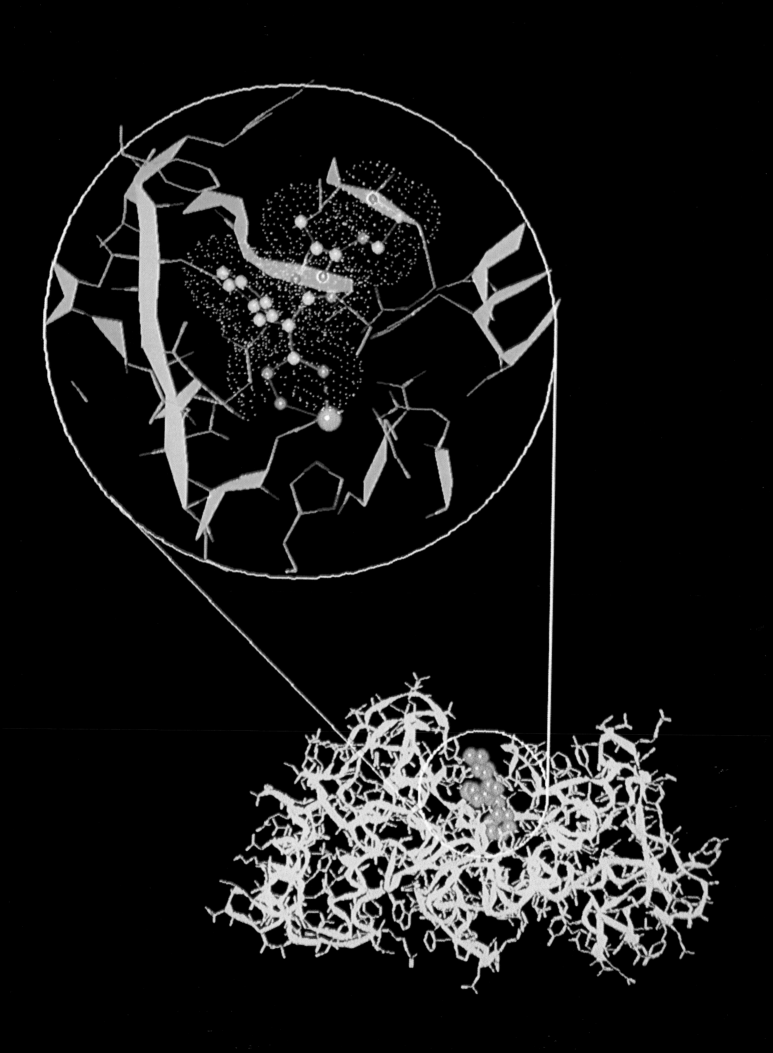

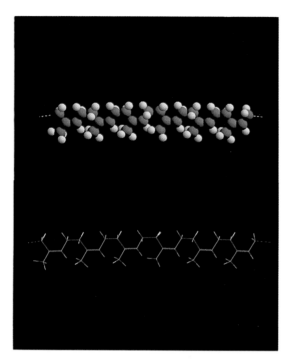

Polysaccharides, as the polymers of sugars are called, are, like protein, diverse in their uses. In addition to structural components, living organisms also use them as energy storage compounds – starch in plants and the similar glycogen in animals. The blood-clotting factor, heparin, is also primarily a polysaccharide. As with amino acids, sugars also occur naturally in smaller molecules, oligosaccharides. Many naturally-occurring gums are oligosaccharides.

The simple sugars from which many polysaccharides derive are composed only of atoms of carbon, hydrogen and oxygen. A few naturally-occurring substances, such as rubber, are composed solely of carbon and hydrogen. Natural rubber is a polymer of the simple molecule isoprene, one of a family of compounds called unsaturated hydrocarbons. The simplest of this class of molecules is ethylene, a compound of considerable importance industrially (♦ page 133). Ethylene also occurs naturally. It is produced by plants such as apples and tomatoes during ripening.

Fossil hydrocarbons

Although most of the ethylene produced today by industry is considered to be synthetic, it ultimately derives from natural sources – the fossil hydrocarbon deposits which play such a key role in the world economy. The best known fossil hydrocarbons are oil, natural gas and coal. These have formed over millions of years from the remains of living organisms. Under the geological conditions prevailing, the oxygen and nitrogen have tended to disappear from the complex molecules of life, to give mixtures of hydrocarbons.

The way in which fossil hydrocarbon deposits occur depends on their geological history. In addition to the readily exploitable deposits, there are many others, such as the huge expanses of tar sands in central Canada and the oil shales in many other parts of the world where the organic materials are intimately mixed in with inorganic materials.

▼ *The Earth's crust contains large deposits of hydrocarbons. In some places such as Trinidad, there is even a lake of molten pitch. Hydrocarbons are compounds of carbon and hydrogen, the simplest of which is methane gas (CH_4). Oil is largely long-chain hydrocarbons, while coal and pitch contain ring compounds as well.*

▲ *Most hydrocarbons are liquids or waxes, but they can also be polymers such as polyethylene and polypropylene, which have chains of hundreds of thousands of carbon atoms. Natural latex rubber is a hydrocarbon polymer. This is best seen in the stick diagram where the green rods represent the chain of carbon atoms.*

Synthetic Compounds

14

The earliest synthetic compounds...Perkin and aniline dyes...The uses of synthetic compounds...Planning a new molecule...Reactions in synthetic chemistry... PERSPECTIVE...Heat and chemical rections...The 19th-century chemist's laboratory...Designing molecules by computer

Chemical knowledge is based on experiments carried out in the laboratory. It is here that natural compounds are analyzed and their behavior studied, and where new compounds are first synthesized.

More than six million different chemical compounds have now been recorded. The great majority of these are organic (substances based on carbon). Most are synthetic, in that they did not exist until they were made in a laboratory. Much of the history of chemistry, however, has been concerned with the study of chemical substances which occur naturally. These were analyzed in laboratories to discover the elements of which they were made, how many atoms of each were present in a molecule and how those atoms were joined together.

Compounds were first synthesized as a means of checking the structures of natural compounds postulated on the basis of analytical findings. Both by accident and design, this led to the synthesis of new molecules. In an attempt to synthesize the naturally-occurring anti-malarial quinine, on the basis of what was then believed to be its structure, the British chemist, William Henry Perkin (1839-1907) oxidized aniline in 1856 and accidentally produced mauveine, the first synthetic dyestuff. This breakthrough led to the rapid development of a new branch of chemical industry.

Heat and chemical reactions

Heat has been used as an energy source in the chemical laboratory and its precursors since the dawn of scientific experiment. Heat is not only important in separating mixtures by distillation. It can also reduce the time needed for an experiment, because the rate of a reaction approximately doubles with each 10°C rise in temperature. On the other hand, too much heat can lead to decomposition of the reaction mixture, sometimes explosively. Consequently, controllable heat sources have been sought.

The earliest treatises on alchemical experimentation include instructions for heating substances. Mild heat could be supplied by surrounding a vessel with horse manure. The water bath, in which a vessel is immersed in boiling water, has also been used for centuries. The temperature at which the water boils places an upper limit on the amount of heat to which the reaction is exposed. This type of bath is still sometimes called a bain marie, after its reputed inventor, Maria the Jewess. According to a 3rd or 4th century work by the Greek alchemist Zosimos, she also invented the ambix or alembic, the forerunner of the still.

A common source of heat in the laboratory, introduced in the 19th century, is the Bunsen burner. Invented by the German chemist Robert Bunsen (1811-1899), it provides a gas flame of variable hotness. The heat of the flame depends on the amount of air that is mixed with the gas before it burns.

Many laboratories today use electric hotplates or heating mantles. These can be controled, like an electric hotplate on a domestic stove, to provide the required degree of heat.

◄ *William Henry Perkin (1839-1907) was the first person to make a synthetic dye, which he did when he was a student at the Royal College of Chemistry in London in 1856. He was trying to make synthetic quinine but the result was merely a sticky, black mess. When, however, he worked with aniline, a by-product from coal tar, he found that his product dissolved in alcohol to give a beautiful purple solution. Perkin, still only 18, left college and with his father and brother began to manufacture the new dye, which he called mauveine. The response of British manufacturers was less than enthusiastic, but once the color became the rage of Paris its success was assured. Called mauve, it dominated fashion for a decade. Following the success of mauve came a range of other synthetic dyes and soon plants to produce them were operating in Britain, France and Germany. Today chemical dyes color everything around us.*

Chemistry in the 19th Century

▲ ▶ *Justus von Liebig and his laboratory at the University of Giessen in 1842. Liebig was one of the fathers of modern chemistry, and to his laboratory came chemists from all over Europe and America. He developed methods for the exact analysis of organic compounds and is best known for introducing the Liebig condenser, although this lithograph shows the old-fashioned retort still in use (right center).*

The father of modern chemistry

One of the great practical chemists of the 19th century was Justus von Liebig (1803-1873). Born in Darmstadt, Germany, he became interested in chemistry through his father's coloring materials business. When castigated at school for his lack of diligence and asked what he intended to become, he replied that he would be a chemist. At this, the whole school "broke into an uncontrollable fit of laughter, for no one at that time had any idea that chemistry was a thing that could be studied".

After studying chemistry in Germany, he went to Paris for further practical study. From there, he went to Giessen, where he was professor of chemistry for 26 years, during which time he trained many chemists in the techniques of practical chemistry. Among his more famous students were August Kekulé (1829-1896), discoverer of the structure of benzene, and August von Hofmann (1818-1892), subsequently the first professor of chemistry at the Royal College of Chemistry in England. Liebig also made considerable research contributions in the field of organic chemistry, particularly through his development of techniques of organic analysis.

In the 1850s, Liebig moved to Munich, where he confined himself to lecturing rather than practical instruction. In later life also, he devoted himself to more broadly based problems, such as the use of chemicals in agriculture (◆ page 145).

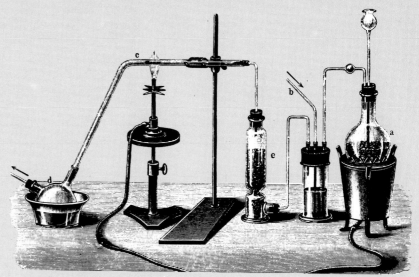

A 19th-century laboratory

In 1827, the British scientist Michael Faraday (1791-1867) published a 600-page volume, "Chemical manipulation; being instructions to students of chemistry on the method of performing experiments of demonstration or of research, with accuracy and success". In this, he described many processes familiar to experimental chemists today – weighing, distillation, precipitation, filtration, crystallization and many more. But the most important item in the laboratory as he described it was the general furnace. "Its use," he wrote, "is partly domestic, partly chemical; for it has to warm and air the place, occasionally to heat water, as well as to supply the means of raising a crucible to ignition, or of affording a high temperature to flasks through the agency of a sand bath."

The development of the voltaic pile as a conveniently available source of electricity also had a profound effect on chemical research. Faraday's book contains a long section on the construction of batteries and their use in experiments (◊ page 56). Important contributions to chemical theory came from this addition to the laboratory.

Through the efforts of Faraday and his contemporaries, chemistry became one of the exact sciences during the 19th century. In part, this was due to the development of an experimental technique which led from a laboratory dominated by a furnace to one in which many pieces of apparatus still in use today were to be found – the Liebig condenser, the polarimeter (device for measuring optical activity, used to study molecular configurations), knife-edge balances for accurate weighing of small amounts of chemicals, and the spectroscope (◊ page 94).

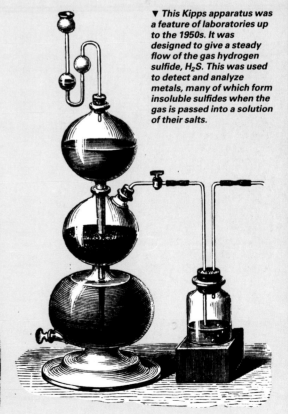

▼ This Kipps apparatus was a feature of laboratories up to the 1950s. It was designed to give a steady flow of the gas hydrogen sulfide, H_2S. This was used to detect and analyze metals, many of which form insoluble sulfides when the gas is passed into a solution of their salts.

◄ Sulfur trioxide, SO_3, dissolves in water to form sulfuric acid. In 1831 SO_3 was made from sulfur dioxide, SO_2, by passing this gas and air through a red hot platinum tube. Such simple apparatus is typical of that found in many 19th-century laboratories.

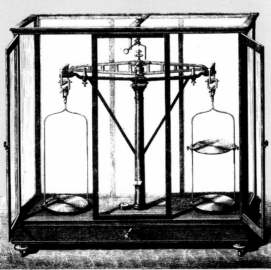

► The Scottish chemist Joseph Black (1728-1799) was the first to recognize the importance of weighing in chemical research, and through his influence chemistry was put on a sound footing. Analytical scales were accurate to within a milligram.

Chemistry Today

The modern laboratory

The key to practical chemistry today is electricity. Although the processes carried out in laboratories are basically the same as they have been for many years, laboratories are now cleaner, more efficient places, as a result of electrification. To heat substances, for example, synthetic chemists use electric hotplates or thermostatically-controlled heating units in which reaction flasks are designed to sit, rather than Bunsen burners. To concentrate solutions, rotary evaporators are used in which a flask of solution is turned continually by an electric motor so that only a portion of its contents is exposed to heat at any one time, thus making it easier to handle sensitive materials.

Other services may also be on tap: ordinary and deionized water (used in making aqueous solutions for analytical and physical measurements), vacuum lines for carrying out distillations under reduced pressure or for aiding filtration, and possibly one or more types of gas such as nitrogen.

▼ The white coats remain but modern chemistry laboratories are different from the traditional image of benches, Bunsen burners and bottles. Here materials are being analyzed for content in an automated distillation apparatus, from which the data is processed by computer. The linking of laboratory equipment to computers is becoming a regular feature, especially in analytical laboratories. The storage and access to information on chemicals allows easy identification of unknown compounds and components in mixtures.

► This maze of glass tubes, taps and bulbs, is called a vacuum line. To handle volatile substances and gases that are highly toxic, or react with oxygen or water vapor, it is necessary to manipulate them within the confines of such all-glass apparatus. As its name implies, a vacuum line can be completely evacuated before the dangerous or air-sensitive materials are introduced. Once in place they can then be safely analyzed, purified and reacted, operations otherwise done in the open laboratory.

▶ *Chemical analysis consists of identification and exact measurement. Over eight million chemical compounds are known (◊ page 88), and identification of one is made possible by comparing, for example, the mass or infrared spectrum of an unknown substance with those on record. Often it is only tiny impurities in food and drink, blood and urine, air and water, that are of interest. The exact determination of the amount of a contaminant is the purpose of this type of analysis. Here, samples of river water are being collected for pollution tests. The chemist may have to measure amounts of material that are present only in parts per million, or even parts per billion.*

Balances and spectrometers

In the analytical laboratory the importance of electricity and electronics is most apparent. Balances, used for weighing to levels of accuracy previously unknown, and spectroscopic equipment are electrically powered. The advent of microchip electronics has also added considerably to the sophistication available to the late 20th-century analyst. It has also brought quick analysis within the reach of all chemists.

In the early 1960s, for example, most infrared spectrometers were large instruments operated by specialist technicians. Today they are desktop machines readily useable by any chemist to provide an idea of how a reaction is proceeding, for example. Similarly, other types of spectroscopy – ultraviolet, NMR, electron spin resonance – and mass spectrometry are now treated far more as commonplace tools for the research chemist than they were a few years ago.

As well as increasing the availability of everyday instrumentation, the electronics revolution has led to ever more sophistication for advanced spectroscopy to assist the work of the physical chemist. The develoment of new types of magnet has pushed forward the frontiers of NMR spectroscopy, greatly expanding its value as a research tool. Computers also play an important role, enabling chemists to study the possible interactions of as-yet unsynthesized molecules. When an apparently interesting molecule is discovered, other computer programs can help to plan synthetic routes to it.

A more mundane, but nonetheless important change in the modern laboratory is an increased awareness of the potential hazards both from chemical substances themselves and from the reactions they undergo. Modern chemists operate under safety rules that were unknown in the past. Containment areas, such as fume cupboards, are far more efficient than they once were, and chemists are more often required to wear protective clothing, such as rubber gloves and eye protectors, when dealing with hazardous materials or with new compounds whose hazards are not yet known.

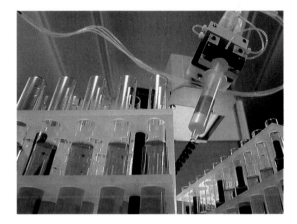

▲ *The tedious repetition of the same analytical process was a job once carried out by laboratory technicians. Today such batch sampling and testing can be automated. Robot analyzers carry out more tests for more components, and with accuracy; and they never get bored!*

▼ *Accurate weighing has always been an essential part of chemical research and analysis. Weighing to a tenth of a milligram used to involve tiny weights and tweezers, and was very time-consuming. Today's electronic balances are quick and give a digital readout of the weight.*

Computer techniques can help scientists to identify chemical carcinogens

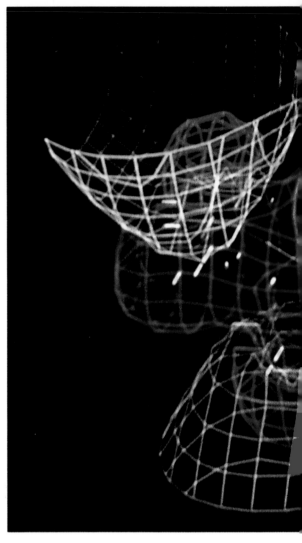

◄▼ *Molecular modeling techniques can be used to study virtually any type of chemical structure, and special styling may highlight particular structural features. In the protein (below) the helices – secondary structures of the backbone – are shown as cylinders and the sheets as arrows. In inorganic structures it is important to show coordination about different atoms, and the coordination polyhedra of copper aluminum borate are shown left.*

Making new molecules by computer

For many years, chemists have used ball-and-stick models to help to visualize the 3-D shape of chemical structures. This is important because many of the chemical and biological properties of molecules are crucially dependent on their shape. Now, however, chemists are using computers, not only to study the shape of existing molecules, but also to help them design new compounds. Computer-aided molecular modeling systems allow the chemist to build large and complicated structures very quickly and display them on the screen using computer-graphic techniques. Once built, a molecule can be moved around the screen and twisted or modified (an oxygen atom replaced by a sulfur atom, for example). A variety of calculations can be performed, from simple geometry calculations (such as the distance between the atoms) to highly complex quantum-mechanical computations. The chemist can determine the energy of a particular structure in a given conformation, or plot out the distribution of electron density around a molecule – information that may be crucial to understanding the way in which it will react with other molecules. With such a powerful tool at their fingertips, researchers are now able to identify potentially useful drugs, pesticides or catalysts without moving from their computer terminals.

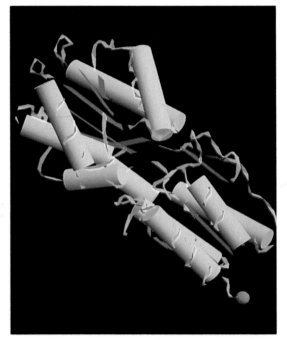

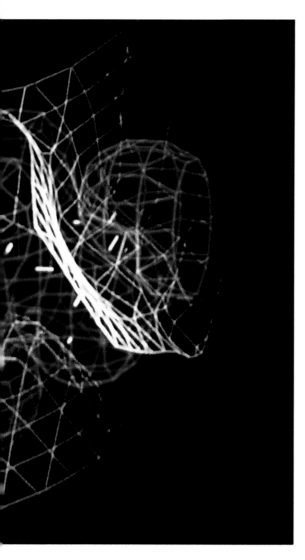

Most laboratory synthesis is now aimed at making entirely new molecules. Much of it is done in the hope of finding molecules with properties which will make them useful commercially. However, some chemical syntheses are also carried out simply in order to improve basic chemical understanding.

When the group of elements in the right-hand column in the Periodic Table – helium, neon, argon, krypton, xenon and radon – were isolated around the beginning of the 20th century (◀ page 75), they were called the "inert" or "noble" gases, for their unreactivity.

In the 1930s, the United States chemist Linus Pauling (b.1901) suggested that the inert gases might form compounds with highly reactive elements such as the halogens. Attempts to make such compounds failed. It was not until 1962 that Neil Bartlett (b.1933) reacted platinum hexafluoride with xenon and produced an orange crystalline compound, xenon hexafluoroplatinate. Other chemists looked again at the reaction between the two elements xenon and fluorine and almost immediately produced xenon fluoride. Since that time, a number of other compounds of xenon, krypton and radon have been made and what were formerly known as the "inert" gases are now called the "rare" gases. The synthesis of these compounds has improved understanding of the nature of chemical bonding.

Tracer techniques, using the carbon isotope C-14, were used extensively in the late 1940s and 1950s by the United States chemist Melvin Calvin (b.1911) to work out the chemical steps involved in plant photosynthesis. They are today used not only in elucidating natural biochemical pathways, but also in studies on the fate of potential pharmaceutical compounds in the body (◀ page 225).

▲ ► *The energy map of the histamine molecule (right) as it is twisted about two bonds shows the angle of twist plotted about horizontal and vertical axes, with colored contours indicating the energy of the molecule as the angles change. In the electrostatic potential map of the amino acid glycine (above) areas of positive potential energy are shown red, and negative blue. Such maps can show which part of a molecule is likely to be reactive.*

Computer modeling – a case study
Scientists in Britain have used computer modeling techniques to predict whether particular chemicals are carcinogenic, and have shown that this may depend on whether the chemical can fit the active site of an enzyme (◀ page 110).

All animal cells have two related families of enzymes: the cytochromes p448 and p450. The cytochromes p450 eliminate chemical carcinogens from the cell, by catalyzing a reaction to convert them into a product that will react with other molecules in the cell and form a harmless soluble compound that can be excreted. But if a carcinogen comes into contact with p448, a compound results which can act on DNA to cause cancer.

The active sites of cytochromes p448 are flat and planar, so only flat, thin molecules can react with them, whereas the molecules that react with p450 are "bent". By using computer modeling techniques to study the shapes of over 200 compounds, a team at Surrey University in Britain has predicted their carcinogenicity with 95 percent accuracy. Further work in this field could lead to the design of molecules to fit the active sites of cytochromes p448, thus blocking any reaction with carcinogens.

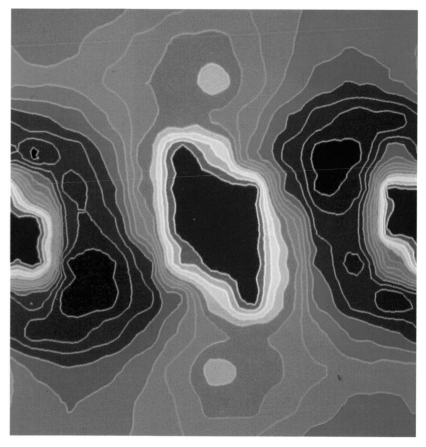

Most new compounds made in the laboratory are complex organic molecules. In general, the chemist decides the structure of the molecule that is required and then works back from that, through a series of reactions, to a readily available starting material. The basic decision on which structure to make may depend on various factors. In looking for new drugs, for example, a molecule with a particular chemical activity and a particular shape will probably be sought, so that it can interact exactly with a specific enzyme or biological receptor in the body (◀ page 108). In a situation like this, the chemist may study computer-generated shapes of different molecules before choosing which to make.

Types of synthetic reaction
Each step in the reaction sequence usually only changes one part of a molecule. Thus, a carbon double bond may be broken and a different atom added by a single bond to each of the two carbon atoms (◀ page 86). This is an addition reaction.

Another common type of reaction is a substitution reaction, in which one atom or group of atoms in a molecule is removed and replaced by a different group. Other types of reaction commonly encountered in organic chemistry are elimination reactions, in which part of a molecule is removed; and rearrangements, in which the basic skeleton of the molecule changes shape.

From the record of experiments which have been conducted in the past, a "literature" of chemistry has built up. The chemist can use this not only to discover the type of reaction that will be needed to effect a change in a certain part of a molecule, but also the likely yield and side products, and an indication of how fast the reaction will proceed.

While each reaction step is intended to change only one part of the molecule, it is important to ensure that other parts are not inadvertently changed at the same time. This can easily happen in complex molecules which contain several chemically reactive groups. In some cases, additional steps have to be added to the reaction sequence to protect one part of the molecule while another is being changed. Usually, any such protection adds two steps to the sequence: one to modify an active group by adding a protective function and then, at some later stage, removing that function.

Much of the known information about chemical reactions has now been incorporated into computer programs, so that a potential synthetic scheme to produce a new molecule can be worked out rapidly. The computer may provide alternative routes, from which the chemist has to choose one to try. This choice will be based not only on factors such as availability of raw materials, but also on the relative ease of carrying out each stage under laboratory conditions. Some reactions may work well only at high pressures, for example. While this need not be a problem on an industrial scale, many laboratories do not have the equipment to carry it out.

Synthetic experiments are often carried out in laboratory glassware which would still be recognised by a late 19th century chemist. In recent years, however, there has been an increase in the development of automated syntheses, particularly for molecules, such as proteins, where the synthesis involves repetition of a cycle of a few identical steps. Thus in 1987 chemists developed a machine for closely monitoring peptide synthesis. This means that a series of chemical reactions in which amino acids are strung together sequentially while dangling from a polymer can now be fully automated.

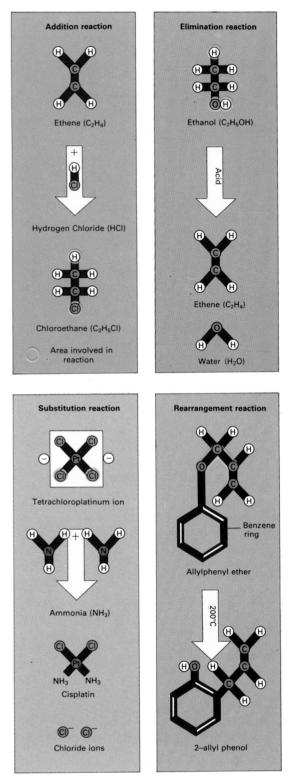

▲ There are many ways in which chemists can change one substance into another. In an addition reaction new atoms are added to molecules with double bonds. Conversely atoms can be removed to create a double bond by an elimination reaction. In a substitution reaction atoms in a molecule are replaced by other atoms, or even by other molecules. The anticancer drug cisplatin is made this way. Finally a molecule can be induced to do a rearrangement of its own atoms, say by heating it. In these and other ways over ten million different molecules have been made!

*Moving from the laboratory to the large-scale plant...
Designing for safety, efficiency and economy...Reaction
vessels...Distillation equipment...PERSPECTIVE...Making
effective use of raw materials and byproducts...
Controlling the plant...Unit operations...Distillation*

In planning how to make a compound commercially, the industrial chemist has to consider a range of factors. First is the availability of raw materials. If a large sum is to be invested in manufacturing plant, it is essential that the raw material for the manufacturing process can be obtained in predetermined quantities regularly, without great fluctuations in price. In some cases, where different routes are possible for synthesizing the product, it may be more practical to use an expensive starting material, the supply of which is reliable, rather than a cheaper raw material for which availability may be uncertain.

A second consideration is ease of handling. It is usually easier to deal with liquids and gases on a large scale rather than solids. For a chemical reaction to take place satisfactorily, it may be necessary to dissolve solids, or to remove impurities. On a large scale, these processes can be expensive.

Choice of process will also depend on the value of the by-products, which may be positive or negative. If a by-product has to be disposed of in a specific way – because of its pollution potential or toxicity, for example – this can add significantly to the process costs. On the other hand, if the by-product can be sold profitably, then the overall process becomes more economical.

Finding a profitable production method

When selecting a suitable method for large-scale production of a chemical, it is necessary to take into account many economic factors, including the availability of raw materials and the possibility of marketing by-products of the main reaction. A good example is vinyl chloride, for which production methods have changed over the years to reflect changing economic realities.

Vinyl chloride ($H_2C=CHCl$), the starting material for polyvinyl chloride (PVC: ♦ page 142) used to be made from acetylene (C_2H_2) and hydrogen chloride (HCl). Acetylene at the time was made via calcium carbide from coke and limestone. This process was energy-intensive, and it ran into competition with production of ethylene from oil. For a number of processes, ethylene (C_2H_4) could be used instead of acetylene.

To make vinyl chloride from ethylene, it had to be treated with chlorine, producing hydrogen chloride as a byproduct. This was unwanted, as hydrogen chloride was not in short supply at that time. This problem was solved for a time by using a mixed process, in which the hydrogen chloride produced in making vinyl chloride from ethylene was used to make more vinyl chloride from acetylene.

In the long term, even this proved uneconomic. The problem of excess hydrogen chloride production was solved by the development of an oxychlorination process, in which a mixture of hydrogen chloride, ethylene and oxygen generates vinyl chloride and water.

A further problem with vinyl chloride concerned contamination of the air. In the 1970s longterm workers in PVC factories were observed to be liable to contract cancer of the liver, and measures had to be brought in to control worker exposure.

◄ *Most crude oil is refined into fuels and burned, but a small fraction ends up as feed-stock for petrochemical plants like this (left). There it is turned into the plastics, drugs, dyes, detergents and fibers upon which the quality of modern life depends.*

In designing a large-scale plant, the chemical engineer has to attend to details that would pass unnoticed in the laboratory

Saving energy costs

The energy balance of a chemical process is important, as energy is an important cost in the chemical industry. If some parts of a chemical process generate energy, then it is most efficient to run these in a way that conserves the energy (by creating steam, for example) so that it can be used elsewhere in the process (or the plant as a whole) where energy is required.

Safety is also important. A reaction which can be carried out safely in the laboratory may not be suitable for scaling up if it uses inflammable or toxic solvents, for example. While these may be controllable on a small scale, the problems associated with their large-scale use can mean such a substantial investment in safety equipment that the process would be uneconomic (◗ pages 234-235). It is also necessary to bear in mind simple geometry. The heat produced by an exothermic reaction in a small flask may dissipate readily through the walls of the flask but when the reaction is scaled up, with increasing vessel size, the surface-to-volume ratio decreases. In a larger "flask", there is relatively less surface area to dissipate heat. If this is not taken into account, dangerous conditions can easily be created.

Once a chemical product has been identified, development scientists will study all these problems with a view to devising the most efficient process. This is called "optimization". It is used not only for newly discovered substances, but also for those which have been in production for many years.

In getting from the laboratory to production in an industrial plant,

▲ *Energy losses from a chemical plant are revealed by this aerial photograph taken with a heat sensitive infrared camera. "Hot spots", such as chimneys and some pipes, are white. The coolest parts, including storage tanks, show up as blue.*

▼ *Heated vessels and distillation apparatus waste a lot of energy if they are poorly lagged and insulated. New materials can reduce heat loss dramatically, and although this adds to the initial cost of a plant, the investment is repaid over its working life.*

▼ *Time is money and very expensive money when it has to be borrowed. The nine major stages from original idea to actual saleable product are shown below. The average time taken for each stage is shown in the left hand side of each box, with the total number of months to reach the end of that stage at the right hand side. For a typical plant the cost can be hundreds of millions of dollars, and the time before it comes onstream will normally be over five years. Even when a plant has started to produce the chemical for which it was designed, it may be another five years before it has paid for itself. If raw material prices, demand and international currencies change in the meantime it may be several more years before the plant shows a return on the investment. But the world's chemical industries continue to flourish with annual sales exceeding $500 billion.*

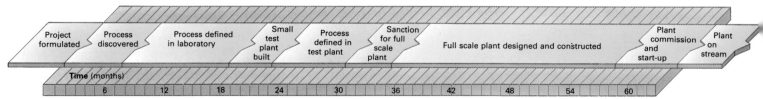

Project formulated | Process discovered | Process defined in laboratory | Small test plant built | Process defined in test plant | Sanction for full scale plant | Full scale plant designed and constructed | Plant commission and start-up | Plant on stream

Time (months)
6 12 18 24 30 36 42 48 54 60

▲ ▼ *Constant monitoring of all stages of a chemical process, to ensure smooth running and quality control, can now be carried out using sensors which relay information to microcomputers and central control. Even processes that we imagine are heavily dependent on human assessment, such as the production of flavors (below), now use data processors at shop-floor level. New plants are protected by automatic early-warning systems. Strategically placed detectors along the boundary fences can even give warning of any loss of chemicals to the environment.*

the skills of both chemists and engineers are needed. A chemical plant dedicated to the production of a single substance can use conditions of temperature and pressure not normally found in a laboratory. Chemical studies in the laboratory may indicate that the best way to scale up a reaction is to use high temperatures and pressures. These, in turn, will require that different materials are used in the construction of the plant than are used in laboratory apparatus.

An additional factor to be taken into account is that a plant may be expected to run, day in and day out, producing the same chemical, for many years. This does not happen in the laboratory. Consequently, the stresses placed on the materials used in the full-scale plant are different. The type of materials which will best resist corrosion by the reactants and products need to be selected. If the wrong materials are chosen, the plant may become hazardous.

Because glass is inert towards many chemical compounds, it is used extensively in small-scale plant. Many of the reactors used for batch production of chemicals are glass-lined metal. This not only provides a corrosion-resistant surface, but also one which is relatively easy to clean when the product is changed. Much large scale chemical plant is made from mild or stainless steel. In some cases, steel clad with titanium is used. The choice of material is usually based on the corrosivity of the reagents, solvents and products with which it will come into contact. Often corrosion is not wholly avoidable and the "lifetime" of different plant items has to be taken into consideration in establishing the economics of a process.

Controlling the processes

Another major difference between the laboratory and the plant is in the degree of automation and control. Although modern laboratories are becoming more and more automated (◀ page 116), the chemist is likely to monitor the progress of the reaction personally and make adjustments according to the observations. Most large-scale plants are now run with continuous processing rather than preparing the reactions in separate batches, and such plants are now computer-controlled. Computerization is increasingly found even in batch production. Critical parameters such as temperature are measured by sensors in the process streams or the reactor vessel itself, and the information fed back to a computer. These sensors may be thermometers or pressure gauges, or may be flow meters; in some large plants radioisotope monitoring systems (◀ page 234) may be introduced to check for the proper flow of materials through the system.

The computer in turn can operate various valves to adjust these parameters to a set pattern. If the temperature within the reactor exceeds a certain level, for example, the computer may increase the flow of coolant through the reactor jacket.

With the rapid development in recent years of microprocessors, individual items of plant are now being fitted with their own controllers which keep conditions within that sector to the parameters set, and only feed back important information to the central computer controlling the process. Distributed processing, as this is usually called, is likely to increase rapidly during the next few years, bringing with it further improvements in plant operation and reliability.

When scaling up from the laboratory bench to industrial plant, the skills of the engineer as well as those of the chemist are required

Chemical reactors

The key to any industrial-scale chemical reaction is the reactor. In most laboratory reactions, starting materials are placed in a reaction vessel – often a glass flask. After reaction has taken place, the material is removed and purified. This is called a "batch process", because the product is made in individual batches. Many processes for making complex chemicals are also carried out in batches.

In a typical batch process, a reactor may be a metal vessel with glass-lined walls. It may be fitted with a mechanical stirrer and entry ports through which reagents are added. It may also have a metal jacket through which either steam or cold water can be circulated, according to whether the reaction needs to be heated or cooled. The vessel usually also has ports which can be fitted with sensors to determine conditions such as temperature and pH.

General-purpose chemical plant of this type can be used to make a variety of substances. In the pharmaceutical industry, for example, where only small amounts of active ingredients are required, the same reaction vessel may be used for several products, each made at a different time of year. The reactors have to be cleaned thoroughly between products. Small-scale batch production is labor-intensive, despite the introduction of computer-control and automation.

Most processes for the production of very large quantities of chemicals are carried out continuously. Frequently, the yield of product obtained by a single passage of raw materials through a reactor is very low. But if the unreacted starting materials are recycled, the overall yield can be high. This depends on the ease of separation of the product from starting materials and any by-products. If the desired product can be separated easily, then the yield from the reactor is less important than in the case of a product which is difficult – and thus costly – to separate.

Continuous reactors are of two main types: tubular reactors and tank reactors. A tubular reactor can be used for reactions which involve gases and liquids. They are particularly effective for carrying out reactions which take place very quickly and where rapid heat transfer is necessary, either to enhance product formation or to keep down the amount of byproducts.

The reactants mix as they enter the tube, but then pass through it usually without further mixing with incoming reactants. A mixture of reactants and products emerges from the other end. If the reaction needs to be carried out at high temperature, then the tubular reactor may be surrounded by a furnace, as in the case of naphtha-cracking to produce ethylene. Or it may be cooled, as in the production of ethylene oxide from ethylene and oxygen, if high temperatures would favor undesirable products (in this case, carbon dioxide and water).

A variant on the tubular reactor, used for some heterogeneously catalyzed reactions, is a fluidized bed reactor. A fluidized bed is a solid made up of small particles which can be agitated so that they behave like a fluid. A common means of agitation is to pass gas upwards through the bed. Consequently, where gaseous reactants are involved, they can be used to fluidize the catalyst bed as well as to react while passing through it.

A tank reactor is more akin to the laboratory flask. The most common type is the continuous stirred tank reactor. Starting materials are added continuously and product is removed at the same rate. This type of reactor is most suited to reactions in which the product can be separated easily from the starting materials.

▲ *Just as glass apparatus is used for experiments in the laboratory, so glass may be preferred when a process is scaled up for batch processing. Glass is almost the perfect material for reactors: it is inert to nearly all chemicals (hydrofluoric acid and molten sodium hydroxide are notable exceptions), is heat-resistant and transparent.*

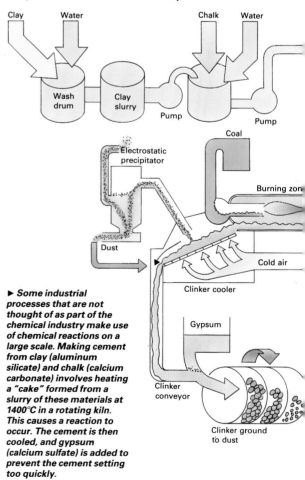

▶ *Some industrial processes that are not thought of as part of the chemical industry make use of chemical reactions on a large scale. Making cement from clay (aluminum silicate) and chalk (calcium carbonate) involves heating a "cake" formed from a slurry of these materials at 1400°C in a rotating kiln. This causes a reaction to occur. The cement is then cooled, and gypsum (calcium sulfate) is added to prevent the cement setting too quickly.*

Unit operations

A key concept used by chemical engineers in designing plant is that of "unit operations". This breaks down the steps involved in the conversion of starting material to product and finds the best way to handle each step. Unit operations can be classified in a variety of different ways. One is to look at them in terms of each stage of a process that is involved.

First, preparation equipment. This includes grinding equipment, for reducing the size of solid particles so that they will dissolve easily, and compressors for achieving high pressures for some gas reactions. This categorization is imperfect, in that some of the operations may be used at more than one stage of a process. If the final product is a solid, it may need to be ground to a particular size before it is ready for use. However, the operation of grinding is carried out according to the same criteria wherever it is used in a process.

Second, reaction equipment. In addition to conventional reactors, this also includes kilns for preparation of solids at high temperature, such as quicklime (calcium oxide) and cement.

Third, separation equipment. This includes apparatus for filtration, distillation and liquefaction as well as absorption towers. It also includes equipment such as electrostatic precipitators.

Fourth, transfer equipment. This includes flow controllers, pumps and fans for mass transfer and heat exchangers for heat transfer.

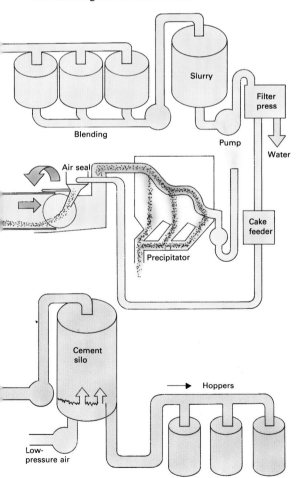

◄ ► Centrifugal force is used to speed up the separation of solids and liquids. In these industrial filter drum centrifuges (above) the spinning action forces the liquid through the holes in the drum, leaving the solid behind. The process resembles the spin cycle of the domestic clothes washing machine.

Industrial centrifuge

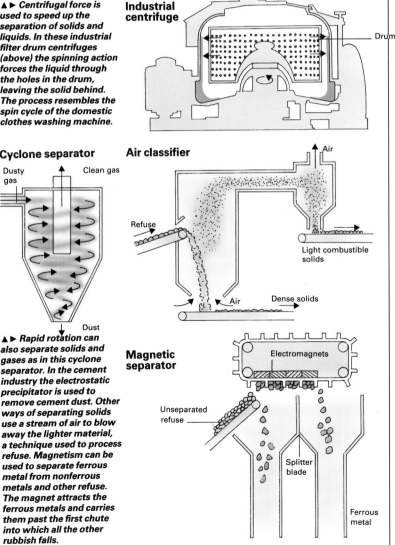

Cyclone separator

Air classifier

Magnetic separator

◄ ► Rapid rotation can also separate solids and gases as in this cyclone separator. In the cement industry the electrostatic precipitator is used to remove cement dust. Other ways of separating solids use a stream of air to blow away the lighter material, a technique used to process refuse. Magnetism can be used to separate ferrous metal from nonferrous metals and other refuse. The magnet attracts the ferrous metals and carries them past the first chute into which all the other rubbish falls.

Distillation

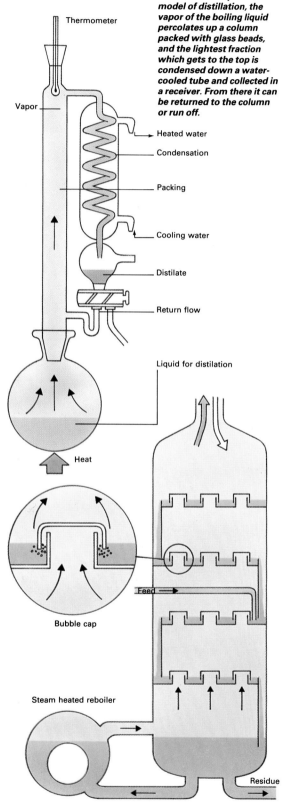

◀ In a laboratory-scale model of distillation, the vapor of the boiling liquid percolates up a column packed with glass beads, and the lightest fraction which gets to the top is condensed down a water-cooled tube and collected in a receiver. From there it can be returned to the column or run off.

Thermometer

Vapor

Heated water

Condensation

Packing

Cooling water

Distilate

Return flow

Liquid for distillation

Heat

Bubble cap

Steam heated reboiler

Feed

Residue

▲ Petroleum refining uses a bubble tower to separate the various fractions of hydrocarbons into petrol, paraffin, diesel oil, etc. The heated vapor rises up the tower and bubbles through a series of capped vents that are designed to ensure intimate contact between the lighter fraction distilling up the tower and the heavier fraction trickling down.

Distillation

One of the earliest chemical processes and still one of the most important is distillation. This separates mixtures on the basis of differences in boiling point between the components. It was probably known in classical times and was certainly used by Arab chemists in the middle ages for the production of perfumes. Later, in western hands, the distillation of alcohol – which involved the additional step of having a condenser to convert the vapor produced back to liquid – gave rise to the production of spirits such as whisky, brandy, gin and vodka.

Distillation is important in the chemical industry today, not only in the separation of products from mixtures and in the recovery of solvents, but also as an integral part of some reactions. The production of some organic esters, for example, involves boiling together an acid and an alcohol and removing the ester and water by distillation as they are formed. As esterification is an equilibrium reaction, continuous removal of products pushes the equilibrium in their favor, thus increasing yields. Distillation processes in commercial use range in temperature from the very low, used for separation of air into its component gases, to 1500-2000°C in the purification of some metals.

As the temperature of a pure liquid is raised, its vapor pressure increases. When its vapor pressure reaches that of the surrounding air, the liquid boils. That is, it turns to a gas. If this gas is cooled, it condenses back to a liquid. If a mixture of liquids with widely differing boiling points is heated, then the more volatile of the two vaporizes first and can be separated from the less volatile by condensing its vapors so that they are collected in a separate vessel from the one containing the boiling mixture.

Many commonly found mixtures of liquids do not have widely differing boiling points, even at atmospheric pressure. If distillations are carried out under reduced pressure, which reduces the boiling points and thus saves energy, the differences are even less. At any temperature, however, there is generally a difference between the composition of the liquid and the composition of the vapor in equilibrium with it, the vapor being richer in the more volatile component. Consequently, when the mixture is distilled, the first condensate will be richer in the more volatile component than the starting mixture.

If this initial condensate is itself distilled, the first fraction of condensate from it will be richer still in the more volatile component. This has led to fractional distillation, in which a distillation column is effectively made up of large numbers of miniature distillation steps. In an industrial process, such as the separation of related organic compounds, the starting material may be fed continuously into the middle of the distillation column. As material rises up the column, it becomes richer in the more volatile components. Conversely, as it falls down the column, below the entry point, it becomes richer in less volatile components. In this way, it is possible to separate mixtures of compounds which have boiling points relatively close together.

Another useful aspect of distillation is the behavior of liquids which do not mix with one another, such as oils and water. These will distill together at temperatures below the temperature of the lower boiling component. The composition of the distillate is related to the molecular weights of the different components. Consequently, steam distillation can be used to obtain relatively high molecular weight organic compounds in a pure form. This is particularly useful for obtaining perfume ingredients, the molecules of which decompose at temperatures below the boiling point of the pure substance.

Chemicals in Bulk

16

The importance of the chemical industry...Bulk chemical production...Chemicals from petroleum...PERSPECTIVE... Important organic chemicals...Sulfuric acid production ...Ammonia and nitric acid...Caustic soda...Sodium carbonate...Ethylene

The chemical industry is now one of the most important large-scale industrial activities in the world, yet only a small number of compounds make up the bulk of its output. For many years, the leading chemical has been sulfuric acid (H_2SO_4).

Currently, about one-third of the world's annual production of sulfuric acid is used in fertilizer manufacture (◆ pages 146-147). Another third divides almost equally between the manufacture of other chemicals and of paints and pigments (◆ pages 169-174). About 10 percent of annual production is used to manufacture detergents (◆ pages 175-180). Other uses include "pickling" of metals to remove surface oxide, acid for car batteries and oil refining.

The pattern of use varies widely between different countries. In the United States, for example, where more than 40 million tonnes of the acid is produced annually, 70 percent is used in fertilizer production. Much of this is for the conversion of phosphate rock to phosphoric acid, 85 percent of which goes into fertilizers.

The world's second most important chemical, in bulk terms, is ammonia (NH_3). Annual production worldwide exceeds 60 million tonnes, of which about three-quarters is for fertilizers. Ammonia is also used in the manufacture of other simple nitrogen compounds, such as urea $CO(NH_2)_2$, used as animal feed supplements.

Ammonia is also used to manufacture nitric acid (HNO_3), some of which is also used in fertilizers. The acid is also important in the explosives and dyestuffs industries. Both nitric acid and ammonia play key roles in many organic syntheses that require the introduction of nitrogen atoms into a molecule. Acrylonitrile, from which acrylic fibers and plastics are made (◆ pages 140-141) is synthesized industrially by treatment of propylene with ammonia and oxygen.

The alkali sodium hydroxide (NaOH), also known as caustic soda, is used in many manufacturing processes, including those for soap, paper and detergents. It is also important in the manufacture of many complex chemical substances, including dyes and pharmaceuticals.

For a number of applications, sodium carbonate (Na_2CO_3) can replace sodium hydroxide. However, the major reason for the scale of production of sodium carbonate is its use as an ingredient in glass. About half the sodium carbonate produced goes into soda-lime-silica glass, the most common form of this commodity.

The lime used in soda-lime-silica glass is limestone, or calcium carbonate ($CaCO_3$). Calcium hydroxide – $Ca(OH)_2$ – which is also called lime or "slaked lime", is another compound produced in very large quantities by the world's chemical industry. It is made by calcining limestone, to produce calcium oxide ("quicklime") and then adding water. It is a weakly alkaline substance which is used extensively in metallurgical processes, as a flux in steelmaking for example. It is also widely used in pollution control and wastewater treatment.

Large-scale organic chemicals

Most of the chemical compounds produced industrially today are organic molecules, based on chains of carbon atoms. The key building block for many of these is ethylene (C_2H_4), a two-carbon compound with a double bond that enables it to undergo a very large number of reactions. About 40 million tonnes of ethylene is produced worldwide each year from oil and natural gas. Slightly less than 50 percent of this is turned into polythene (◆ pages 136-137). The remainder is converted into a wide range of more complex organic chemicals.

Among the basic products obtained from ethylene are ethyl alcohol (C_2H_5OH) for industrial use; ethylene oxide; and halogenated products, which are used as dry-cleaning fluids, fire extinguishers and for degreasing metals. Ethylene dichloride is commercially the most important of these halogenated products, used primarily to produce vinyl chloride, the monomer from which polyvinyl chloride (PVC) is made (◆ page 138).

Ethylene oxide is a highly reactive chemical which is used to make many other chemicals, notably ethylene glycol (CH_2OHCH_2OH). This is used in automobile antifreeze, but it is also an ingredient in making polyester fibers and film.

Among the top 20 chemicals in volume terms are several other organic molecules. Notable among these is benzene (C_6H_6), the parent compound of the aromatic hydrocarbons, from which many useful substances are derived. Large amounts of benzene and ethylene are consumed in the manufacture of ethylbenzene, which in turn is dehydrogenated to styrene, the monomer from which polystyrene is made.

Methanol (CH_3OH), another of the top 20 chemicals, is used widely in the manufacture of other chemicals, such as urea and phenol. The importance of methanol may increase in future, because it can be made from natural gas, and partly also because it can be used to make synthetic gasoline and also as "food" for microorganisms which can be grown for animal feeds.

TOP U.S. CHEMICALS, 1980

Chemical	Production (Million tonnes)	Major uses
Sulfuric acid	40	Fertilizer, petroleum refining, synthesis of other chemicals
Ammonia	19	Fertilizer, explosives, production of other chemicals
Lime	18	Cement manufacture, water purification
Oxygen	17	Steel manufacture, welding, paper manufacture, water purification
Nitrogen	17	Ammonia manufacture
Ethylene	14	Polyethylene plastic manufacture
Sodium hydroxide	11	Chemicals, paper, aluminum, petroleum, soaps and detergents
Chlorine	11	Chlorinated hydrocarbon manufacture, pulp and paper, water treatment

A large plant producing sodium hydroxide uses as much electricity as a town of 300,000 people

Sodium hydroxide

Sodium hydroxide is produced by the electrolysis of solutions of common salt ($\blacklozenge$ page 60); the other major product of this process is chlorine. The chloralkali industry, as it is called, has existed for well over a century and its need to be located near sources of salt has been an important factor in the location and the development of some of the world's historic industrial chemical centers, such as northwest England.

When an electric current passes through aqueous sodium chloride, chlorine gas evolves at the anode. This gas is saturated with water and is very corrosive. Consequently, it is cooled to condense out most of the water and is then "scrubbed" with sulfuric acid to give a product suitable for sale.

Sodium is discharged at the cathode, but reacts immediately with water to reform sodium ions and release hydrogen gas and hydroxide ions. The hydrogen gas is sold as a commercial product. Its uses include hydrogenation of liquid fats to make them solid ($\blacklozenge$ page 152) and reaction with chlorine to make hydrochloric acid.

As the reaction proceeds, a mixed solution of sodium chloride and hydroxide forms. Sodium hydroxide reacts with chlorine to form sodium hypochlorite. To prevent this happening, porous diaphragms are included in the reaction cells to separate the anode and cathode compartments.

Brine, containing about 25 percent by weight of sodium chloride, flows through the cell from the anode end to the cathode end and then out. By this stage it contains about 10 percent sodium hydroxide and 15 percent sodium chloride. Concentration in large evaporators gives a 50 percent sodium hydroxide solution, from which most of the sodium chloride crystallizes out and is removed by filtration. The remaining solution contains only about one percent sodium chloride and, for many uses, can be sold without further purification.

An alternative method uses a mercury cathode. As sodium atoms discharge on this, they amalgamate with the mercury, thus preventing them from releasing hydrogen in the cell. The cathode flows continuously out of the cell into a reactor where it is mixed with water over activated carbon to dissolve out the sodium. The mercury then flows back into the electrolysis cell. The water from the second reactor is converted into a 50 percent sodium hydroxide solution without the need for evaporation. The product is also salt-free. On the other hand, two reactors are needed rather than one.

The mercury cell process has been falling out of favor in recent years, primarily on environmental grounds. Traces of mercury may enter the wastewater from the plant and, under certain conditions, may be converted to highly toxic organomercury compounds. This has led to further developments of the diaphragm cell. The traditional asbestos diaphragm is being replaced by synthetic polymer diaphragms which are impermeable to chloride ions. The liquid that emerges from the cathode compartment is a salt-free 40 percent sodium hydroxide solution.

▲ *Sodium hydroxide is produced by electrolysis of brine and water ($\blacklozenge$ page 60), using diaphragm cells which ensure that the chlorine (one of the by-products of the process) and sodium hydroxide are kept separate.*

▼ *Chloralkali plants (those producing chlorine and sodium hydroxide by electrolysis) are generally sited near good supplies of natural salt, good transport by road and rail and a cheap supply of electricity. Energy costs are a significant element in the production of caustic soda: the power consumption of a cell installation is comparable to that of a city of 300,000 people.*

Uses of sodium hydroxide

- Miscellaneous
- Manufacture of chemicals
- Soaps and detergents
- Neutralization
- Alumina
- Pulp and paper
- Rayon and acetate fibers

Uses of sodium carbonate

- Oils, fats and waxes
- Miscellaneous
- Heavy chemicals
- Fine chemicals
- Food and drink
- Dyes and colors
- Textiles

Sodium hydroxide manufacture

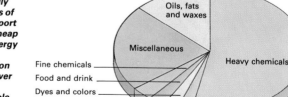

Electricity

Brine (NaCl in solution)

Electrolysis cells

Chlorine

Hydrogen

Sodium hydroxide

Sodium carbonate

Like sodium hydroxide, sodium carbonate has been produced on a large scale for more than a century. The starting material for synthetic sodium carbonate is sodium chloride, obtainable from brine. The overall reaction is the double decomposition of sodium chloride and calcium carbonate to give sodium carbonate and calcium chloride. However, to achieve this apparently simple interchange, it is necessary to use a number of steps, because the reverse reaction is favored thermodynamically. The modern method of manufacture was first made feasible by the French chemist Ernest Solvay (1838-1922). His process, introduced in 1863, displaced the earlier Leblanc process, which had operated since the French Revolution, and which had used sodium sulfate as a starting material.

In the modern version of the Solvay process, water is injected into underground salt deposits to form brine solution. Sodium hydroxide or carbonate are added to this to precipitate any contaminating calcium and magnesium ions which would otherwise form scale on the plant, thus reducing its efficiency. The salt solution is then treated with ammonia and the ammoniacal brine passed down tall towers, through which carbon dioxide is passed upwards. Plates built into the towers break up the gas stream and ensure efficient mixing of solution and gas.

The ammonia and carbon dioxide form ammonium carbonate and the latter reacts with sodium chloride to form sodium bicarbonate ($NaHCO_3$), which precipitates and is removed by filtration. It is converted to sodium carbonate by calcination. The result is a low density product called light soda ash. The ammonium chloride formed is treated with calcium hydroxide, releasing ammonia for reuse and forming calcium chloride, the other end product of the simple double decomposition.

Much of the demand for sodium carbonate, particularly for glass manufacture, is for the denser heavy ash. This is made by treating light ash with water to produce crystals of sodium carbonate monohydrate: calcined they produce heavy ash.

Synthetic sodium carbonate is no longer produced in the USA. There it has suffered competition from natural deposits of trona ore. Trona is sodium sesquicarbonate, a combination of carbonate and bicarbonate. It is dissolved, in some cases by the same type of solution mining as is used elsewhere for brine production from underground deposits. The trona deposits having been formed by evaporation of ancient seas, the solution contains impurities, which are removed by filtration and by passage over carbon. This removes any organic impurities which would discolor the product.

The solution is evaporated to give sodium carbonate monohydrate which is dried in rotary evaporators. Because the process is so simple, a major cost in buying sodium carbonate extracted from trona is its transportation. The European soda industry still uses mainly the Solvay process as this gives a product which is cheaper than that imported from the United States.

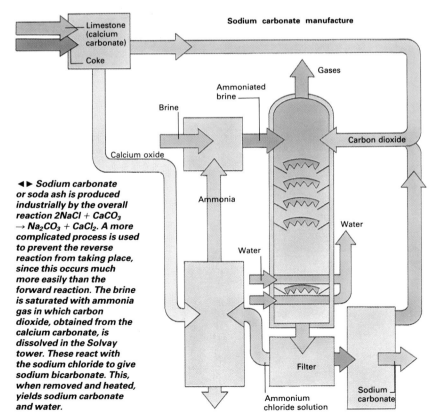

Sodium carbonate manufacture

◀▶ Sodium carbonate or soda ash is produced industrially by the overall reaction $2NaCl + CaCO_3 \rightarrow Na_2CO_3 + CaCl_2$. A more complicated process is used to prevent the reverse reaction from taking place, since this occurs much more easily than the forward reaction. The brine is saturated with ammonia gas in which carbon dioxide, obtained from the calcium carbonate, is dissolved in the Solvay tower. These react with the sodium chloride to give sodium bicarbonate. This, when removed and heated, yields sodium carbonate and water.

Sulfuric acid is one of the cheapest manufactured chemicals available

Sulfuric acid

Sulfuric acid is made primarily by burning molten sulfur in dry air to produce dilute gaseous sulfur dioxide. The dioxide reacts over a catalyst with oxygen to form sulfur trioxide which, on combining with water, gives sulfuric acid. In about 40 percent of production, the sulfur dioxide comes from roasting metallic sulfides rather than from elemental sulfur. However, in recent years, the need to remove impurities from the process has led to the dominance of elemental sulfur. This is obtained both from natural sulfur deposits and from desulfurization of crude oil and "sour" gas (natural gas which contains hydrogen sulfide).

The catalytic process has existed for more than a century, but it has changed during that time to optimize efficiency and reduce its pollutant effects. This has meant considerable effort to control temperatures at each stage of the process. Thus, in a modern sulfuric acid plant, the gas emerging from the sulfur burner may be as hot as 1,000°C. Before passing into the converter, which contains vanadium oxide catalyst, it is cooled to just over 400°C. This is because the reaction between sulfur dioxide and oxygen releases heat; consequently, the higher the temperature, the more the reverse reaction (breakdown of sulfur trioxide to dioxide and oxygen) is favored. This temperature effect has to be balanced against the decrease in catalyst effectiveness as the temperature drops, and about 420°C is the optimum.

The converter is a cylinder containing several layers of vanadium oxide, usually in the form of pellets which also contain silica (as a support) and potassium sulfate (as a promoter). During its passage over the first layer of catalyst, about two-thirds of the sulfur dioxide is oxidized. As this produces heat, the gas is then passed through a boiler, where the heat generates steam which can supply energy to other stages of the process. The gas then passes back into the converter, where it meets the second layer of catalyst. After going through the boiler again, it passes over the third catalyst layer. By this time, about 95 percent of the dioxide has been converted to trioxide.

In the most modern plants, designed to meet recent environmental legislation, the gas is then cooled and passed through a tower containing sulfuric acid. This absorbs the sulfur trioxide. The residual gas, which is now at about 80°C, is reheated to 420°C and passed over the fourth layer of catalyst. Removal of the sulfur trioxide before this fourth contact with the catalyst shifts the reaction equilibrium in favor of sulfur trioxide, so that the overall conversion achieved is greater than 99.5 percent. The gas from the fourth passage is similarly contacted with sulfuric acid to remove sulfur trioxide, after which residual gas can be discharged to the atmosphere.

The final stage is also exothermic, but it does not produce heat at a usable temperature. However, steam generated in the earlier exothermic stages can be sold as energy. In a modern plant, the sale of this energy can be enough to cover all running costs apart from the cost of the sulfur used. As a consequence, sulfuric acid is one of the cheapest manufactured chemicals available.

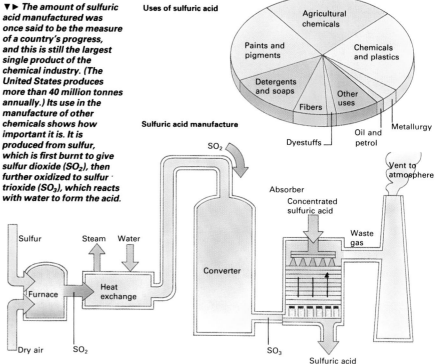

▼ ▶ **The amount of sulfuric acid manufactured was once said to be the measure of a country's progress, and this is still the largest single product of the chemical industry. (The United States produces more than 40 million tonnes annually.) Its use in the manufacture of other chemicals shows how important it is. It is produced from sulfur, which is first burnt to give sulfur dioxide (SO_2), then further oxidized to sulfur trioxide (SO_3), which reacts with water to form the acid.**

Uses of sulfuric acid

- Agricultural chemicals
- Chemicals and plastics
- Paints and pigments
- Detergents and soaps
- Other uses
- Fibers
- Dyestuffs
- Oil and petrol
- Metallurgy

Sulfuric acid manufacture

Sulfur · Steam · Water · Furnace · Heat exchange · SO_2 · Converter · SO_3 · Absorber · Concentrated sulfuric acid · Waste gas · Vent to atmosphere · Dry air · SO_2 · Sulfuric acid

Ammonia and nitric acid

The scale of ammonia production depends almost entirely on its use in fertilizers. It was fear of a world shortage of nitrogen fertilizers – then obtained mainly from nitrate deposits in Chile – in the early years of this century which led the German chemist Fritz Haber (1868-1934) to develop the process named after him and still used today. Ammonia manufacture is still made by the passage of nitrogen and hydrogen gases over an iron catalyst at high temperature and pressure.

One molecule of nitrogen combines with three molecules of hydrogen to give two molecules of ammonia. Like most reactions, this is reversible. Because the reaction uses four molecules of starting material to produce two molecules of product, product formation is favored by high pressures. On the other hand, as the reaction is exothermic, high temperatures favor the back reaction. Unfortunately, no catalyst has been found for industrial use which is active at low temperatures. Consequently, a temperature of about 400°C is used and the process is run in such a way that unreacted hydrogen and nitrogen are recycled through the synthesis converter.

In a modern ammonia plant, hydrogen is generated in steam reformers, either from natural gas or higher hydrocarbons. Where natural gas is used, it is first treated to remove impurities – notably sulfur, which poisons many catalysts. The pure natural gas is effectively methane, which is reacted over a nickel catalyst with steam at high temperature. This gives a mixture of hydrogen and carbon monoxide and dioxide.

In a second reaction, air is added which provides nitrogen for the final stage of the process. The oxygen in this air burns in the hydrogen to give steam. This oxidizes the carbon monoxide to carbon dioxide and regenerates hydrogen. The carbon dioxide is then removed from the gas stream by absorption into alkaline solution.

After further purification, the mixture of hydrogen and nitrogen is compressed and passed through the ammonia converter. On leaving the converter, it contains about 15 percent ammonia. Refrigeration condenses the ammonia to a liquid and the unreacted nitrogen and hydrogen are returned to the reaction vessel. Before this happens, a small percentage of the gas stream is removed. This is to prevent a buildup of rare gases (such as argon), which comprise about one percent of air, and methane which is still present in tiny amounts in the mixture entering the converter.

Large quantities of ammonia are converted into nitric acid by oxidation. The overall reaction is for one molecule of ammonia and two of oxygen to give one molecule each of nitric acid and water. The process is achieved industrially using platinum or platinum-rhodium gauze as a catalyst.

In the first step, ammonia is oxidized to nitric oxide (NO) and water. Further reaction between the oxide and oxygen gives nitrogen dioxide (NO$_2$). This then reacts with water to give nitric acid and regenerates part of the nitric oxide. The exhaust gas from the process always contains some nitric oxide and nitrogen dioxide, which form a brown plume that is visible as evidence of pollution.

▲ ▼ ▶ **The major use of nitrogen compounds is in fertilizers. In plants like the one above, nitrogen from the atmosphere is reacted at high temperatures and pressures with hydrogen gas (H$_2$), obtained from natural gas (CH$_4$). The product is ammonia (NH$_3$), used in liquid or solid form.**

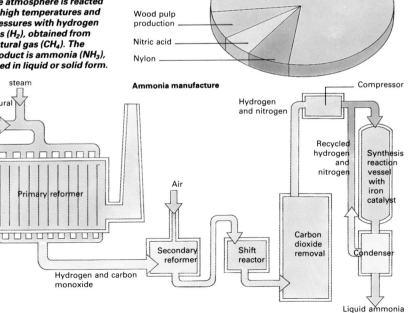

Uses of ammonia

Fertilizer

Wood pulp production

Nitric acid

Nylon

Ammonia manufacture

Chemicals derived from petroleum

More than 90 percent of the organic chemicals used today have a petrochemical origin: the starting materials for their manufacture are crude oil or natural gas. From these are made a number of chemical building blocks, of which the most important is ethylene.

Higher alkenes, notably propylene (C_3H_6) and butadiene (C_4H_6), are also important in the manufacture – often by multistep processes – of the organic chemicals which go into plastics, paints, dry-cleaning solvents, antifreezes, detergents, synthetic rubber and fibers, and many other materials which are important in industrialized societies.

Fifty years ago, the majority of organic chemicals were derived from coal. The growth in demand for automobile fuels, which are obtainable from only a part of the total crude oil, led to cheap petroleum fractions being available in massive quantities.

When crude oil is distilled, it is divided into a number of fractions, each having a different boiling range. As the boiling point rises, so does the average molecular weight of the hydrocarbon molecules in the fraction. Thus, the first fraction, with a boiling-point of up to 25°C, contains molecules with between one and four carbon atoms. In the boiling range between approximately 20 and 200°C is the mixture of hydrocarbons from which gasoline is obtained (4-12 carbon atoms). At higher temperatures, larger molecules are distilled for kerosene, diesel oil and lubricating oil. Excess quantities of any fraction can be "cracked" to give smaller molecules, to increase the amount of feedstock for gasoline and for the chemical industry.

If oil supplies become scarce and thus more expensive in future, it is possible that the chemical industry – and the automobile industry – may depend once more on coal, for known coal reserves worldwide greatly exceed known oil reserves. A 21st-century coal-based chemical industry would probably start with the conversion of coal and steam to synthesis gas – a mixture of carbon monoxide and hydrogen. This could be reformed to methanol, which could become the major feedstock for the chemical industry and also, through the use of zeolite catalysts, the major source of automobile fuel.

Ethylene (C_2H_4)

Ethylene is produced from a variety of hydrocarbon sources, usually in conjunction with other unsaturated hydrocarbons, notably propylene.

In the USA in recent years, ethylene has been made mostly from its saturated analog, ethane, obtained from natural gas. Unlike European natural gas, which is almost pure methane, many of the natural gas deposits in the USA contain usable quantities of higher hydrocarbons, such as ethane and propane. Most European ethylene has been obtained by steam reforming of naphtha, a mixture of hydrocarbons obtained from distillation of crude oil but not usable as petrol.

Vaporized naphtha is mixed with about half its own weight of steam and passed through coiled tubes in a furnace which heats the reaction zone to about 900°C. When the gas mixture emerges from the furnace it is cooled in a heat exchanger, which generates steam for use in the process.

After cooling, the reaction product is separated into a gaseous phase and a liquid phase. The gas phase is compressed and cooled further, so that all its components, apart from methane and hydrogen, liquefy. The liquid phase is then transferred to another separator, where the temperature is adjusted so that the molecules containing two carbon atoms can be removed as a gas, while the heavier hydrocarbons remain liquid. The C_2 fraction is selectively hydrogenated and the ethylene and ethane are then separated by adjusting the temperature, so that the ethylene comes off as a gas.

Producers of the basic unsaturated hydrocarbons have had to become more flexible in their source of raw materials. At the same time, changes in demand for the different hydrocarbons also affect which feedstocks are most desirable.

An alternative future source of both ethylene and propylene could be methane. Thermal cracking tends to reduce the size of the molecules in a reaction mixture. Consequently, methane has not until now been very useful as an industrial chemical. However, it can be converted to methanol, from which ethylene, propylene and a number of other hydrocarbons can be built up, using zeolite catalysts.

Zeolites are complex inorganic molecules which have pores of specific sizes in their crystal structures. This means that they can be used as "molecular sieves" to separate out mixtures of hydrocarbons or as catalysts where the pore size influences the nature of the product.

As oil supplies become scarcer, the production of petrol and unsaturated hydrocarbons for the chemical industry from methanol by zeolites may become increasingly important. The development of the petrochemicals industry has taken place almost exclusively since World War II. When polythene manufacture started in the 1930s, ethylene was made by dehydration of ethanol (C_2H_5OH). Today, ethanol for industrial uses is made largely from ethylene. However, if oil and natural gas supplies do run out, it would be possible to produce ethylene again from ethanol made by fermentation from carbohydrate-containing raw materials.

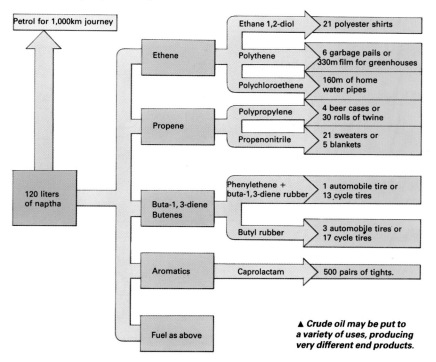

▲ Crude oil may be put to a variety of uses, producing very different end products.

17

Plastics

Plastics and polymers...Thermoforming and thermosetting...Polythene...Acrylic, Perspex...Nylon and synthetic fibers...Bakelite...PERSPECTIVE...Low and high density polythene...Silicone and silicon

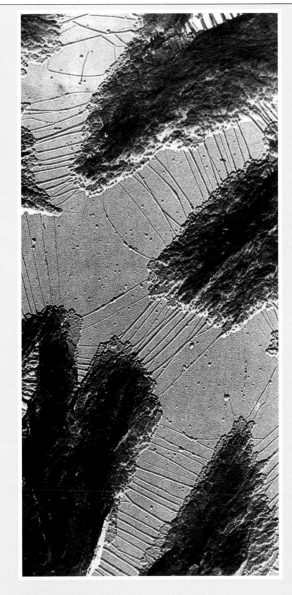

Plastics are polymers, or very large molecules. Some plastics were discovered more than a century ago, but the growth in their development and use has taken place only in the past 50 years. It was largely after the work of the German chemist Hermann Staudinger (1881-1965) in the 1920s that scientists first became convinced that polymers did exist in nature and could also be made in the laboratory.

Polymers consist of small, repeating units derived usually from simple organic molecules or monomers. They also have the particular property of plasticity – that is, the ability to be shaped or molded at some stage of their manufacture. Plastics are divided into two major classes: thermoplastics and thermosets. Thermoplastics can be repeatedly heated and cooled, becoming plastic when hot and solid when cold. Thermosets do not have this reversibility. On heating, reactions take place which produce complex linkages between the polymers so that they are no longer plastic. Thermosets are often called resins and are used to make materials, such as kitchen worktops, that are resistant to heat and chemical action.

Familiar polymers are formed by addition or condensation. In addition polymerization, the monomers contain at least one double bond between a pair of carbon atoms (♦ page 86). During polymerization, the double bonds break to give a single bond and spare combining capacity on the two carbon atoms, which enable them to link with other molecules to create a long chain. In condensation polymerization, bonds between monomer molecules are formed with the elimination of a small molecule, such as water.

▲ *Polythene is the simplest of all polymers, being constructed of long chains of carbon atoms matted together. Each carbon has two hydrogen atoms attached. In chemical shorthand this is written $(CH_2)_n$. When a sample of polythene is examined under a very high magnification, the long chains are evident.*

◄ *The idea of giant molecules, or polymers, was first put forward by Hermann Staudinger in 1926. In 1953 his work on polymerization gained him a Nobel Prize. Chemists had worked with polymers long before, without realizing that they were dealing with chains of atoms made up of the same few links endlessly repeated.*

The Derivation of Plastics

▶ ▼ *Plastics are polymers made up from a small number of raw materials, many of which derive from hydrocarbons. Polyethylene (far right) is a simple molecule, an assembly of ethylene monomers. Polypropylene (center) has the same backbone, but with a methyl (CH_3) group on every alternate carbon atom. Nylon 66 (right) has a larger repeating unit, with amide groups (NCHO) along its chain.*

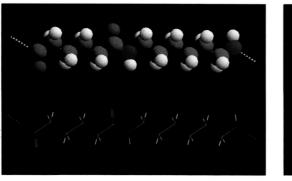
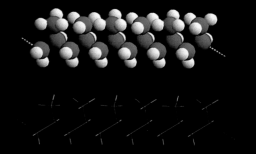

Derivation of some of the more important chemicals used in the manufacture of plastics

Derivation of polymers

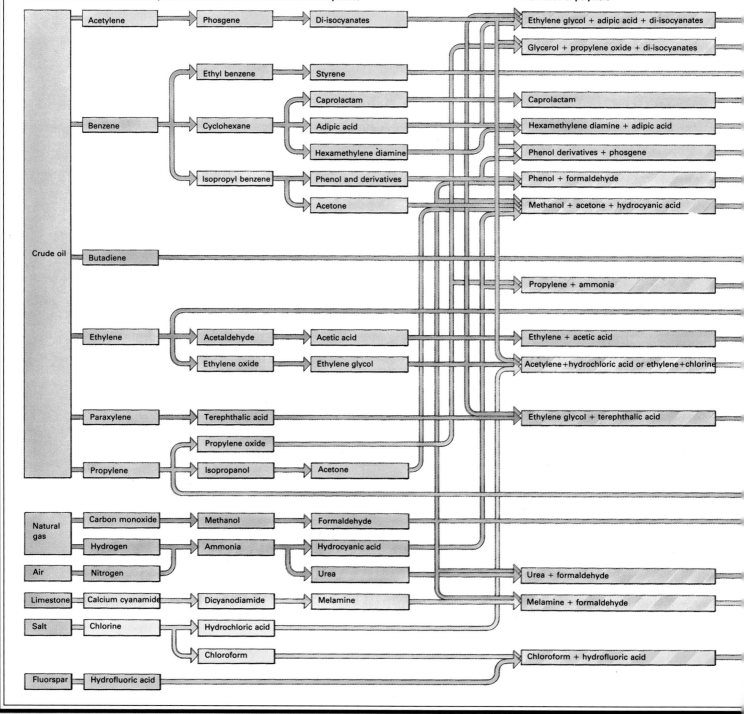

Crude oil	Acetylene	Phosgene	Di-isocyanates	Ethylene glycol + adipic acid + di-isocyanates
				Glycerol + propylene oxide + di-isocyanates
	Benzene	Ethyl benzene	Styrene	
		Cyclohexane	Caprolactam	Caprolactam
			Adipic acid	Hexamethylene diamine + adipic acid
			Hexamethylene diamine	Phenol derivatives + phosgene
		Isopropyl benzene	Phenol and derivatives	Phenol + formaldehyde
			Acetone	Methanol + acetone + hydrocyanic acid
	Butadiene			
				Propylene + ammonia
	Ethylene	Acetaldehyde	Acetic acid	Ethylene + acetic acid
		Ethylene oxide	Ethylene glycol	Acetylene + hydrochloric acid or ethylene + chlorine
	Paraxylene	Terephthalic acid		Ethylene glycol + terephthalic acid
	Propylene	Propylene oxide		
		Isopropanol	Acetone	
Natural gas	Carbon monoxide	Methanol	Formaldehyde	
	Hydrogen	Ammonia	Hydrocyanic acid	
Air	Nitrogen		Urea	Urea + formaldehyde
Limestone	Calcium cyanamide	Dicyanodiamide	Melamine	Melamine + formaldehyde
Salt	Chlorine	Hydrochloric acid		
		Chloroform		Chloroform + hydrofluoric acid
Fluorspar	Hydrofluoric acid			

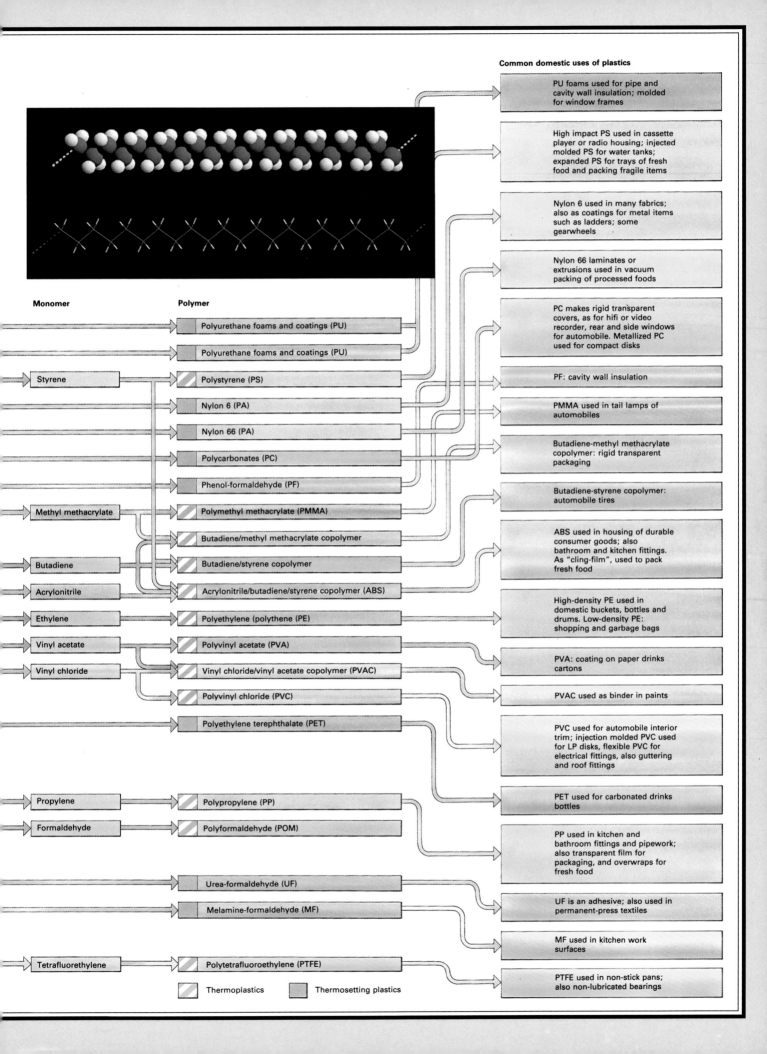

Common domestic uses of plastics

PU foams used for pipe and cavity wall insulation; molded for window frames

High impact PS used in cassette player or radio housing; injected molded PS for water tanks; expanded PS for trays of fresh food and packing fragile items

Nylon 6 used in many fabrics; also as coatings for metal items such as ladders; some gearwheels

Nylon 66 laminates or extrusions used in vacuum packing of processed foods

PC makes rigid transparent covers, as for hifi or video recorder, rear and side windows for automobile. Metallized PC used for compact disks

PF: cavity wall insulation

PMMA used in tail lamps of automobiles

Butadiene-methyl methacrylate copolymer: rigid transparent packaging

Butadiene-styrene copolymer: automobile tires

ABS used in housing of durable consumer goods; also bathroom and kitchen fittings. As "cling-film", used to pack fresh food

High-density PE used in domestic buckets, bottles and drums. Low-density PE: shopping and garbage bags

PVA: coating on paper drinks cartons

PVAC used as binder in paints

PVC used for automobile interior trim; injection molded PVC used for LP disks, flexible PVC for electrical fittings, also guttering and roof fittings

PET used for carbonated drinks bottles

PP used in kitchen and bathroom fittings and pipework; also transparent film for packaging, and overwraps for fresh food

UF is an adhesive; also used in permanent-press textiles

MF used in kitchen work surfaces

PTFE used in non-stick pans; also non-lubricated bearings

Monomer

Styrene

Methyl methacrylate

Butadiene

Acrylonitrile

Ethylene

Vinyl acetate

Vinyl chloride

Propylene

Formaldehyde

Tetrafluorethylene

Polymer

Polyurethane foams and coatings (PU)

Polyurethane foams and coatings (PU)

Polystyrene (PS)

Nylon 6 (PA)

Nylon 66 (PA)

Polycarbonates (PC)

Phenol-formaldehyde (PF)

Polymethyl methacrylate (PMMA)

Butadiene/methyl methacrylate copolymer

Butadiene/styrene copolymer

Acrylonitrile/butadiene/styrene copolymer (ABS)

Polyethylene (polythene) (PE)

Polyvinyl acetate (PVA)

Vinyl chloride/vinyl acetate copolymer (PVAC)

Polyvinyl chloride (PVC)

Polyethylene terephthalate (PET)

Polypropylene (PP)

Polyformaldehyde (POM)

Urea-formaldehyde (UF)

Melamine-formaldehyde (MF)

Polytetrafluoroethylene (PTFE)

Thermoplastics   Thermosetting plastics

Polystyrene, polythene and polypropylene are all based on pure hydrocarbons

◄ Polythene, here shown during manufacture, was discovered in 1933 by polymerizing the gas ethylene (C_2H_4) at high temperatures and pressures – a process that often resulted in explosions. In addition, the polymer did not pack easily, and was of low density. The discovery by Natta of a low pressure process for making polythene revolutionized the industry in the mid-1950s. The secret of Natta's method was the use of metal catalysts discovered by the German chemist Karl Ziegler. Natta and Ziegler shared the 1963 Nobel Prize for chemistry for their work.

► Rubber is a polymer of isoprene, which is obtained by tapping the rubber tree, or can be produced synthetically. Natural latex only found wide use after the discovery in 1839, by Charles Goodyear, that it could be made stronger by heating with sulfur (vulcanization). This results in sulfur atoms cross-linking the polymer chains, which thereby imparts the required stiffness needed for its principal use in tires.

▼ High-density polythene (HDPE) consists of straight chains of CH_2 units that pack more closely than low-density (LDPE) to give a higher density polymer. This form of polythene is rigid enough for the plastic kitchenware that is today found in every home.

Rubber, natural and synthetic

Natural rubber is made from an unsaturated hydrocarbon monomer, isoprene (C_5H_8), which contains two double bonds separated by a single bond. When it polymerizes, the double bonds break so that the end carbons can link to other molecules of monomer and a double bond forms between the two carbon atoms which were single-bonded.

Today, "synthetic" rubber can be made using Ziegler-Natta catalysts. However, such rubber has to compete with other synthetic rubber made from different monomers. Butadiene (C_4H_6) resembles isoprene in its four-carbon backbone with alternating double bonds, but it lacks a methyl side chain. It can also be polymerized to produce a rubbery material. The major rubber produced nowadays is a copolymer of styrene and butadiene.

Different types of polythene

Originally, polythene was made by a high-pressure process. The chain-lengthening process was complicated by reactions which caused chain branching, and the molecules did not pack closely. Polythene made in this way is low-density (LDPE).

Subsequently, a low-pressure process was developed. The molecules pack together more easily, to give a high-density product (HDPE). LDPE is now used primarily for making plastic films, which are made into objects such as carrier bags and refuse sacks; HDPE is used mainly for objects made by injection- and blow-molding (◆ page 141).

A new type of polythene, linear low-density polythene (LLDPE), has been made by a low-pressure process which produces unbranched chains. However, a small percentage of a second monomer, such as hex-1-ene is added. Hex-1-ene is a chain of six carbon atoms with a double bond between the two at one end. When incorporated into a growing polythene chain, the remaining four carbon atoms stick out to one side, to give a branch of defined length. This reduces the ease of molecular packing, giving a lower density product, but with better controlled properties than LDPE.

▶ *Polymers can be modified by having a variety of atoms attached to their backbone chain. Polypropylene has a methyl group (CH_3) attached to alternate carbon atoms. This has different properties depending upon the way in which these methyl groups arrange themselves. If they are randomly orientated (upper image) the polymer is atactic and is rather soft; if they are all pointing in the same direction (lower) the polymer is isotactic and much harder. Isotactic polypropylene, obtained by special catalysts, is the form used commercially.*

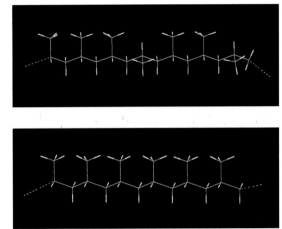

The commonest thermoplastics are polythene and polypropylene, derived from simple, unsaturated hydrocarbons (containing a carbon-carbon double bond). Ethylene comprises two carbon atoms linked by a double bond (◆ page 87). Polythene is made by joining many molecules of ethylene together on a chain.

The processes by which growing polymer chains terminate are partly random. Consequently, a sample of polythene contains molecules of slightly differing lengths, although all with the same average composition. The molecular weight of a polymer is, therefore, an average. Commonly, polythene comprises molecular chains of between 2,000 and 10,000 monomer units.

For many years, the random nature of polymerization processes made it impossible to produce a satisfactory plastic from propylene, the homolog of ethylene in which one hydrogen atom is replaced by a methyl group (◆ page 89). The methyl groups occurred as randomly orientated branchlets emerging from the linear chain.

This problem was overcome by the Italian chemist Giulio Natta (1903-1979) using catalysts originated by the German chemist Karl Ziegler (1898-1973) for the low-pressure polythene process. These metallic catalysts control the polymerization reaction, making it stereospecific (◆ page 109). As a result, it became possible to make isotactic polypropylene, in which all the methyl groups are to one side of the polymer chain. Because of these groups, polypropylene has a high softening point so it can be used to make articles which have to undergo steam sterilization in use.

The disordered, or atactic, polypropylene, which was all that could be made before Ziegler-Natta catalysts were developed, is also a minor byproduct of isotactic polypropylene manufacture. Its structure makes it rubbery in appearance, and it is also somewhat tacky. After being unwanted for some years, it is used to make asphalt road surfaces more stable and, in conjunction with minerals such as limestone, can be used in the backing of carpet tiles for its nonslip character.

Polymers have been prepared from other unsaturated hydrocarbons but the only one of major importance is styrene. This is an analog of ethylene, in which one hydrogen atom has been replaced by a benzene ring (◆ page 83). The benzene ring has a number of special characteristics so that polystyrene is a quite different polymer to polythene and polypropylene. All three are based on pure hydrocarbons, and therefore do share some properties such as being electric insulators. However, while polythene and polypropylene are soft and translucent, polystyrene is hard and transparent.

Acrylic and Perspex can both be tolerated by the body, and are used in spare parts surgery

One of the best known plastics is polyvinyl chloride (PVC). The monomer for this, vinyl chloride, is an analog of ethylene in which one hydrogen atom is replaced by a chlorine atom. The latter is considerably bulkier than a hydrogen atom, and this means that the chlorine atoms which stick out from the hydrocarbon chain in PVC are like small side chains, as with the methyl group in polypropylene. However, because chlorine is chemically different from a methyl group, the resultant polymer is unlike polypropylene. PVC was originally prepared in the 1870s. Its development as a commercial plastic came much later because, paradoxically, pure PVC is not plastic: it required the incorporation of suitable low molecular-weight substances as plasticizers before it could be made workable in a plastic sense.

Like the pure hydrocarbon polymers to which it is related, PVC is an addition polymer. But most synthetic polymers which have achieved prominence in the clothes trade have been condensation polymers, like the natural materials which they have tended to replace.

Two other analogs of ethylene which have been used successfully by the plastics industry are tetrafluoroethylene and methyl methacrylate. In the former, all the hydrogens of ethylene have been substituted by fluorine atoms. Polytetrafluoroethylene (PTFE) is a very inert, heat resistant material. This has made it useful in fabricating parts, such as valves for chemical plants. Its most familiar use, however, is as a coating on nonstick pans for cooking.

Polymethyl methacrylate, better known as Perspex, is a hard, transparent material. It became commercially available at the beginning of World War II, during which it was used extensively in aircraft canopies. Because it is also weather resistant, it is now used in outdoor advertising signs, motorcycle windshields and street light fittings.

A related monomer, acrylonitrile, gives a polymer which is not useful as a plastic. However, polyacrylonitrile can be dissolved in some organic solvents and spun from solution as acrylic fiber for wide use in textiles for clothing and carpets. Polyacrylonitrile is also important in composite materials production. When carbonized under controlled conditions, it produces carbon fibers (page 100), which can be used to strengthen thermosetting resins.

▲ ▶ *One of the most successful polymers has been polyvinyl chloride, PVC. By incorporating into the PVC other compounds as plasticizers, to make it supple, and stabilizers, to protect it from strong sunlight, it can be used almost anywhere. Here it is lining a reservoir built into the crater of an extinct volcano on Tenerife.*

▲ *Traditionally false eyes were made of glass; today they are made of acrylic polymer, which is not only more comfortable to wear but less prone to breaking if dropped. The presence of the artificial eye encourages the tear duct to work and generate antiseptic fluid to protect the eye socket from infection. The eyes can be machined to individual requirements.*

◀ *The Perspex squash court enables the sport to be enjoyed by spectators. Perspex is polymethyl methacrylate, first used in World War II for the canopies of Spitfire aircraft. This polymer is as clear as glass without being brittle. Even when fragments of Perspex penetrated wounds, the body tolerated it, a discovery that led to its use in cataract surgery.*

The first synthetic fiber – rayon – was made from cellulose derived from wood

Synthetic fibers

For thousands of years, natural polymeric fibers have been known. Silk is a protein fiber, while cotton is composed of cellulose. As early as 1884, the French chemist the Comte de Chardonnay (1839-1924) succeeded in making fibers from cellulose nitrate, a substance obtained by treating wood cellulose with nitric acid. Because this "artificial silk" reflected light so well, it was called "rayon". Its inflammability made it undesirable for textile use. However, various other rayons were developed by modifying cellulose with other chemicals.

The American chemist Wallace Carothers (1896-1937) decided to try to make synthetic fibrous materials by reacting together monomers which could form the same kind of bond found in natural proteins. The original nylon was a copolymer of adipic acid (an organic molecule containing six carbon atoms and having a carboxylic acid function – COOH – at each end of the hydrocarbon chain) and hexamethylenediamine, which contains six carbon atoms with an amino group – NH_2 – at each end of the chain.

By using these two molecules, Carothers found he could make a long-chain polymer with hydrocarbon groupings of six carbon atoms linked by amide bonds. This product is now known as nylon 66, to indicate its monomeric composition. An innovation introduced by German chemists soon after Carothers' original discovery was to use a single monomer with an amino group at one end and a carboxylic acid group at the other. The most common form of nylon produced in this way is nylon 6, made from caprolactam.

Carothers originally tried to make fiber-forming polymers by esterification. An ester is the condensation product of a carboxylic acid and an alcohol (OH). Carothers abandoned this work because the products could not be melt-spun into suitable fibers. Nevertheless, his ideas were taken up by others. In 1941, two British scientists, Rex Whinfield and J.T. Dickson, discovered polyethylene terephthalate, a polyester in which the carboxylic acid part of the molecule involves a benzene ring. This confers additional stability on the polymer which was absent from Carothers' polyesters.

Polyesters and nylons have subsequently proved very successful as textile fabrics. However, they have both also shown their worth as plastics. Nylons are used to make small solid objects, such as gear wheels, by injection-molding, while a major development in the use of polyester has been the PET (polyethylene terephthalate) bottle, a plastic bottle which can be produced very cheaply.

Originally, PET bottles could only be molded with rounded ends and black bases had to be glued onto them. More recently, ways have been found to mold the bottles with almost flat bases without introducing stresses that would make them unable to withstand the pressure generated by carbonated drinks.

As the relationship between molecular structure and properties of macromolecules has become better understood, new thermoplastics have been designed to work under more extreme conditions. These so-called engineering plastics are much more resistant to high temperatures than ordinary thermoplastics. They are generally based on monomers that contain benzene rings, but which often also contain other elements, such as sulfur and oxygen. Polyethersulfones (PES) and polyetheretherketones (PEEK) are better-known examples of this type of material. As well as being used in oil-well drilling and airplane construction, they can be used to make lightweight composites with carbon fiber (page 100).

▼ *The United States chemist Wallace Carothers became head of research into organic chemistry for the E.I. du Pont de Nemours company in 1928, where he discovered the first synthetic rubber. Later work on the chemical properties of polyamides resulted in the discovery of nylon. He produced the first nylon threads in 1935 by melting the polymer and forcing it through spinnerets.*

▲ *A row of plastic cups come off the production line after being shaped by the vacuum-forming process, in which a male mold forces the soft plastic into the female and a vacuum from below assists the molding operation. The plastic is cut, the mold opened, and the cup ejected by compressed air. Such methods of shaping objects are fully automated: a production line may produce a million or more units a day.*

Vacuum forming

Press
Plastic sheet

Mold

Vacuum

◀ In the vacuum molding of plastic articles, the sheet of molten plastic is passed over a mold and heated to make it soft. The air in the mold is evacuated from below, sucking the plastic sheet to the sides of the mold, where it is cooled to set in shape.

Blow molding

Air blown

Mold

Plastic tube

◀ In the blow molding technique, a length of hot plastic tube is put into the mold and sealed at the bottom. Air is blown into the tube from above, forcing the soft plastic to take the shape of the mold. The mold is then opened and the formed product drops out. This technique may be used for products such as bottles and flasks.

Injection molding

Plastic granules   Hydraulic ram   Mold

Heating element

◀▶ Injection molding is used for solid objects and objects that require a careful finish on both surfaces. Plastic is introduced to the mold in the form of granules, which are heated as they are forced into the mold by a hydraulic ram. The object forms as the plastic cools, and the mold is opened to release the product.

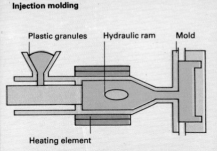

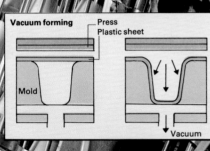

Thermosetting polymers

As thermoplastics become used in more and more applications, they are encroaching on the traditional markets of thermosetting materials. Thermosetting polymers, or resins as they are also called, are condensation polymers that form complex three-dimensional networks. Often, the starting materials are partially polymerized to give solids which can be ground into powders and molded. Once the powder has filled the mold, further polymerization is induced, either by heating or adding a small amount of a substance which acts as an initiator. This causes extensive cross-linking, giving a hard product that will not lose its shape on heating.

One of the first commercial plastics, Bakelite, introduced more than 60 years ago, is a resin made from phenol and formaldehyde. It is still used in applications such as automobile distributor heads, where good electric insulation and heat resistance are more important than appearance. For applications where decorative finishes are preferred, such as kitchen worktops, it has been superseded by resins made from formaldehyde (H.CHO) and either urea ($CO(NH_2)_2$) or melamine ($C_3H_6N_6$).

Complete polymerization of a thermoset is sometimes stopped by adding an inhibitor. Such "inhibited resins" can be used as pastes, with further polymerization being induced by an additive which counteracts the inhibitor. The time taken to complete polymerization – or "curing" as it is also called – can be controlled so that the paste remains workable while it is shaped. This type of formulation is often used to build up complex shapes, such as boat hulls, strengthened with glass or other fibers.

The resins used in these composites are often polyesters. They differ from the thermoplastic polyesters in that one of the monomers includes a carbon-carbon double bond which provides a site for cross-linking reactions.

▲ Glass-fiber reinforced resin plastics are often used for objects such as these helicopter blades in which strength and lightness are of equal importance.

▼ Bakelite, the first commercially-produced plastic, was used for many domestic electrical and luxury goods from the 1920s. These items date from the late 1940s.

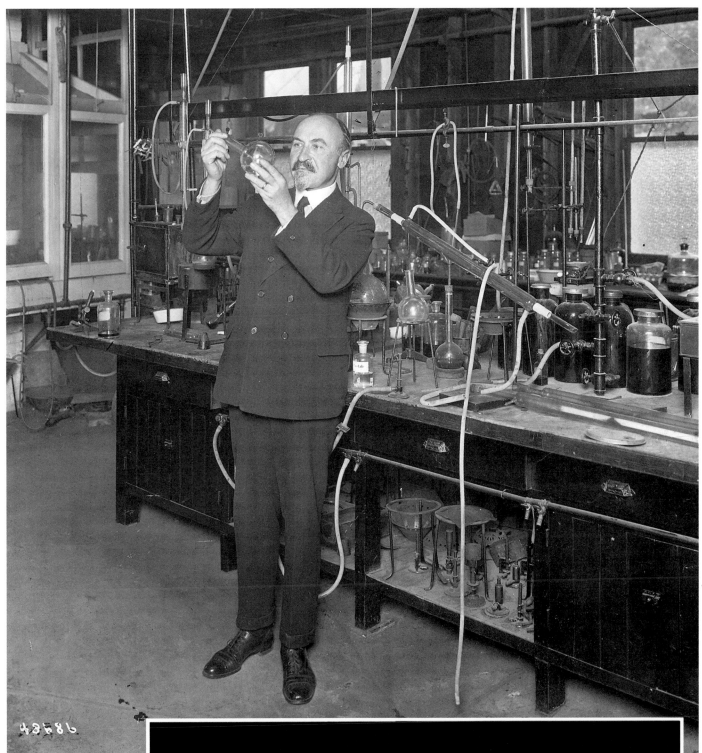

43586

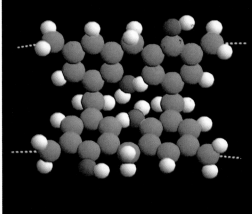

▲ The Belgian-born chemist Leo Hendrik Baekeland (1863-1944) emigrated to the United States in 1899 and developed a photographic paper that could be developed in artificial light. He began exploring phenol-formaldhyde resins, which yielded Bakelite, while researching a substitute for the varnish shellac. His discovery provided the foundation for the modern plastics industry.

► The chemical structure of the Bakelite monomer.

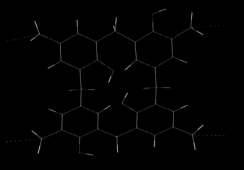

▲ A researcher working on the qualities and applications of new polymers tests the setting qualities of an "improved resin" in the laboratory.

▼ Wiring as used for defense purposes. In the event of a fire, the silicone in the insulation melts and turns to silica, with good insulating properties.

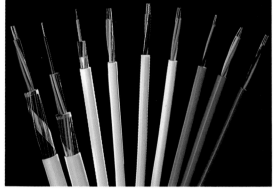

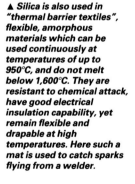

▲ Silica is also used in "thermal barrier textiles", flexible, amorphous materials which can be used continuously at temperatures of up to 950°C, and do not melt below 1,600°C. They are resistant to chemical attack, have good electrical insulation capability, yet remain flexible and drapable at high temperatures. Here such a mat is used to catch sparks flying from a welder.

The uses of silicon

In the Periodic Table, silicon falls below carbon in group 14. Consequently, silicon shows similarities to carbon but is different in some respects. It has been suggested that, on planets such as Mars, life might have evolved based on natural polymers of silicon rather than the carbon-based polymers on which terrestrial life depends. However, silicon-silicon bonds are much less stable than carbon-carbon bonds and double bonding between silicon atoms does not normally occur.

Macromolecules can be formed from alternating silicon and oxygen atoms. One of the commonest substances on Earth, silica (the main ingredient of sand) is an inorganic polymer of silicon and oxygen. It is also the basis of one of the plastics with which we are most familiar: glass. Window glass is made by fusing silica with several minor ingredients which reduce the melting point of the mixture and make the melt easier to process.

Glass is thermoplastic, and the shape is not wholly fixed. At ambient temperatures, glass undergoes plastic deformation, albeit very slowly. Measurements made on very old samples of glass show that the bottom of a pane is thicker than the top as a result of the slow flow of the glass under

the effect of gravity (◀ page 36).

Silica is a three-dimensional polymer, because each silicon atom can bond to four other atoms, and the links to adjacent oxygen atoms in a silicon-oxygen chain only take up two of these valencies. A major step in the chemistry of synthetic polymers was the use of the remaining valencies to bond with organic side chains, such as methyl groups.

As a result, it has been possible to produce liquid silicone, silicone grease and silicone rubber. Silicone plastics have a number of advantages over their organic counterparts. In a fire, for example, a silicone is reconverted to silica. As a result, silicone rubber has been used as an insulating material for electric wiring in defense applications.

Silicone materials are used in medicine and cosmetic surgery. Being so unlike physiological materials, they produce less adverse reactions from the immune system than do organic materials.

Silicones are also effective water repellents and have been used to treat many different materials, including paper, textiles and masonry.

An everyday application of silicones is as an additive in polishes and waxes. The lubricating effect of the silicone on the wax in the polish makes it possible to obtain a good finish with less effort.

Agrochemicals

18

*Natural fertilizers...Nitrogen and the soil... Phosphates
and potassium fertilizers...Insecticides and other
pesticides...Herbicides and selective weedkillers...
PERSPECTIVE...Chemical warfare between plants...
Modern methods of spraying*

Since humans took up agriculture about 10,000 years ago, continued
efforts have been made to improve crop yields. Long before anything
scientific was known about soil fertility, crop rotation was used to
help keep soils balanced by ensuring the minerals in the soil were not
depleted by a single crop. Animal manure was first used as fertilizer
centuries ago, while the search for substances to keep pests from
damaging crops has a history going back to classical times.

Attempts at a scientific understanding of soil fertility began
seriously in the mid-18th century. However, the age of agrochemistry
did not begin in earnest for nearly another hundred years. In 1837,
the British Association for the Advancement of Science com-
missioned the German chemist Justus von Liebig (1803-1873) to study
soil fertility. He pointed out that fertile soils were rich in the elements
nitrogen, phosphorus and potassium, and also contained calcium and
magnesium. Unfortunately, he made an error in assuming that plants
could obtain all their nitrogen requirements from atmospheric nitrogen and that only mineral elements were essential to soil fertility.

▲ Animal manure is rich in
all the nutrients that plants
need and has a high organic
content which helps to
condition the soil. The most
important elements are
nitrogen, phosphorus and
potassium, which need to
be added each year to
improve yields. Trace
elements are also required
but these are rarely in short
supply in the soil.

▼ Phosphate is found in all
living things, yet it is easily
lost from soil or chemically
bonded to metals in the soil
so that it cannot be tapped
by plant roots. An early
source for fertilizers was
guano. This phosphate,
from bird droppings, lies
several meters thick on
islands off the coast of Peru.
The birds live off fish whose
bones are rich in phosphate.

Warfare among the plants

*It has been known for many years that plants have
defense mechanisms against predators. Chemists
have begun to study plants to see whether some
contain substances which will keep others in check.*

*Proponents of chemical warfare between plants
call the subject allelopathy. Identified as
allelopathic chemicals are a number of simple
organic acids which are known to inhibit seed
germination. It is possible that the "tartness" of
underripe fruit is to prevent premature germination
of the seeds. Other chemicals identified as
allelopathic encourage seed germination.*

*Supporters of allelopathy argue that plants may
release such substances into the soil, either to
prevent germination of seeds that are close to them
or to trigger "suicidal" germination (that is
germination of the seed at a time of year when the
seedling will not be able to survive).*

*Juglone, a substance produced by walnut trees
which has been shown to have insecticidal
properties, may also be allelopathic. Tomatoes and
alfalfa wilt if exposed to it. Salicylic acid, produced
by willow trees, affects the ability of some plants to
retain essential metal ions. This is probably why
little will grow in proximity to willow trees.*

*It has been argued that there is no evidence that
these chemicals will permeate the soil sufficiently
to create a protective barrier for the plant that
produces them. If more potential allelochemicals
can be identified, they might provide ideas for
synthetic analogs which could be herbicides.*

A hectare crop of barley takes 110kg of nitrogen, 14kg of phosphorus and 56kg of potassium from the soil

Nitrogen fertilizers

Fertilizer production is now a major part of the chemical industry. The fixation of nitrogen to ammonia is the second largest-scale chemical manufacture in the world, while sulfuric acid (◀ page 130) is only larger because it too plays a key role in fertilizer production.

Nitrogen is used for protein production. It is absorbed by plant roots as the nitrate ion, although ammonium ions can also be used as fertilizer as they are oxidized to nitrate by soil bacteria. Phosphorus stimulates root development and is also important in plant ripening, while potassium plays a key role in photosynthesis.

Nitrogen is supplied to soils in a variety of chemical forms. In some places, ammonia is injected directly into soils. Where this can be done, it is an effective way of supplying the element, because it comprises more than 80 percent of the ammonia molecule. Two other forms in which nitrogen is commonly added to soil are urea (46 percent nitrogen) and ammonium nitrate (35 percent nitrogen).

Urea is manufactured by compressing carbon dioxide and ammonia to high pressures. An exothermic reaction (◀ page 95) converts the gases to ammonium carbamate, most of which dehydrates to give urea and water. Unreacted gases can be removed and recycled, while the urea solution is evaporated to give solid product. More than a quarter of the world's ammonia production is used to make urea and about three-quarters of the product is for fertilizer use. Half the remainder is also used agrochemically, as an animal feed additive.

Ammonium nitrate is made by passing gaseous ammonia into nitric acid solution. This generates considerable heat, which is used to evaporate the water present in the acid and that formed during the reaction. This process must be carried out under carefully controlled conditions as the nitrate is potentially explosive. The molten product is sprayed down the inside of a high tower, called a prilling tower. It breaks into droplets which solidify during their fall to the base of the tower, giving a product in pellet form.

Potassium and phosphates

Compound fertilizers, which contain nitrogen, phosphorus and potassium, often have the first two elements in the form of ammonium nitrate and ammonium phosphate. These are made by neutralizing a mixture of the acids with ammonia. The solution is concentrated, mixed with potassium chloride and then granulated.

The earliest phosphorus fertilizers derived from bone meal, and in 1841 a process of manufacturing "superphosphates", in which the phosphate was made more readily soluble, was developed. The starting material for phosphate fertilizers is now phosphate rock, of which about 100 million tonnes are mined annually. The rock varies in composition according to its place of origin, but is primarily insoluble calcium phosphate. This is made either by mixing phosphate rock with sufficient sulfuric acid to convert two-thirds of the calcium present to calcium sulfate. The mixture is allowed to mature and is used without further processing. This "superphosphate" contains about 10 percent phosphorus. "Triple superphosphate", which contains about 25 percent phosphorus, is made by treating phosphate rock with phosphoric acid, so that the product is not diluted with calcium sulfate.

Most of the phosphoric acid used in fertilizers is obtained by the "wet process", in which phosphate rock is treated with excess sulfuric acid. The insoluble calcium sulfate is removed by filtration. The acid

solution still contains many of the impurities present in the rock. Nevertheless, it can usually be used without further purification.

Potassium is obtained from natural deposits of potassium chloride. Frequently such deposits are mixtures of metal chlorides. Sylvinite, for example, is a mixture of potassium and sodium chlorides, which is separated by a "hot leach" process. Potassium chloride is much more soluble in hot water than in cold, while the solubility of sodium chloride changes little with temperature. Sylvinite is crushed and mixed with concentrated brine (sodium chloride solution) and then heated. As the brine is already saturated with sodium chloride, only the potassium chloride dissolves. The mixture is then filtered out and the filtrate cooled, so that potassium chloride precipitates out.

◄ *Hydroponics is the cultivation of plants by feeding them a solution of inorganic nutrients. Here the roots of tomato plants are wrapped and solution pumped through.*

► *Withhold an essential nutrient from a growing plant and it suffers. The condition known as blossom end rot affects tomatoes and is caused by a deficiency of calcium.*

▼ *A rich source of nitrogen is ammonium nitrate. This is made as granules by spraying the molten chemical, and allowing the droplets to fall down tall "prilling" towers.*

Crop protection is a major use of chemicals in agriculture

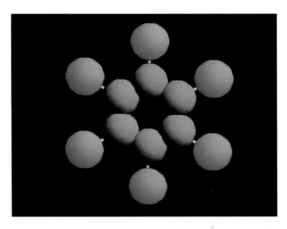

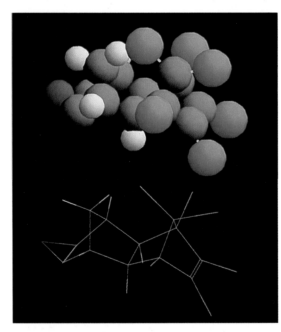

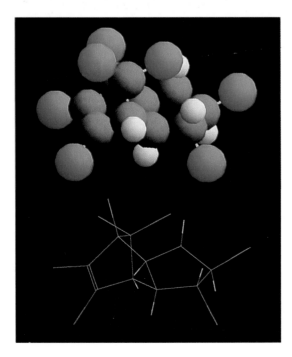

◄ *The compound DDT had been discovered in 1873 but it was not tested as an insecticide for another 60 years, when it was tried successfully as a mothproofing agent. Once the potency of such chlorinated compounds was realized, chemists made several others. Benzene hexachloride (top) – the carbon atoms are green, the chlorine atoms are blue – was found to be effective and safe enough to use against head lice. Dieldrin (center) was widely used to combat crop-eating insects. Unfortunately, these kinds of insecticides persist in the environment for many years and so enter the food chain. Fish and birds were found to have high levels of chlorinated insecticides in their bodies. Some birds of prey were brought near to extinction, and DDT was even found in humans. In addition DDT-resisting strains of insects were starting to evolve. Consequently in the 1970s several countries banned the widespread use of persistent insecticides such as DDT. Other chlorinated insecticides have also been banned on environmental grounds: these include the compound chlordane (bottom).*

Pest control

Insect control has a very long history, but again much of the chemical progress has been in the past half century. Sulfur compounds were used to control insects more than 4,000 years ago, while the natural insecticidal properties of some plants have also been exploited for many years. Thus dried flowers of *Chrysanthemum cinerariafolum* (pyrethrum) were an article of commerce in the 17th century as was derris dust, the powdered root of the derris plant. World demand for pyrethrum flowers is now in excess of 25,000 tonnes annually, most of which comes from Kenya, Tanzania and Ecuador.

Attempts in the 1930s to discover new mothproofing agents for textiles led the Swiss chemist Paul Müller (1899-1965) to discover the insecticidal properties of diphenyldichlorotriphenylethane. Better known as DDT, this compound was synthesized first in 1873, but its effects had gone unnoticed. It rapidly went into use as an insecticide and was soon followed by the discovery of other chlorinated hydro-carbon insecticides, such as benzene hexachloride – first synthesized by Michael Faraday in 1825. It was not until some years later, how-ever, that it was discovered that one isomer in particular, of the six possible forms, was far more effective than the others. Known as lindane, it is still used in some parts of the world to treat head lice.

Other chlorinated hydrocarbon insecticides discovered during the 1940s included dieldrin and chlordane. The effect of introducing chlorine atoms into the chemical structures was to make compounds which could not easily be broken down by soil microorganisms. Consequently, the organochlorines are known as persistent insecti-

◄ *Human attempts to control the ravages of insect pests were largely ineffective until the present century. The first insecticide to be used on a large scale was DDT in the 1940s. Some areas of the tropics were made virtually mosquito-free and malaria-free by DDT. Nearly 2 million tonnes of DDT have been used throughout the world since then.*

Better spraying through electricity

In recent years, considerable attention has been paid to developing better means of applying pesticides. This cuts down harmful environmental effects and also reduces costs. Low-volume spraying techniques can now spread as little as 2.5 liters of pesticide solution over one hectare of crop. Even this figure can be reduced several-fold by using a spray method that relies on static electricity.

In conventional spraying, the solution is broken up into droplets mechanically. In the Electrodyn system, the spray is created by applying 25,000v across the outlet nozzle of a container of pesticide. The pesticide is dissolved in an oil-based solvent rather than the aqueous systems more often used.

The solution breaks up into very fine, electrically charged droplets. These repel one another and consequently spread out widely as they leave the nozzle. At the same time, they are attracted to the nearest "earth", such as the plants over which they have been sprayed. There is the added advantage that both tops and undersides of leaves are covered with the spray. In this way one liter of insecticide will cover two hectares of crops.

Despite the high voltage across the nozzle, a hand-held Electodyn sprayer operates on four 1.5v batteries. The system was first tried out in 1981 on cotton and was found to increase farmers' profits by up to 50 percent.

▲ *The Electrodyn sprayer, invented by Dr Ron Coffee, gives each tiny drop of insecticide an electric charge. This causes the spray to be attracted to all parts of the plant, including the underside of leaves, and consequently none is wasted.*

cides. Their persistence, which at first was seen as an advantage, later meant they fell out of favor on environmental grounds.

At about the same time as the organochlorines were introduced, research into the toxicity of organophosphorus compounds was under way. These are poisons which kill by interfering with the chemistry of nerve transmission. Some are stockpiled as potential chemical weapons ("nerve gases"). Others have found widespread application as insecticides and, given the way in which these substances are metabolized, they are far more lethal to insects than to mammals. Some of these compounds are absorbed by growing plants and then ingested by insects which feed on the plants. Insecticides which operate in this way are called systemic. Unlike the organochlorine pesticides, organophosphorus pesticides break down fairly rapidly.

A third major class of insecticides are the carbamates. These also interfere with nerve transmission. The carbamate structure occurs in the natural, poisonous alkaloid physostigmine. It was this that led to the investigation of this grouping as a component of insecticides.

More recently, pyrethroids have developed as an important new class of insecticides. These are synthetic compounds based on the chemical structures found in the naturally occurring pyrethrins, the active ingredients of pyrethrum flowers. The natural compounds break down rapidly in the presence of sunlight, and much of the synthetic work has been aimed at modifying parts of the molecule to confer greater stability on it. Much of the research was carried out in the 1960s and 1970s at the Rothamsted Experimental Station, Britain, founded by John Bennett Lawes.

▲ ▼ *Herbicides (weedkillers) are used to increase food production, such as in giving a young oil palm enough room to establish itself (above). Some herbicides overstimulate the plant's growth system, such as the phenoxyacetic acid derivatives (1,2). Paraquat (4) is a contact herbicide, which fatally interferes with a plant's metabolism. However, Simazine (3) is a selective weedkiller used to protect corn.*

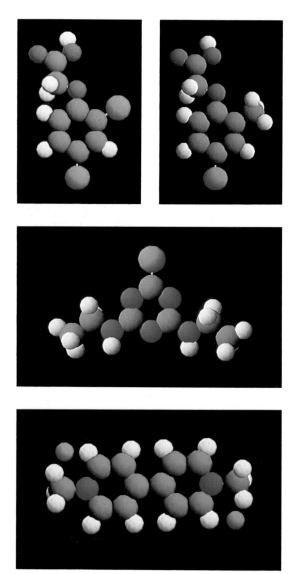

Some of the microorganisms which attack plants cannot be seen by the naked eye, yet they can be devastating in their effect. The great famine in Ireland in the 1840s was caused by a fungus which blighted potatoes. A number of fungicides have been developed. Many are organic molecules which also contain a metal atom, such as the thiocarbamates zineb (zinc containing) and maneb (manganese containing).

Other pests which need controlling if crops are to be protected are nematodes (tiny worms which attack the roots of plants) and rodents. Nematodes are controlled by soil fumigation with organohalide compounds. Major poisons used against rodents are anticoagulants, which prevent blood clotting and cause death from internal bleeding. A key compound used for this purpose is warfarin. Many of the early pesticides are now less effective, however. Thus, not only are there warfarin-resistant rats, there are also weeds which have developed the ability to break down the common triazine herbicides. Consequently, chemical warfare against agricultural pests continues.

Controlling weeds

The chemistry of weed control has advanced dramatically in the past half century. Herbicides are used to prevent weeds from competing with crops, to remove noxious weeds from grazing lands and to control growth of vegetation on roadsides and railroad tracks. Some inorganic chemicals for weed control were introduced in the late 19th and early 20th century. Copper sulfate was the first, followed by ferrous sulfate and sodium chlorate. These compounds kill plants indiscriminately, so they had to be applied carefully.

The modern era of herbicide chemistry came into being through the discovery of natural growth-promoting substances in plants in the 1930s. Once these had been identified, synthetic analogs were made, some of which were found not only to be effective as growth promoters but, at higher concentrations, to kill some plants. This action was sufficiently selective for some of the compounds to be introduced as herbicides. Subsequently, there has been a continued search for selective weedkillers which can be used with different crops. The older herbicides tend to kill broad-leaved plants, while leaving grasses (a class which includes cereal crops) unharmed. They are, therefore, no use where the crop is broad-leaved, like soya beans and sugar beet, and the major weeds grasslike.

Herbicidal activity is usually discovered by screening large numbers of compounds to find those which have some herbicidal effect. The hit-and-miss nature of this activity stems from the relatively poor state of knowledge of plant biochemistry. Once a candidate has been spotted, the molecule is modified in an attempt to enhance its activity against weeds while reducing any toxic effect on crop plants. Thus, in 1956, the triazine herbicide simazine was introduced for controlling weeds growing among corn. The corn was unaffected. Subsequently this was shown to be because it possesses an enzyme which breaks down the herbicide. In some cases it has been possible to find very selective compounds. Carbyne is a carbamate herbicide which kills wild oats but does not harm wheat, although they are both cereals.

Another important herbicide discovery was paraquat. This kills plants by interfering with their photosynthetic mechanisms. It is not selective, but can be used in place of plowing to clear fields ready for sowing. Unlike some other herbicides, which persist in soil, paraquat is inactivated on contact with soil. This is because of the molecule's shape which makes it strongly absorbed by clay particles.

Chemistry and Food

19

Preservatives and colorings in food...Artificial flavorings...Emulsifiers and thickening agents... PERSPECTIVE...Additives in fast food...Margarine manufacture...Alcohol and malt whisky...Synthetic foods and synthetic sweeteners

Nearly all food bought in industrialized countries has been processed in some way. This is obvious for products such as bread and cheese, which bear no resemblance to the grain and milk from which they come. But even whole fruit and vegetables, such as apples and tomatoes, are often treated with ripening agents and preservatives before they reach the consumer. Much food processing is physical or biological, but chemistry also plays a part in maintaining and extending the variety and palatability of the modern diet.

Much of the chemistry of food processing centers on additives. These may be natural or synthetic substances which are added to food for one of a number of reasons. Their use goes back to ancient times. The Roman naturalist Pliny the Younger, who lived in the 1st century AD, recorded the addition of various substances to wine to make it taste better. One of the most common additives was chalk (calcium carbonate), which would have neutralized excess acid and made the wine less sharp to the taste.

The use of preservatives
The use of nitrites and nitrates in brine for curing meat also has a long history. This mixture prevents the growth of the bacterium *Clostridium botulinum*, which produces a toxin that can cause severe food poisoning and even death. At the same time the mineral salts and hemoglobin in the meat react chemically to form nitrosohemoglobin, so the process imparts a pink color to pigmeat, which it converts to ham or bacon.

Microbial growth is a major reason why food goes bad, and this may be retarded by the use of refrigeration. But many substances which occur naturally in food are chemically unstable. On exposure to air, fats oxidize to give unpleasant tasting and smelling products which we associate with rancidity. The shelf-life of such materials can be increased by adding antioxidants – substances which help to prevent this chemical reaction from taking place. These may be naturally occurring substances such as ascorbic acid (vitamin C), or synthetic, such as butylated hydroxyanisole (BHA) and butylated hydroxytoluene (BHT).

Oxidation of fats is catalyzed by metal ions, some of which also occur naturally in foods. A way to prevent oxidation is therefore to add sequestering agents, substances which grab onto the metal atoms and prevent them from acting as catalysts. Citric acid, found in lemon juice, is frequently used as a sequestrant, as is phosphoric acid.

As the shelf-life of foods is increased by the use of preservatives, the likelihood increases that natural pigments will fade with exposure to light or air. Consequently, a range of artificial coloring agents, many derived from coal-tar dyes, are frequently added in order to keep products such as strawberry jam looking fresh.

Fast food – junk food?
"Let's go out for some fast food." In a hamburger you will probably have sodium or calcium polyphosphate (stabilizer), sodium glutamate (flavor enhancer) and some synthetic flavoring. Make it a cheeseburger with a slice of processed cheese and you add the following to your meal: sorbic acid (mold inhibitor), capsanthin, annatto or carotene (yellow colorants), sodium citrate (emulsifier, stabilizer and enhancer of the effect of antioxidants) and possibly more polyphosphate.

Have some instant mashed potato and you will also eat butylated hydroxyanisole or butylated hydroxytoluene and sodium metabisulfite (antioxidants). Prefer crisps? Then you get the antioxidants, plus possibly some sodium 5 – ribonucleotide (flavor enhancer). Baked beans to complete the main course – plus guar gum or modified starch (thickener) and possibly some sodium bicarbonate (to reduce acidity).

For dessert, how about apple pie? If it is prepackaged, it will include modified starch (thickener), potassium sorbate (a salt of sorbic acid) and also give you some citric acid (prevents discoloration) and malic acid (flavoring).

Some of these materials occur naturally in food. Citric acid occurs in citrus fruits and malic acid in apples, for example. No health risks are known for most additives. However, some may cause digestive disturbances and others are thought to induce allergic reactions.

▲ *Not what Nature intended but almost as good?*

The discovery of margarine in 1869 was the result of a competition run by the Emperor Napoleon III to find a butter substitute

▶ *Margarine is made increasingly from vegetable fats such as palm oil, groundnut oil, or sunflower oil. These are first purified to remove impurities such as carbohydrates, phospholipids and resins, and then neutralized: the fatty acids combine with caustic soda to form soaps, which are run off (*◆ *page 175). If the oil used is unsaturated, it is made more stable and given a more solid consistency by the hydrogenation process. Unwanted odors are then removed from the oil by steam distillation. If necessary, different oils are blended together to achieve the required consistency, and colorings, vitamins and an emulsifier added, to complete the oil blend.*

▶ *In the hydrogenation process, unsaturated compounds are treated with hydrogen in the presence of a catalyst to saturate the carbon-carbon double bonds. The single carbon-carbon bond is much more stable and raises the melting point.*

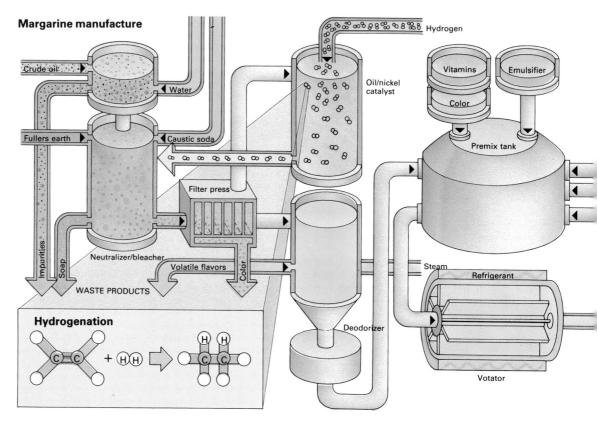

Margarine manufacture

Hydrogenation

▲ *The taste of the food we eat is increasingly the product of the art and science of the food chemist. This study combines a knowledge of the way in which the body recognizes tastes (still a poorly understood subject), and attempts to enhance the flavors in the food offered to the consumer. The chemist may be involved in ensuring that any processing techniques used do not harm the natural flavors of the fresh food.*

Food and flavorings

The complex flavors of natural foods can be analyzed by techniques such as gas chromatography. The constituents in the sample are separated by preparing it in a gaseous state, injecting it into a carrier gas and passing this along a tube coated with a liquid. The various components of the sample dissolve at different rates in this liquid, and therefore arrive at the end of the tube at different rates. Many fruits have dozens of volatile substances contributing to their flavor, although one or two usually predominate. Many of these substances are esters or aldehydes which can be made in the laboratory. They can then be used to make food products, such as fruit-flavored candies or desserts, which resemble the natural fruit flavors. It is possible to categorize chemical flavors as imitation, artificial and synthetic. Synthetic flavors do not imitate any natural substances, whereas artificial flavorings comprise natural products in proportions not found in nature. Imitation flavors incorporate at least some artificial products.

Emulsifiers

Much margarine today contains 4-*cis*-heptenal, a substance which contributes to the creamy taste of butter. Margarine was patented in 1869 by the Frenchman Hippolyte Mège-Mouriès (1817-1880). His discovery came as a result of a competition run by the Emperor Napoleon III to find a substitute for butter. Mège-Mouriès's product was made substantially from beef tallow and lard but, in 1902, Wilhelm Normann discovered that vegetable oils could be chemically modified by treatment with hydrogen gas to give solid fats. Most margarine now consists primarily of hydrogenated vegetable oils. The hydrogen saturates carbon-carbon double bonds, which occur more frequently in plant triglycerides than in those from animals, thus raising their melting points.

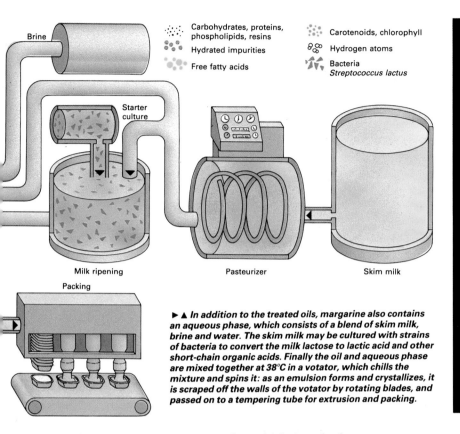

Brine

Carbohydrates, proteins, phospholipids, resins

Hydrated impurities

Free fatty acids

Carotenoids, chlorophyll

Hydrogen atoms

Bacteria
Streptococcus lactus

Starter culture

Milk ripening

Pasteurizer

Skim milk

Packing

▶ ▲ *In addition to the treated oils, margarine also contains an aqueous phase, which consists of a blend of skim milk, brine and water. The skim milk may be cultured with strains of bacteria to convert the milk lactose to lactic acid and other short-chain organic acids. Finally the oil and aqueous phase are mixed together at 38°C in a votator, which chills the mixture and spins it: as an emulsion forms and crystallizes, it is scraped off the walls of the votator by rotating blades, and passed on to a tempering tube for extrusion and packing.*

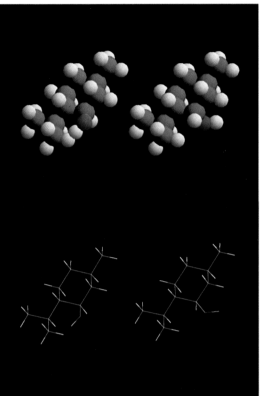

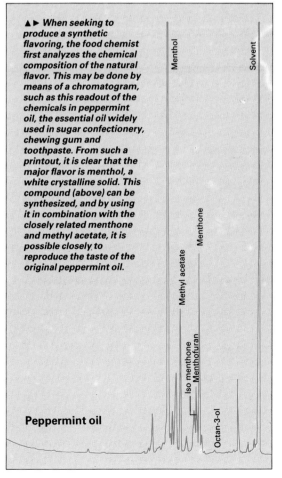

▲ ▶ *When seeking to produce a synthetic flavoring, the food chemist first analyzes the chemical composition of the natural flavor. This may be done by means of a chromatogram, such as this readout of the chemicals in peppermint oil, the essential oil widely used in sugar confectionery, chewing gum and toothpaste. From such a printout, it is clear that the major flavor is menthol, a white crystalline solid. This compound (above) can be synthesized, and by using it in combination with the closely related menthone and methyl acetate, it is possible closely to reproduce the taste of the original peppermint oil.*

Menthol

Solvent

Menthone

Methyl acetate

Iso menthone

Menthofuran

Octan-3-ol

Peppermint oil

Although margarine and butter, for which it substitutes, are composed mainly of fats, they also contain other solids and some water. Water and fatty substances do not ordinarily mix together. However, many foods are stable emulsions of tiny particles of fat suspended in an aqueous medium or, as with butter, vice versa. Emulsifying agents work in the same way as household detergents (◆ page 177). One part of the molecule has an affinity for non-polar materials (fats) while the rest has an affinity for polar substances, like water. The emulsifier thus provides an interface between the incompatible substances so that they remain evenly mixed.

An important natural emulsifier is lecithin. This occurs plentifully in egg yolks and accounts for the widespread traditional use of eggs in preparing foods such as mayonnaise. Lecithin can also be obtained from other natural sources and many mass-produced foods incorporate soya bean lecithin as a stabilizer. Thickening agents are another type of additive widely used. These increase the viscosity of products such as soups and "instant" milk puddings. If a soup is made by traditional methods, it may thicken naturally from the slow production of jelly-like materials from breakdown of collagen in meat tissue, or it may be thickened by the addition of starch (often corn-flour). "Convenience" foods contain other thickening agents, because unmodified starch gives a gel which is not sufficiently stable to undergo many processing operations. Often alginates, obtained from seaweed, are used instead. These are complex carbohydrates similar to pectin, the substance in fruits which thickens jam.

Another food additive which occurs naturally in some foods is monosodium glutamate. The sodium salt of one of the essential amino acids, this substance is a flavor enhancer. It brings out the flavor of meats, making them taste stronger, but it is not clear how this happens. It is used widely in oriental cooking.

Chemistry and Spirits – Making Malt Whisky

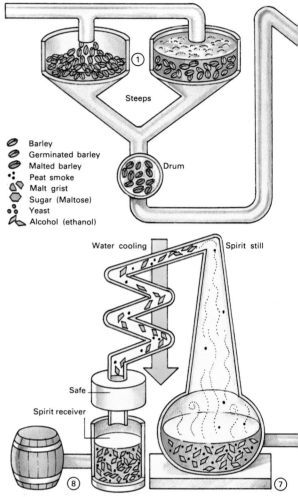

Barley
Germinated barley
Malted barley
Peat smoke
Malt grist
Sugar (Maltose)
Yeast
Alcohol (ethanol)

Steeps

Drum

Water cooling   Spirit still

Safe

Spirit receiver

Alcoholic beverages

The production of beverages containing ethanol (C_2H_5OH) as an intoxicant goes back thousands of years. The simplest form of alcoholic beverage is one in which naturally-occurring sugars from plant material are fermented by yeast to give a weak alcohol solution, as is the case with beer (made from cereals) and wine (made from grapes).

Stronger drinks can be made by distillation of beer, wine or a similar weakly alcoholic mixture. This increases the proportion of alcohol to give spirits such as vodka, brandy and whisky.

The production of alcoholic beverages is an important part of the food industry. While some drinks are still made by traditional methods, others are produced in huge manufacturing plants.

The flavor of a particular alcoholic drink depends on the mixture of trace volatile components which it contains. These may be derived from the main material used in its manufacture, or from an additive, such as hops in beer and herbs and berries in gin. These components react slowly with the alcohol or with one another, so that the taste changes with time. In many wines, traces of fatty acids form esters as the wine matures. Beverages may also vary in flavor through the way they are stored. When drinks are matured in wooden casks, naturally occurring chemicals leach from the wood and react with components of the beverage.

▲ Turning the carbohydrate store of plants into alcohol (ethanol) was one of the earliest chemical processes to be discovered. Today it is the basis of the beer, wine and spirits industries. The key step is the fermentation of plant sugars by yeasts which produces the ethanol and the gas, carbon dioxide.

▼ Every drop of whisky that comes from the stills has to be accurately recorded by a spirit safe, like the one below, before it is put into bond. There it can rest for up to 12 years before it is considered fully mature. Laws governing the use of stills are enforced by the Customs and Excise.

▶ In these beautiful copper stills the dilute solution from the fermented grain that contains the alcohol is heated. The ethanol that boils off is condensed and collected, but yet contains some water. It is then piped to a second still to undergo a final distillation to give the pure spirit.

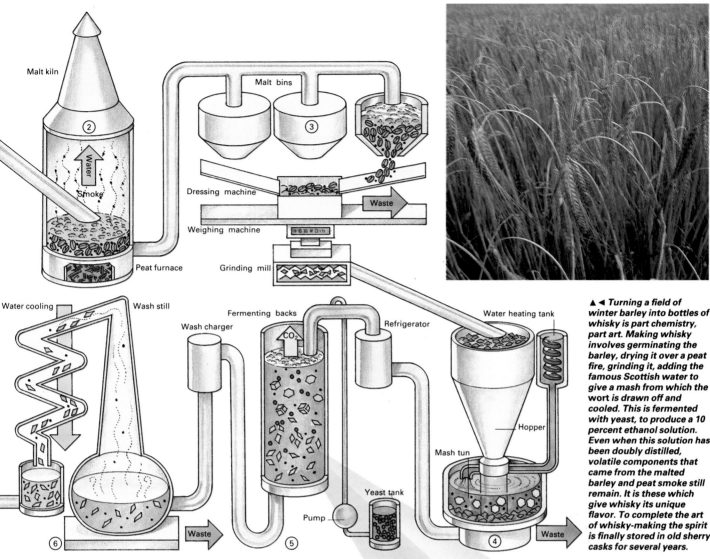

Malt kiln

② Water Smoke

Malt bins

③

Dressing machine

Weighing machine

Waste

Peat furnace

Grinding mill

Water cooling   Wash still

Wash charger

Fermenting backs   Refrigerator

CO_2

Water heating tank

Hopper

Mash tun

Yeast tank

Pump

Waste

⑥   ⑤   ④   Waste

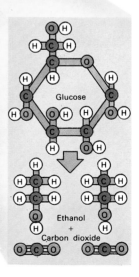

▲ ◄ *Turning a field of winter barley into bottles of whisky is part chemistry, part art. Making whisky involves germinating the barley, drying it over a peat fire, grinding it, adding the famous Scottish water to give a mash from which the wort is drawn off and cooled. This is fermented with yeast, to produce a 10 percent ethanol solution. Even when this solution has been doubly distilled, volatile components that came from the malted barley and peat smoke still remain. It is these which give whisky its unique flavor. To complete the art of whisky-making the spirit is finally stored in old sherry casks for several years.*

▼ *Even the chemistry of whisky manufacture is mysterious. The long chains of starch in the malt break down into individual glucose rings in steeping. The reaction that turns glucose into ethanol and carbon dioxide is carried out by the yeast enzymes in the fermenting backs.*

Glucose

Ethanol + Carbon dioxide

Scotch whisky

Scotch whisky, one of the world's most famous alcoholic beverages, is of two kinds: malt and grain. Malt whisky is made solely from malted barley, while grain whisky also uses unmalted barley and corn.

The first stage in malt whisky production is malting. Soaked barley is allowed to germinate under carefully controlled conditions. This breaks down some of the starch in the barley to soluble carbohydrate. At the end of the malting process, the barley is dried. This drying may be carried out using peat fires, which impart a smoky taste to the finished product.

The dried barley is ground and mixed with hot water. The mixture is then filtered to give a sugary liquid called wort. After cooling, yeast is added and fermentation takes place over a period of about 48 hours. The weakly alcoholic liquid so produced is then distilled twice.

The first distillation separates the alcohol from the unfermentable material and the yeast residues, but still produces only a weakly alcoholic solution, called low wines. This is distilled a second time. When the alcoholic strength reaches an appropriate level, the distillate is run off into a spirit receiver. The raw spirit is then filled into oak casks and allowed to mature. In the case of a malt whisky, this process may go on for more than 10 years.

Synthetic food

Although much of today's food is processed, most of it still comes from animal or vegetable sources. Meat substitutes may be made from soya beans, for example. Even in the case of high-technology processes, such as the production of microbial protein for animal feed from methanol or hydrocarbons, living organisms make up the foodstuff.

It is possible, however, to make entirely synthetic food components. During World War II, the Germans, faced with severe food shortages, produced fats from mineral oil. Oil hydrocarbons can be separated into fractions of different molecular weights. If an appropriate fraction is oxidized, it gives carboxylic acids of about the same chain lengths as those found commonly in triglycerides. These acids can be esterified with glycerol (which can also be made from oil) to give fats and oils which are completely synthetic. However, metabolism of the synthetic materials is not the same as that of natural fats.

No satisfactory synthesis of carbohydrate polymers has been achieved, although the Russian chemist, Alexander Butlerov (1828-1886), discovered that formaldehyde could be polymerized to give a mixture of sugar-like compounds. Subsequent researchers have developed this reaction to produce and separate glucose, the monomer from which both starch and cellulose are made.

The problem so far has been to find a way to polymerize this in the precisely-ordered sequences found in nature and on which the properties of the polysaccharides depend. In nature, this is done enzymically. Although chemists have developed highly specific catalysts in recent years, these have been for the production of new materials such as polypropylene rather than for synthetic carbohydrates, which would be unlikely to compete on economic grounds with carbohydrates from plant sources.

Although not yet approved for human consumption, various sucrose polyesters have been made, using the same fatty acids for esterification as occur in triglycerides. Because the enzymes which break down triglycerides cannot cope with the larger molecules based on sucrose rather than glycerol, these materials can act like fats in their contribution to food composition and texture but cannot be digested. Similarly, interest is now growing in producing sugar in which the component monosaccharides have the "wrong" biological conformation. The taste receptors on the tongue are not stereospecific, so they still detect this compound as sweet, but the enzyme which digests sucrose cannot digest this isomer.

There are already approved artificial sweeteners, such as saccharin and aspartame. However, these are much sweeter than sugar. Consequently, they are only needed in much smaller quantities. While this does not matter in sweetening drinks such as tea and coffee, in much food manufacture, such as cakes and pastries, sugar serves not only as a sweetener, but also as a bulking agent. This is why a non-digestible sugar could be a major contribution of chemistry to food processing.

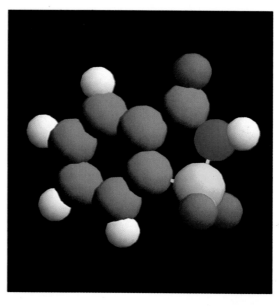

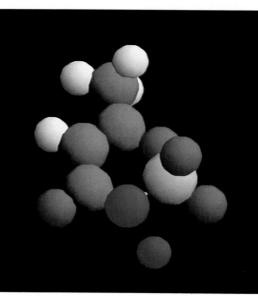

▲ In 1861 the Russian chemist Alexander Mikhailovich Butlerov reported that a sugar-like compound could be made in the laboratory from the simple chemical, formaldehyde. This discovery showed that the chemical production of food was possible. Yet making of carbohydrates, like sugar, by purely chemical methods still eludes modern chemists. That has to be left to Nature's chemists, the plants!

◄ Chemists have, nevertheless, had considerable impact on other aspects of food. Sweetness is a taste sensation that we find very pleasant. In nature the sweetest sugar is fructose (1), which is found in honey and is produced from sucrose (sugar). Unfortunately these natural sweeteners are also energy rich and we may not want to take in the extra calories. Artificial sweeteners are a way round this dilemma. Hundreds of times sweeter than sugar or fructose, they can be entirely calorie-free.

The first commercial sweetener was saccharin (2), discovered accidentally in 1879 by Ira Remsen and Constantin Fahlberg at the Johns Hopkins University in the USA. This molecule is sweet but has a slightly bitter aftertaste. Even so it has been widely used. In 1967 Karl Claus of the Hoechst chemical company discovered acesulfam (3), chemically a very similar molecule to saccharin and nearly as sweet, but without the bitter aftertaste.

*The earliest natural drugs...Designing and manufacturing modern drugs...The first synthetic drugs – morphine and heroin...Safer alternatives...Modern painkillers...Sulfa drugs and antibiotics...Natural and synthetic hormones...*PERSPECTIVE*...Biological and chemical drugs...Anesthetics old and new...Antiviral drugs*

Until the later years of the 19th century, the only chemicals effective in medicine were those obtained from natural sources such as plants. Even by 1932, there were only 32 synthetic drugs available, whereas now there are thousands.

Probably the oldest known effective drug is opium. Its extraction from the poppy, *Papaver somniferum*, and use as a pain-killer were well established in classical Greece. Other plant drugs came into use as a result of observations made during the conquest of the Americas in the 16th century. Both quinine, the antimalarial from cinchona bark, and ipecacuanha, for treating dysentery, were introduced into western Europe during the 17th century.

It was not until early in the 19th century, with the isolation of morphine from opium, and its identification as an alkaline substance, that scientific pharmacy began. Chemists searched for other alkaline substances in plants, giving them the general name alkaloid. Through the formation of salts with acids, it became possible to purify these substances and study their effects more accurately.

In 1819 quinine was isolated from cinchona bark and shown to be its active principle. Within 10 years, a factory in France was producing just under four tonnes of quinine sulfate a year.

Introducing new drugs

The development of a new drug is a lengthy and complex operation. After the therapeutic target is identified, and the approach chosen, a variety of possible active substances may be extracted or synthesized on the laboratory scale for evaluation before the most likely 20 or 30 products are identified. After this stage, preclinical testing may take up to three years, with biological and pharmaceutical tests, and animal trials undertaken to test for toxicity. At the same time, chemical testing is done to analyze specific activity, stability and degradation products. Radioisotope labeling (◗ page 225) may be used to track the movement of the drugs in the body. In the process the number of possible candidates will have been whittled down to between five and ten. All this data is written up and passed to the appropriate regulatory body with an application for authority to begin clinical trials. These may take a further five years, beginning with tests on healthy volunteers to discover the highest tolerable and smallest effective doses, and to record the intended and side effects of the various doses. After these have been completed, the first controlled trials on the efficacy of the drug on a patient are begun, and if these are successful a larger-scale therapeutic trial may be undertaken. By this time the final product has been selected, and, while clinical trials continue, tests for longterm toxicity and carcinogenicity are undertaken, and plans drawn up for large-scale production. The new drug is then registered with the health authorities, and launched onto the marketplace.

Biological and chemical drugs

Natural substances – animal, vegetable and mineral – have been used for thousands of years in the treatment of disease. Today, doctors still use substances obtained from biological sources, such as the heart drug digitalis (from foxgloves) and the hormone insulin, even when the chemical structure of the active molecule is known. Most pharmaceutical agents are, however, synthetic chemicals.

The major use for biological materials in medicine is as vaccines, which are used to prevent disease. They do this by stimulating the body's immune system to make antibodies which recognize and attack specific pathogenic organisms. The recognition is based on one or more parts of the molecular structure, called antigenic sites.

The vaccine is either an attenuated (weakened) form of the disease organism, as in the case of oral polio vaccine, or a particular antigenic component of the organism. In either case, the crude vaccine is produced from a large-scale culture of the disease organism. The organism is grown under special conditions in a nutritive broth, in much the same way as yeast is grown in carbohydrate solution to make alcoholic drinks.

Many antibiotics are also produced by microbial culture (◗ page 160). But the antibiotic is isolated from the culture and purified to give a single type of molecule. Most modern vaccines contain a mixture of molecules; sometimes this can lead to adverse reactions. With recent advances in biotechnology, it may soon be possible to produce vaccines by genetic engineering which consist of a single, defined chemical component. If that is done successfully, the already blurred distinction between biologicals and chemicals in medicine will become even less clear.

▲ *In the 19th century pharmacists made up medicines from raw materials such as mineral salts and plant extracts. Skill and knowledge were required to prepare the various formulations, though few of their remedies were effective.*

The American public spends more than a quarter of a billion dollars each year on pain reducers

The earliest synthetic drugs

As the 19th century progressed, many more pure substances became available and were tested for their therapeutic effects. At the time, while the theory that germs were the causative agents of disease was still in its infancy, these tests were carried out without any scientific basis. Nevertheless, some useful drugs were discovered. A number came from minor modifications of known substances.

Thus, in 1898, diacetylmorphine was introduced. This is obtained by acetylating the two hydroxyl groups which occur in morphine. Its originators hoped that this modification would retain the painkilling properties of morphine while reducing the unwanted side effects, such as respiratory depression. The German chemical company Bayer marketed this new drug, but after four years it was found that it was highly addictive. Heroin, as the new substance had been named, is still one of the most widely used drugs of abuse.

Nevertheless, the principle of modifying chemical structures to enhance some aspects of a substance's activities while reducing others is sound. Codeine is a naturally occurring analog of morphine in which one of the hydroxyl groups has been replaced by a methoxyl group (♦ page 159). It has been used as a painkiller for many decades. In addition, it acts as a cough depressant.

After the chemical structure of morphine was worked out in the early 1920s, various attempts were made to synthesize similar structures. One of these was levorphanol, which was more potent and less sedating than morphine. Like morphine, it is a laevorotatory optical isomer (◀ page 109). The dextrorotatory isomer has no painkilling activity, but its methoxy-analog, dextromethorphan, shows codeine's cough-suppressant activity. This substance is still widely used in cough medicines.

Aspirin and other analgesics

Another naturally occurring substance to be studied in the 19th century for its physiological effects was salicylic acid. Salicylic acid is the active ingredient in willow bark, which was chewed to relieve the symptoms of malaria in the mid-18th century. Although effective in suppressing fevers it was very unpalatable. Just before the turn of the century, Bayer introduced acetylsalicylic acid, which had the same effect but was more palatable. They gave it the trade name aspirin.

The American public spends more than a quarter of a billion dollars each year on over-the-counter pain reducers. These belong to the general class of analgesics, which help relieve pain without affecting the individual's consciousness. Of these aspirin is probably the most widely used drug in the world today, despite the existence of many competitive analgesics. At the same time as aspirin's antipyretic effects were observed, acetanilide was found to behave similarly. An analog of this, phenacetin, was discovered soon after and became a popular painkiller for many years. However, its use was subsequently restricted because of its damaging effects on the kidneys. Also, it was discovered in the 1950s that in the body phenacetin breaks down into a slightly different molecule, paracetamol. This is now widely used as an alternative painkiller to aspirin.

Paracetamol contains a hydroxyl group and acetylsalicylic acid a carboxylic acid group. These two groups can be used to form a single substance, the ester benorylate, which breaks down in the body to a mixture of paracetamol and aspirin. In addition, many other pain-killing drugs have been introduced, such as ibuprofen.

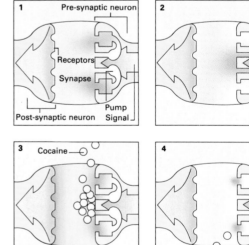

▲ *Cocaine affects neurons in the brain and leads to a feeling of wellbeing (a "high"). The synapse is the tiny gap between nerve fibers, where an incoming electrical signal releases molecules which cross the gap and trigger off the receptors at the other side. Cocaine blocks their return so the molecules continually fire the receptors. Intense mental stimulation results – until the body removes the cocaine.*

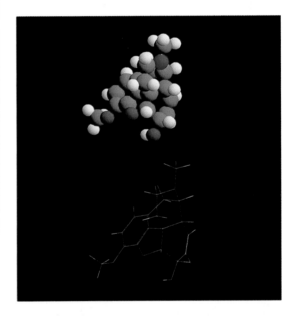

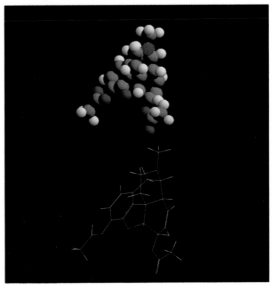

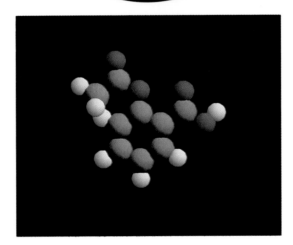

◄▲ *Cultivation of the white-flowered opium poppy is officially discouraged, but demand is such that it continues, as here in Thailand. To harvest the raw opium the seed pods are scored and the material that oozes from the cuts is scraped off.*

◄► *Morphine, codeine and heroin are all chemically related. The differences are in the groups of atoms at the bottom left of each molecule. Morphine (top right) is the simplest with an hydroxy group (OH), codeine (center) has a methoxy group (OCH₃), and heroin (bottom) has an acetate group (OCOCH₃), with another acetate group bottom right. All are excellent painkillers, and are used as such, but all are addictive. For relief from mild pain such as headaches there are safer, nonaddictive, substances such as aspirin (left).*

The chemistry of antibiotics

A major attack on bacterial infections came through the discovery of a substance which affects bacteria but not mammals. It was known in the 19th century that microorganisms could compete with one another, and it was thought that some might produce substances which harmed others.

The key observation came in 1928, when the Scottish bacteriologist Alexander Fleming (1881-1955) noticed the effect of a *Penicillium* mold on the growth of staphylococci bacteria. This was followed up in Britain by Ernst Chain (1906-1979) and Howard Florey (1889-1968), and by 1945 penicillin was manufactured on a large scale.

The development of penicillin manufacture was a triumph for the microbiologist; the chemist did not play a significant role until the 1950s. It was in the late 1950s that the key part of the penicillin molecule was identified as 6-aminopenicillamic acid (6-apa). Once 6-apa had been identified, it was discovered that it occurred on its own in the fermentation broth. By changing conditions, it was possible to produce much more 6-apa and use this to produce a wide range of semisynthetic penicillins. This made it possible to overcome two major drawbacks of benzylpenicillin, the form first produced. Benzylpenicillin is not acid-stable, so is not effective orally. By substituting a different group for the benzyl group, it was possible to make the molecule more resistant to acid and thus introduce oral penicillins.

The effectiveness of penicillin depends on its retaining its beta-lactam ring intact. A number of infective organisms are resistant to natural penicillin because they produce an enzyme which opens this ring. By adding other substituents to the molecule, which placed bulky groups close to the ring, scientists were able to make penicillins which could not be acted upon by this enzyme.

In the past 40 years, many other antibiotics have been discovered. Most have come originally from molds, but in many cases chemical modification has improved their performance. Generally, the basic chemical structures are so complex that production of the raw material by fermentation is more economic than chemical synthesis.

◄ *Alexander Fleming (1881-1955) made his momentous observation at St Mary's Hospital, London in 1928. Colonies of staphylococci on a Petri dish had not grown around a blue mold that had accidentally contaminated his culture (inset). The mold was Penicillium notatum and he realized that it was releasing a chemical lethal to the bacteria. He called the chemical penicillin.*

► *Penicillin works by interfering with the outer membrane of a bacterium. The effect of this can be seen on staphylococcus (the bacteria responsible for boils and abscesses). The lower bacterium is unaffected, and is about to divide, whereas the upper bacterium has come into contact with penicillin. The penicillin has ruptured the bacterium and released its contents.*

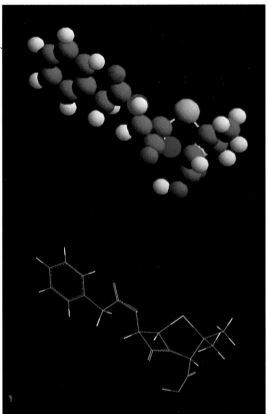

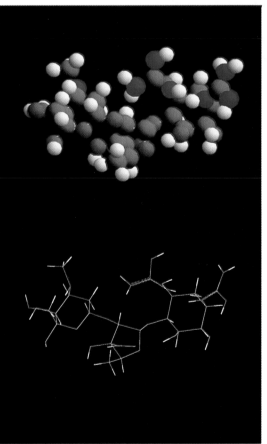

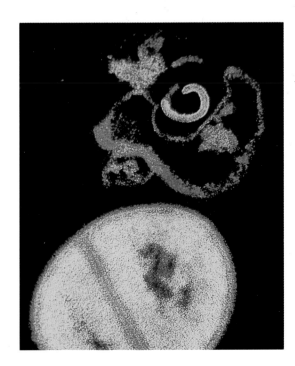

▲ The production and packaging of antibiotics must be carried out under the most sterile conditions to prevent contamination. Here vials of the drug cephalosporin are filled.

► Penicillin (top) is a large molecule that is produced by microbes such as Penicillium notatum. There are several types in which different groups of atoms are attached to a basic double ring system that contains a nitrogen atom (blue) and a sulfur atom (yellow). Some are made naturally others chemically. The second successful antibiotic was streptomycin (bottom), which is a molecule produced by the soil organism Streptomyces griseus. Introduced in 1947, it was very effective against tuberculosis. It works by interfering with the bacteria's ability to produce certain vital proteins.

Of all the substances that chemists have produced, the medical drugs have been the most appreciated

Early anesthetics

The use of general anesthetics has its origins in a now discredited concept for the treatment of illness with different gases. "Pneumatic medicine", as this was known, postulated that different ailments would be relieved by inhalation of different gases. It arose at the beginning of the 19th century, following the chemical preparation of a number of gases, such as oxygen and carbon dioxide, in a relatively pure form.

One of the gases experimented with, notably at the Pneumatic Institution in Bristol, by the young British chemist Humphrey Davy (1778-1829), was nitrous oxide (N_2O), which became known as "laughing gas" because of its euphoric effects. It was also noticed that, under its influence, people were less aware of pain.

Nitrous oxide inhalation became a popular pastime at parties. In some cases, ether was used as a substitute. It was from observations on the effect of these gases that the idea of their use as anesthetics arose. They were first tried out in the 1840s.

Both had some drawbacks as anesthetics. This led to the introduction of chloroform, which became popular after its use in childbirth by Queen Victoria in 1853.

Anesthetics in the 20th century

The search for different, more effective anesthetics continued. Although its anesthetic effects on rabbits had been observed in the 1870s, ethylene was not introduced as an anesthetic until the 1920s. It continued in use until the 1950s. Another hydrocarbon anesthetic, cyclopropane, was discovered while trying to identify a contaminant that was believed to cause unpleasant side effects from propylene-induced anesthesia.
It was introduced in the 1930s and is still sometimes used.

Both ethylene and cyclopropane caused concern because of their flammability and the consequent possibility of explosions in operating theaters. Hydrocarbons containing a number of halogen atoms in place of part of the hydrogen were investigated during the 1930s as refrigerants (a purpose for which they are still used). Some of these were also shown (like chloroform) to have anesthetic properties, but none was really satisfactory.

During World War II, from work on the purification of uranium for the atomic bomb, considerable experience was gained in the preparation and handling of fluorine-containing compounds. This led to an expansion of fluorine chemistry after 1945, and the development of further halogenated hydrocarbons for anesthetic use. The most successful of these is halothane, which was introduced in the late 1950s and still very widely used.

▶ **The early chemists were not averse to trying out the effects of the new gases they made on themselves. Nitrous oxide, or laughing gas as it became known, was one of their more pleasant discoveries. Its pain-killing properties were used by dentists for over a century. It is still used today as a safe anesthetic. These early demonstrations of gases were rudely mocked by the cartoonist Gilray!**

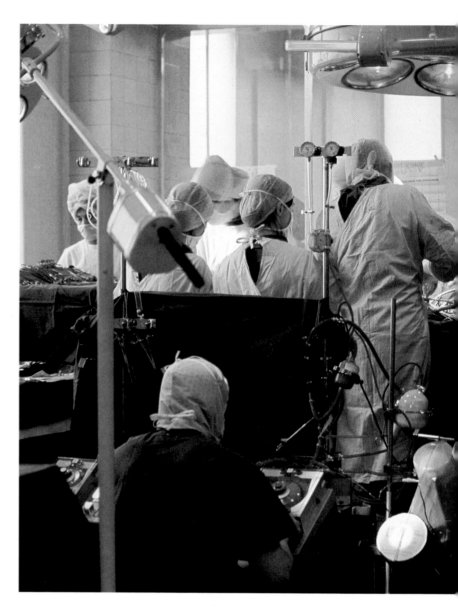

Sulfa drugs

While analgesics may make an ill person more comfortable, they only relieve symptoms. Probably the most significant development in health care during the 20th century has been the introduction of anti-infectives. The search for these became serious after the establishment of the germ theory of disease. An avenue which excited many researchers was the way in which certain synthetic dyes showed an affinity for micro-organisms. The dyes were used to stain the organisms so that they could be studied more easily under the microscope.

In the early 1930s, the German chemist Gerhard Domagk (1895-1964) discovered that a dye, Prontosil red, could cure mice suffering from fatal staphylococcal infections. This dye had a sulfonic acid grouping at one end of the molecule, introduced to make the molecule bind more strongly to textile fibers. Within a year of Domagk's announcement of his discovery – for which he was awarded the 1939 Nobel Prize in medicine – it was discovered that it was not the dye itself which was active, but sulfanilamide, a substance produced from it in the body and containing the sulfonic acid function. This led to the synthesis of many chemical compounds based on sulfanilamide. The best known of these was M&B693, or sulfapyridine, synthesized by the British drug firm, May & Baker. This was the first effective drug against pneumococcal pneumonia and is credited with saving the life of the British leader, Sir Winston Churchill, during World War II.

The effectiveness of the sulfa drugs was later discovered to derive from the competition between the basic sulfanilamide structure and p-aminobenzoic acid. This substance is used by some bacteria in the synthesis of folic acid, an essential growth factor. As mammals canot synthesize folic acid but have to include it in their diet, sulfanilamide does not interfere with mammalian metabolism.

◄ Chemists have led the fight against pain. Nitrous oxide, chloroform and diethyl ether were all in use by the 1850s as general anesthetics for surgery. However, sudden death could sometimes occur and safer alternatives were found such as the less toxic and nonflammable fluorocarbons used today. Chemists have discovered compounds that act as local painkillers for dentistry, sports injuries and minor operations.

◄ ► Sulfanilamide was one of the first medical drugs. Introduced in the late 1930s, it was effective against bacterial infection. Research since then has produced a whole range of drugs that are prescribed as antibiotics, painkillers, sedatives and in treatment of heart disease. Not all have been properly used; the stimulant amphetamine (left) became a street drug in the 1960s. It is still used today but as an appetite depressant for people suffering from obesity.

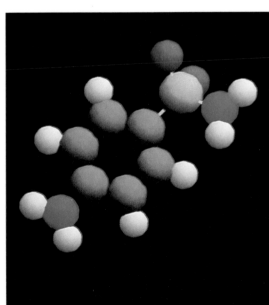

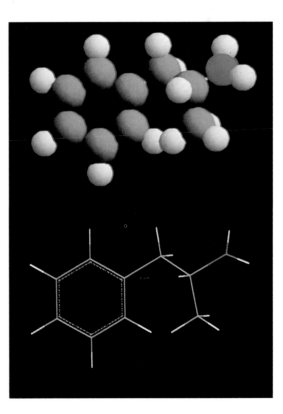

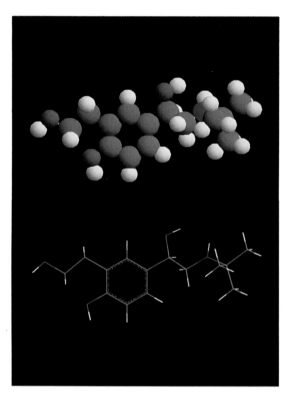

▲ **Even when chemistry cannot cure, it can greatly ease the symptoms. This molecule can restore gentle breathing to those fighting for every breath. Salbutamol (trade name, Ventolin) has helped millions of asthma, bronchitis and allergy sufferers. It is so effective that only a tiny dose of the drug is needed. When inhaled as a fine powder, as little as a ten-thousandth part of a gram (0·0001g) brings instant relief.**

Viruses — a new frontier

Many of the infective diseases which have scourged mankind for centuries are caused by viruses. These have proved virtually impossible to treat, although prevention by vaccination has been effective in keeping some of them at bay.

A virus is not a free-living organism. It contains nucleic acid, often protected by a protein coat, and works by taking over the metabolic machinery of a cell. Attempts to attack viruses chemically have tended also to attack the host cells.

In recent years, it has been found that viruses may have weak spots. Although they use most of the machinery of host cells, they also often need a few enzymes of their own to replicate satisfactorily. It now seems possible to design molecules which will interfere with these enzymes and thus prevent viral replication without damaging the host organism.

The first effective, specific antiviral substance to become available was acyclovir, an analog of guanosine, a nucleotide involved in nucleic acid synthesis. This was discovered in 1974 and found to be active against the herpes family of viruses.

Acyclovir can diffuse in and out of mammalian cells. If it enters a cell infected with herpes virus, it is converted to a phosphorylated form by the viral enzyme thymidine kinase. Once phosphorylated, the acyclovir brings nucleic acid synthesis to a halt. As this synthesis replicates the key part of the virus, viral proliferation is prevented.

The synthesis of hormones

Many diseases are caused by problems arising in the body itself. A number of successful drugs interfere with or mimic the action of some of the body's own chemicals in order to alleviate such problems.

The body contains a number of hormones, small molecules which circulate via the bloodstream and produce physiological effects by linking onto receptors at various sites. The actions mediated by a particular hormone may be quite different, according to the type of cell whose receptor they interact with.

One of the most potent hormones is epinephrin (adrenaline). In the late 19th century, it was discovered that an extract from adrenal glands raised blood pressure. The pure substance named epinephrin was isolated in 1901 and found to have a wide variety of effects, including the conversion of glucagon to glucose in muscles, and the speeding up of heart and breathing rates. The elucidation of its structure led to the synthesis of analogous compounds, as well as their isolation from other sources. For example, the active principle of a herb used in China for 5,000 years was found to be ephedrin. This has similar effects to epinephrin but can be taken orally.

It was used for some time in the treatment of asthma because it had an effect of dilating the bronchi. Subsequently, it was superseded by a synthetic epinephrin analog, isoprenaline. This, in turn, was superseded by other synthetic analogs, designed to overcome one major drawback of natural molecules and their close analogs. Not only does the body have mechanisms for producing substances which exert powerful effects, it also has mechanisms for removing those substances rapidly. Epinephrin is in a class known as catecholamines, because they are related to the aromatic alcohol, catechol. The body also contains enzymes that are capable of destroying catecholamine. Consequently, isoprenaline was only usable for relieving acute attacks of asthma: it was destroyed too rapidly to have a lasting effect. Orciprenaline was based on the idea that a long-acting drug could be made if the structure could interact with the body's catecholamine receptors, but not be attacked by the enzyme. One of the hydroxyl groups on the benzene ring was moved around from the catechol configuration. This achieved the desired effect. The presence of the hydroxyl groups made the molecule more polar than ephedrine, thus preventing it from crossing the blood-brain barrier. So orciprenaline did not cause insomnia, a side effect of ephedrine.

A further development in anti-asthmatic drugs based on this principle was salbutamol. In this case, instead of moving the hydroxyl groups around the aromatic ring, one of them was displaced from the ring by a methyl group. At the same time, the inventors of this molecule changed slightly the hydrocarbon group attached to the nitrogen atom. This had the effect of increasing the molecule's affinity for receptors in the lungs and decreasing affinity for the heart receptors – a failing of isoprenaline had been the strength of its effect on the heart, which led to a number of deaths from its use.

Research in this area led to a class of drugs which block the action of epinephrin. An analog of isoprenaline, dichloroisoprenaline, was found to antagonize the cardiac effects of epinephrin. At the same time, it had some agonistic (epinephrinlike) effects. The British pharmacologist Sir James Black (b.1924) discovered a series of beta-blockers, so-called because they block one type of adrenergic receptor. The most notable, propranolol, introduced in 1964 to reduce anginal pain, was also successful in controlling high blood pressure.

Chemistry and Photography

21

*The structure of photographic film...Silver halides...
Developing the film...Printing the image...Alternative
methods of film processing...PERSPECTIVE...Early
experiments in photography...Color transparency and
negative film...Polaroid and Cibachrome*

Photography relies on the physicist to devise lenses of sufficient
sharpness to focus a sharp image of a subject; and the skills of the
chemist are required to transfer this image onto paper and preserve it.

The energy associated with photons of light (♦ page 195) is sufficient to cause chemical reactions in some substances, and photography is a branch of chemistry that depends on the interaction between light and matter. Most photography is based on the effect of
light on silver salts (halides), even in color films. How the exposed
film is manipulated subsequently produces the range of colors.

Three silver halides – chloride, bromide and iodide – are used in
photographic films and papers. The most commonly used is silver
bromide. The chloride reacts less quickly to light than the bromide
and the iodide more quickly. These halides are used, either alone or in
combination with the bromide, to make film and photographic paper
slower or faster in their response to light.

Early experiments in photography

*The ability of light to decompose silver salts was
discovered early in the 18th century. Then in 1777
the Swedish chemist Karl Wilhelm Scheele (1742-
1786) discovered that it was the blue end of the
visible spectrum which caused this decomposition.*

*Thomas Wedgwood (1771-1805), a son of the
British potter Josiah Wedgwood, first tried to make
use of this by exposing paper and leather
impregnated with silver salts to images in a camera
obscura. These images were too faint to make an
impression on the light-sensitive material, but
Wedgwood was able to produce impressions by
placing the material behind paintings on glass.
Neither he nor the British chemist Sir Humphry
Davy (1778-1829) was able to fix the impressions –
to find a way of removing unreacted silver salt so
that the surface was no longer light-sensitive.*

*In 1839, Henry Fox Talbot (1800-1877) announced
a process which overcame the fixing problem and
converted the originals – which were negative
impressions – into positive images. His material
was made by soaking paper in silver nitrate
solution and converting this to silver chloride by
treating it with sodium chloride solution.*

◄▲ *In 1841 William Henry
Fox Talbot took out a
patent for his method of
printing photographs,
which he called calotypes.
One of the first such
pictures, of Lacock Abbey,
where he lived, was taken
on 24 September 1840 (left).
Talbot's process greatly
improved the negative
image left by light falling on
silver iodide paper. He
developed the negative by
treating it with gallic acid
and silver nitrate solution.
Meanwhile in France Louis-
Jacques-Mande Daguerre
had discovered that a
copper plate coated with
silver iodide and exposed to
light developed an image
that could be fixed by
treatment with mercury
vapor and a solution of
sodium chloride.*

When a film is exposed, each photon causes a chemical change in the halide grains: when it is developed, the effect of this reaction is greatly magnified

In a modern photographic film tiny grains of silver halide are dispersed evenly through a transparent substance (usually gelatin). This emulsion, as it is called, is spread thinly on a solid transparent base of cellulose acetate or polyester.

The halide grains have a crystal structure in which each silver ion is surrounded by six halogen ions and vice versa. However, there are imperfections in the crystal lattice at the surface of each grain and these are where the reaction with light probably takes place.

When a photon strikes a silver halide grain, it knocks an electron off a halide ion. The halogen atom (chlorine, bromine or fluorine) formed combines chemically with the gelatin, while the spare electron joins a silver ion and turns it into an atom of metallic silver.

A single grain of halide contains many million silver ions; each photon striking it converts only one of them to metallic silver. When the film is developed, these metal atoms are used to channel electrons from the developer to other silver atoms in the same grain, thus converting the whole grain to metallic silver and so vastly magnifying the effect of the original photon. The active ingredient in the developer is usually a complex organic chemical which gives up electrons easily by converting to a more highly oxidized molecule.

Grains of silver halide in which no metal atoms are present are not affected by the developer. When development is complete the film is treated with a solution containing thiosulphate ions which dissolve unreacted silver halide. This "fixes" the film so that it is no longer light-sensitive. The parts which have been exposed to most light and so have most silver atoms are blackest, and those which have been exposed to no light are white, giving a negative image of the original scene as taken by the camera.

Printing the photograph

To produce a monochrome positive, sensitized paper is exposed to light shining through the negative. The process of latent image formation occurs again as photons strike silver halide grains in the paper. When this material is developed and fixed, the result is a black-and-white positive image.

Silver salts react with ultraviolet, violet and blue light. To make black-and-white film which responds to light from the whole visible spectrum (panchromatic film), synthetic dye molecules are attached to the silver halide grains. These absorb light energy from the longer wavelengths of the visible spectrum and transfer it to halide ions so that the latent image can form. By attaching appropriate molecules it is possible to sensitize film to infrared wavelengths. Because the halide grains are still sensitive to blue light, infrared film has to be used in conjunction with a filter which absorbs the blue wavelengths before they reach the film. As many objects emit infrared radiation (◀ page 64), infrared film can be used to photograph in the dark.

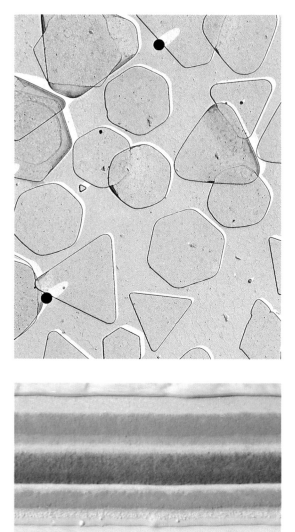

▲ *In this cross-section through color film, the three layers of emulsion are separated by gelatin layers that prevent migration of chemicals between the layers. When exposed and processed each layer registers its complementary color.*

▶ *The chemistry of color photography is also based on silver halides, even though these are only affected by light of short wavelength, the blue and violet end of the spectrum. To make film sensitive to long wavelengths (red) and intermediate wavelengths (green) the silver halide is mixed with synthetic dye molecules which are activated by these colors, and applied as three layers of emulsion to the film. When the film is exposed, each dye captures light from the three parts of the spectrum, and passes it as energy to a neighboring silver halide particle. As light penetrates through the successive layers of emulsion a negative image of each color range emerges. Printing the image requires a color reversal stage.*

Color negative process

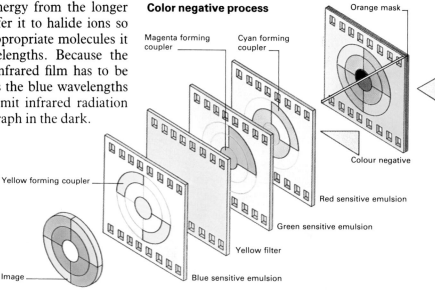

Orange mask

Magenta forming coupler

Cyan forming coupler

Yellow forming coupler

Colour negative

Red sensitive emulsion

Green sensitive emulsion

Yellow filter

Image

Blue sensitive emulsion

 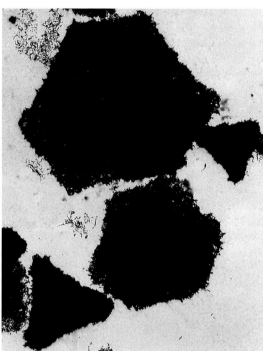

◄ *The chemistry of photography is the chemistry of silver halides. These salts are sensitive to light which converts a little of the silver ion of the salt to metallic silver. This may happen to only one atom of the many million atoms in each grain of the silver halide. The remaining silver ions in the grain are later converted ("reduced") to silver by chemical means during the developing process. Those grains that have not been affected by the light do not develop and are washed away. On the far left is the fine grain structure of a piece of unexposed film. The middle photograph is film that has been exposed, magnified 18,000 times. The clear triangles are silver halide particles and where these have been converted to silver metal we see a black mass. The photograph on the immediate left shows a fully exposed film.*

Paper print process **Color transparency process**

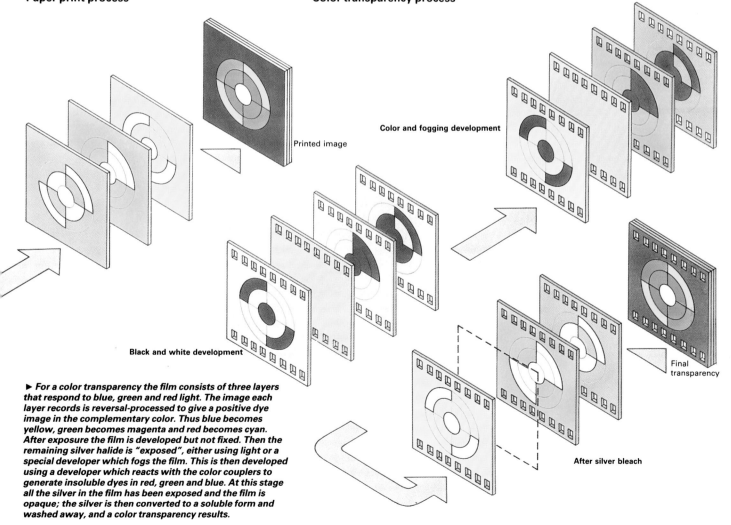

Printed image

Color and fogging development

Black and white development

After silver bleach

Final transparency

► *For a color transparency the film consists of three layers that respond to blue, green and red light. The image each layer records is reversal-processed to give a positive dye image in the complementary color. Thus blue becomes yellow, green becomes magenta and red becomes cyan. After exposure the film is developed but not fixed. Then the remaining silver halide is "exposed", either using light or a special developer which fogs the film. This is then developed using a developer which reacts with the color couplers to generate insoluble dyes in red, green and blue. At this stage all the silver in the film has been exposed and the film is opaque; the silver is then converted to a soluble form and washed away, and a color transparency results.*

Variations on color film development

There are a number of variations on main processes of color film development. In some color film, there are no dye intermediates. These are all added during the processing stages. This type of film is called a non-substantive emulsion and its processing is a complex, multistage process which can only be carried out professionally. All films which can be processed in an amateur's darkroom have substantive emulsions (with dye intermediates built in). In the dye-transfer process, a separate transfer of the dye images from records of the red, green and blue contents of the picture produces the completed print.

Another variant is the Cibachrome process, by which positive prints can be made directly from color transparencies. This works by a dye destruction process. The printing paper has the three primary color dyes coated onto it. When exposed to light, these dyes are destroyed in proportion to the amount and color of light falling on them. Thus, blue light will destroy the red and green dyes, so that the end result is the same balance of colors as in the original transparency.

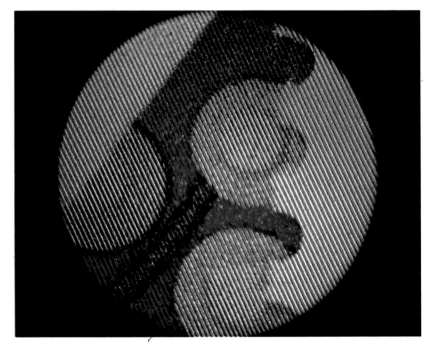

Instant color film

In 1948, the United States inventor Edwin H. Land (b.1909) introduced the Polaroid Land camera which produced instant photographs. The key to this was a film which contained both film (negative) and paper (positive) together with appropriate processing chemicals. Land's invention was expanded to instant color prints in 1963.

In the early 1980s the Polaroid Corporation introduced an instant color transparency film which could be used in any 35mm camera. This differed from conventional film in several ways. The film base, made of polyester, is only 0·074mm thick (compared to about 0.125mm) and it faces the camera lens, rather than the emulsion.

Behind the film base is an additive color screen. This is a repeating pattern of blue, green and red filters, with over 1,000 filters per centimeter. Behind the filter layer, sandwiched between two protective layers, is the positive image receiving layer. A release layer, which is important in the final stage of processing, comes next. Beyond this is the emulsion layer, which contains silver halide, and an antihalation layer to mop up any photons which pass through the emulsion. These seven layers have a total thickness of 0.014mm.

When a photograph is taken with this film, which halide grains are exposed depends on light coming through the tiny filters. Thus, green light will be blocked by the red and blue filters and will only expose silver grains behind green filters.

A mixed processing fluid is applied to the emulsion side of the film and migrates through the different layers. Where silver halide has been exposed, grains are converted to silver. Unexposed halide forms a soluble complex with the processing fluid and migrates into the positive image receiving layer. Here, it too is converted to silver.

After a processing time of about one minute, the film is split apart at the release layer, thus carrying away the silver which was originally exposed. What remains is a transparency. Where green light originally passed through it, there is now a silver layer blocking the blue and red filters, so that only green will show when light shines through it. The same is true for other colors, so that a full color picture is produced.

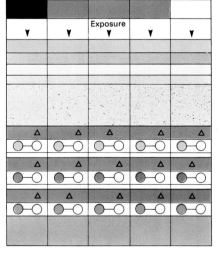

▲ **Emulsions on a Polaroid print under magnification. In 1948 Dr Edwin H. Land's camera made instant photography possible. This used special film containing both negative, positive and the developing chemicals.**

▶ **Polaroid film consists of layers of chemicals sandwiched between a transparent layer and a backing layer. Light passing through the layers activates the silver grains and their associated dyes. Developer is squeezed between the emulsion layers and image-receiving layers as the film leaves the camera, to give a picture in about a minute.**

Exposure

Subject
Plastic layer
Acid layer
Timing layer
Image receiving layer
Developer layer
Migrating dye developer
Blue sensitive layer
Yellow dye developer
Green sensitive layer
Magenta dye developer
Red sensitive layer
Cyan dye developer
Opaque base
△ Unexposed silver halide
▲ Exposed silver halide

Paint, Lacquer and Adhesive

22

The nature of surface coatings...Early paints...Linseed oil and water-based paints...Resins...Acrylics... Powdered paints...Modern pigments...Adhesives... PERSPECTIVE...Restoring old paintings...Color in paint...Painting an automobile

Coatings protect surfaces such as wood and metal from chemical and physical attack, and decorate them. The major products of this area of chemical technology are paints, enamels, lacquers and varnishes; it also includes adhesives and coatings for paper and textiles.

All these products usually contain a *binder* dissolved in an *extender*, which controls the flow properties of the coating and also helps to improve its hardness and resistance to wear. Until recently, the coatings industry was the largest single consumer of organic chemical solvents. Increased environmental concern, coupled with technological advance, has led to an increase in water-based coatings and coatings without extenders, thus reducing solvent consumption.

The binder forms the protective film on the surface being treated. Paints, enamels and lacquers usually also contain pigments, insoluble materials which make the finished coating opaque and colored. Color in varnish comes from soluble dye rather than pigment.

The most common form of binder in a paint is either a natural oil or synthetic material containing a number of carbon-carbon double bonds which oxidize and then crosslink to form a three-dimensional polymer network (◀ page 93). Linseed oil, derived from flax, has been used as a binder in Europe since the 13th century.

▲ As well as protecting, coloring and covering, paint should also adhere to the surface to which it is applied. Today we impose limits on what paint can contain, ruling out such traditional ingredients as the covering agent, white lead. Modern paints use titanium dioxide in its place. Even the organic solvents which once gave paint its characteristic smell are being phased out, and replaced by safer ones. Water-based paints are increasingly used, especially indoors.

◀ The durability of some ancient paintings can be seen in Egyptian tombs, such as this one at Thebes. Protected from the ravages of sunlight and weather, this wall painting remains clear and vivid after 3,500 years. The principles of paint remain the same: the coloring agent and a binder are mixed with a solvent that allows the paint to flow evenly, and evaporates as the paint dries.

Synthetic resins introduced in the 1920s are still very important surface-coating ingredients

Early paints and lacquers

It was not until the 18th century that painting of the exteriors of houses became widespread. It was found that the life of wood and plaster could be extended by coating them with paint made from linseed oil and zinc oxides. The pigment not only provides bulk but also has other advantages. It absorbs ultraviolet radiation (◀ page 66) and thus slows down the photodegradation of the film formed by the oil and also has fungicidal properties. Although linseed oil does not form a very strong film and also yellows on aging (owing to photodegradation), the film is flexible. This is important on surfaces which change shape with age as a flexible film is less likely to crack and peel.

In a lacquer, the binder forms a film by evaporation of solvent, rather than by chemical reaction. Natural resins have been used for this purpose since the 5th century AD in oriental countries. An enamel's binder may form a film either physically or chemically. In general, enamels are dried by heat treatment, which speeds up the drying process.

The development of synthetic binders

Over the past three quarters of a century, there has been a move away from natural film-forming materials to synthetics as binders. This has followed the growth of the plastics industry (◀ pages 133-144). Surface coatings – paints and varnishes – are now important consumers of many different polymers.

Shortly before World War I, Leo Baekeland (1863-1944) invented the thermosetting plastic Bakelite (◀ page 142). This is a highly crosslinked polymer formed from phenol and formaldehyde. Following on from this work, some German chemists discovered that modified phenol-formaldehyde resins could be dissolved in natural drying oils and contributed to film formation. These phenolic resins still occur in some lacquers for specialized uses such as coating chemical plant and food cans. They are also used in printing inks and on surfaces, such as floors, where there is heavy wear.

This discovery was followed by the introduction of other synthetic binders. Urea-formaldehyde and melamine-formaldehyde resins were introduced in the early 1920s. These are widely used in applications such as imparting crease resistance to fabrics and "wet strength" to paper products, and as adhesives.

Urea-formaldehyde and melamine-formaldehyde resins are also still used in paints and lacquers to improve the properties of alkyd resins. These were introduced at the end of the 1920s and are still very important surface-coating ingredients. An alkyd resin is made from a polyfunctional alcohol such as glycerin, a dibasic acid and a fatty acid. The fatty acid is unsaturated and may be derived from a natural source, such as linseed or soya bean oil.

Like linseed oil, alkyds harden by oxidative crosslinking of the unsaturated side chains. Fully-saturated alkyds have been made, but do not dry well. A recent development is the introduction of allyl ether side chains in place of the unsaturated fatty acids. Resins containing these harden very rapidly on treatment with an oxidizing agent, such as hydrogen peroxide, or with electron bombardment or ultraviolet irradiation. These alkyds are used to coat wood – notably furniture – and steel, as well as paper products, such as paperback book jackets. These may also be laminated with plastic, although this is liable to peel off during use.

Studying old paintings

The Adoration of the Magi *(right) by Ludovico Cardi, called Il Cigoli, was painted around 1600. Since then it has been restored several times.*

In the past natural resin varnishes such as mastic and dammar were used to saturate the paint layer; losses were retouched in oil paint. Such additions discolor markedly, and Cigoli's work was obscured by layers of varnish and repaint. To remove these, organic solvents were used. The losses were filled and retouched, then a layer of a synthetic resin varnish was applied which does not discolor.

A cross-section through the paint layers shows the buildup of the paint. The first layer is yellowish-brown, the color deriving in part from a natural ocher (iron oxide) and partly from the oil medium. The second ground is gray, a mixture of lead carbonate with charcoal. The green drapery was laid out with a translucent green verdigris or copper acetate; acetates of copper were made by exposing copper to the vapors of vinegar. Next Cigoli used an impasto containing green (copper resinate), lead-tin yellow and a little lead white. Finally he glazed the drapery using a translucent green paint consisting of just copper resinate.

▼ *The painting had a cracked surface due to an excess of glue applied during previous restoration. The lining canvas had to be removed, the glue scraped away and the painting flattened. A new canvas support was attached to the original with a thin layer of BEVA 371, a heat-activated copolymer resin developed for lining paintings.*

▶ *A cross-section through the paint surface of the green drapery reveals Cardi's technique. Two colored ground layers were applied, then a layer of green copper acetate, a thick impasto and a transparent green surface of copper resinate. The thick brown layer of varnish was removed during cleaning.*

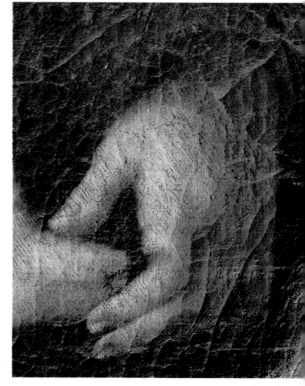

▲ *In its unrestored state, the painting had discolored varnish and the surface was badly cracked.*

▶ *Cardi's* Adoration of the Magi *after restoration.*

◄▲ *Coating an automobile is a multistage process. The stages (above) transform bare metal (lower stripe) using phosphate, cathodic primer, primer surfacer and base coat to the finished clear coat (top stripe). The surface is first prepared by cleaning and then treating in phosphate solution (left) to increase its corrosion resistance. The metal is then coated with primer. Often based on epoxy resins, it is chosen for the good adhesion between metal and paint. Primers are applied by electrodeposition from aqueous solution. The paint binder is an anionic molecule which is attracted to the metal surface by making it the anode in an electric circuit. The walls of the tank containing the paint suspension act as the cathode. A pigmented base coat and clear top coat complete the process.*

Mass production painting: the auto industry

The growth of the automobile industry during the 20th century has led to considerable technological advance in surface coatings for metals. When mass production of automobiles started, linseed oil-based paints still predominated. The slow drying time of these created production-line problems. In 1923, nitrocellulose lacquers were introduced as the first in a series of innovative alternatives to conventional paints for automobiles.

Nitrocellulose was first made in the early 18th century, by treating cotton with nitric acid. In its fully nitrated form, it is highly inflammable (guncotton ♦ page 183). However, partially nitrated material was found to dissolve in a mixture of ethyl ether and ethanol. Collodion, as this solution was called, found several uses in the 19th century, including the formation of protective films on wounds and the preparation of early photographic plates.

In the 1880s, protective coatings based on nitrocellulose dissolved in amyl acetate were introduced. These were used mainly on brass bedsteads, to prevent tarnishing. This was not very successful, because the adhesion between the coating and the metal was poor.

During World War I, nitrocellulose was used widely as a dope, to strengthen fabric for use on airplanes. Shortly afterwards, the American chemical firm du Pont discovered that treatment of nitrocellulose with sodium acetate led to partial breakdown of the polymer chain. This made solutions with a higher solids content possible. It was these which became the first nitrocellulose lacquers, reducing the drying time on automobiles from days to hours. Nevertheless, several coats of lacquer were needed to achieve a satisfactory finish and this then had to be polished for extended periods to achieve a gloss finish.

Nitrocellulose lacquers were succeeded by alkyd- and melamine-based enamels which could be baked on. These had an even higher solids content and therefore needed a smaller number of coats. They also needed less polishing than cellulose lacquers. In the early 1960s acrylic enamels were introduced. These have even higher gloss and are very hard wearing. In recent years, further development of surface coatings for automobiles has been necessary as metal parts have been replaced by plastics.

▼ Acrylic paints are highly versatile, being used in both oil- and water-based paints. They have the advantage that they can be applied by aerosol. They give a light and fairly opaque color that has been used to brighten the urban environment, whether with official sanction or entirely spontaneously.

▲ Murals that enliven many city streets are usually done in emulsions and gloss paints, with good alkali- and acid-resistant qualities. The surface is normally sealed with a stabilizing solution to prime and neutralize it. Non-porous paints, such as acrylics and polyurethane glazes, tend to peel after a few years.

The growth of water-based paints has led to greater complexity in paint manufacture. Many pigments disperse less well in water than in organic solvents. Consequently, special dispersants have to be added. Other additives change the physical characteristics of the paint so that it flows smoothly on surfaces, without dripping from paint brushes or rollers.

As the move away from solvent-based systems continues, there is also growing interest in applying paints as powders. One method is to spray the powder onto a heated surface, where it melts, forms a film and then sets. Both thermoplastic and thermosetting resins can be used for this purpose.

Advances in paint technology
A further major advance in paint technology came in the late 1940s with the introduction of emulsion paints. These contain synthetic polymer as the binder, but rather than being dissolved in an organic solvent, they are emulsified in water. This makes home decorating less messy. More important, it means no loss of expensive and potentially environmentally harmful solvent.

The first composition to be used in emulsion paints was a styrene-butadiene copolymer. Styrene-butadiene synthetic rubber was developed during World War II and is based on a 2:1 ratio of butadiene to styrene; ◀ page 137). The polymer used in emulsion paint, on the other hand, had a 1:2 butadiene-styrene ratio.

The polymer incorporated double bonds, which meant that it could form a film by oxidative crosslinking, like linseed oil. Also like linseed oil, it yellowed on aging. Subsequently, emulsion paints based on polyvinyl acetate and acrylic polymers have come to dominate this sector of the market.

Acrylics, based on polymers of acrylic acid derivatives, are versatile binders, being used in both oil and water-based paints. In many cases, the binder is a copolymer. Originally, copolymers of methyl methacrylate (which, when polymerized on its own forms Perspex; ◀ page 138) and acrylic acid esters were used. To these have now been added a wide range of copolymers with other molecules such as styrene and polyvinyl acetate.

The permanently glossy finish of many modern automobiles is imparted by solvent-based acrylic lacquers. Acrylics also provide water-based gloss paints for woodwork. Water-based acrylics are of interest to the automobile industry in order to avoid the economic and pollution problems which arise from evaporating large quantities of organic solvent. At present, however, they do not impart such a good finish. Acrylics are also used in other types of surface coating, such as floor polishes and finishes for leather.

Polyvinyl acetate (PVA) forms too brittle a film to be used satisfactorily on its own as a paint binder. Initially, it was mixed with plasticizers to make it more flexible. More recently, however, internally plasticized PVAs have been produced. These are copolymers, containing a second monomer such as ethylene which adds flexibility to the polymer chain. The comonomer can also help to stabilize the polymer. The acetate groups in PVA will hydrolyse in alkaline conditions, converting the polymer to soluble polyvinyl alcohol. This limits the applications of PVA in surface coatings. Addition of ethylene or vinyl chloride as a comonomer reduces the sensitivity of the polymer to hydrolysis to a much greater extent than the actual percentage of non-hydrolysable material present.

▲ A contact adhesive bonding formica to metal shows its polymeric nature. Adhesives attach chemically to the surfaces to which they stick, in order to give a permanent bond.

▼ Among natural adhesives is that which holds ocean barnacles firm. Until recently all adhesives were natural polymers such as paste (made from starch) and glue (animal protein).

Sticking together

Many of the same types of molecule used in paint are also used in adhesives: a surface coating needs to have adhesive properties if it is to form a good bond with the surface to which it is applied. An adhesive is a surface coating between two adjacent surfaces.

Adhesion depends on the formation of chemical bonds between the adhesive and the substrate, or on less strong attraction, such as Van der Waals' forces and hydrogen bonding (♦ pages 35, 90). Proximity is very important if these forces are to be effective. Consequently, an adhesive needs good "wetting" properties for surfaces and surfaces to be joined together need to be smooth and clean.

Early adhesives were based on natural polymers such as starch paste and fish glue used in aqueous solution. Adhesion occurred as the water evaporated. Some modern adhesives depend on evaporation of water or an organic solvent. But there are also different types of solventless adhesive.

Hot-melt adhesives are low melting solids which are applied in the molten state and set on cooling. Books are often held together with hot melt adhesives as are the soles and uppers on some shoes.

Two-component adhesives work when a liquid monomer or oligomer (partial polymer) is mixed with a catalyst which induces rapid polymerization to a solid (♦ page 94). No solvent is needed because the starting material has sufficiently low viscosity to spread.

A very strong setting glue, often called "superglue", has been introduced in recent years. This is a cyanoacrylate, polymerization of which is initiated by traces of water. The amount of water needed is very small, so that the glue sets on exposure to most surfaces, which usually have enough water molecules absorbed on them for the reaction to take place.

Natural soaps...Synthetic detergent...The action of detergents...Dry cleaners...Surface active agents in other industries...PERSPECTIVE...Cosmetics old and new...Soap manufacture...The constituents of modern detergents

Soap was one of the earliest chemical products to be manufactured commercially. The discovery of soap may have resulted from observations of the cleansing effects produced when fat and wood ash (which contains alkali) are mixed.

Animal fats and plant oils are primarily triglycerides, esters (◀ page 110) of long-chain carboxylic acids with glycerin. When treated with alkali, the ester linkage is saponified: glycerin is released and metal salts of the fatty acids are formed. If the metal ion in the alkali is potassium or sodium, soluble soaps are produced.

Soap molecules are long and their two ends have different chemical qualities. The carboxylate anion at one end is highly polar (◀ page 90): the body on the other hand is nonpolar. Consequently, one end has an affinity for polar substances such as water (is hydrophilic), and the other for nonpolar substances such as oils (hydrophobic). Soap cleans by two basic mechanisms. First, it reduces the surface tension at the interface between water and other materials such as cloth – in other words it improves the "wettability" of solid surfaces). Second, it can break up oily particles and hinder their reaggregation.

Soap manufacture

Until the development of the chemical industry in the late 18th century, alkalies came from natural soda deposits such as those in Egypt, or from plant ash. In Italy, soapmaking depended on olive oil and ash from barilla, a coastal plant. It was the shortage of barilla during the late 18th century that led to the discovery of the Leblanc process for making synthetic soda (◀ page 129), and created one of the bases of the bulk chemical industry.

Meanwhile, European colonization of other parts of the world was opening up new sources of triglycerides, notably palm and coconut oils, Differences in the composition of oils and fats alter the characteristics of the final product, as does the choice of metal ion. The hardness of a soap is also related to the length of the fatty acid chain.

Control of these variables allows the soapmaker to produce different grades of soap. But the basic process is still the boiling of triglycerides with an alkaline solution. Initially the triglycerides are boiled with weak alkali. As the reaction proceeds, stronger alkali is added until saponification is almost complete. The mixture is then treated with brine to "salt out" the soap.

Two layers form, the lower consisting mainly of water, excess alkali, brine and glycerin. This is removed and the top layer is treated again with alkaline solution to complete saponification. The salting out process is repeated and the two layers are separated. Water is then added to the soap mass until it becomes gelatinous. It is then left to separate into layers. The top layer, containing about 70 percent soap and 30 percent water, is called "neat soap". The other layer, called "nigre", contains impurities, plus about 30 percent soap. It is reprocessed before being discarded.

Other processes used are "cold" and "semiboiled". In these the glycerin produced is not separated but remains in the final product. In the cold process, triglycerides and concentrated alkali are mixed at a temperature just above the melting point of the triglycerides. The mixture is poured into large cooling frames, where it stays for several days. It is then cut into bars.

Soaps made by this method are described as "undersaponified". This means that less alkali than the theoretical maximum is used in order to ensure that no alkali remains in the final product. A soap for personal use which contains more than 0.03 percent free alkali may cause skin irritation.

The "semiboiled" process is similar, except that it is carried out at a higher temperature (70-80°C) so that harder triglycerides can be used.

For the preparation of toilet soaps, neat soap is milled and minced (or plodded, as it is called in the trade). It is then extruded as strips, flakes or pellets, mixed with additives such as colors, perfumes and bactericides, and extruded again in a bar of the appropriate shape.

◀ *Soap was originally made by boiling together animal fat and "lye", the alkali leached out of wood ash. The same chemical reaction is used to make today's soap, although vegetable oils like palm and olive oils are now used in place of animal fat, and sodium hydroxide in place of lye. Apart from toilet soap, other uses of soap such as for washing clothes and scrubbing surfaces, have been superseded by synthetic detergents.*

Cosmetics

▲ Cosmetics have been used since ancient times: witness their discovery in Egyptian tombs of 3500 BC. Toilet boxes, like this one in the British Museum, testify to their continued use down the centuries, as do many Egyptian portraits (◊ page 169).

◄ In the 16th century cosmetics became highly fashionable, with first Queen Elizabeth of England and then many of her courtiers painted with heavily whitened faces.

Ancient cosmetics

Evidence for the use of cosmetics goes back at least 5,000 years. Cosmetics have two major functions. They can be surface coatings – such as lipstick or eyeshadow – designed to enhance or protect a surface (the skin), or they can be cleansing agents.

Early cosmetics included minerals such as malachite, a green copper carbonate used by the Egyptians as eyeshadow, and galena (black lead sulfide) used by them to darken eyebrows. The Egyptians also used red ocher to enhance the color of lips and cheeks, and henna for coloring hair. For cleansing, they had mudpacks containing alum.

The Greeks and Romans used depilatories based on a mixture of orpiment (yellow arsenic sulfide) and lime. On the skin, this forms soluble calcium sulfide which softens hair to a jelly-like consistency.

Galen, a 2nd-century AD Greek physician, published a recipe for cold cream based on beeswax and olive oil into which rose petals were crushed. When cold, this was beaten with water to form a cream. Cold creams have a similar basis today, except that mineral oil often replaces vegetable oil as it does not go rancid.

Modern cosmetics

Most cosmetics are composed of a small number of chemicals. The differences between the many different brands on the market are usually negligible in chemical terms. Face and eye make-up is basically coloring material in a particular type of base. The base used determines various characteristics of the product, such as how long it will last and its ease of application.

Face powders contain three or four key ingredients – an opaque substance, such as zinc or titanium oxide, to cover skin blemishes; mineral talc or zinc or magnesium stearate to provide adhesion and "slip" (for ease of application); an absorptive material, such as kaolin or magnesium carbonate to absorb perspiration. The powder may also contain guanine or mica to give it sheen.

Lipsticks are made from mixtures of oily liquids such as castor oil, waxes such as beeswax, and pigments. Nail polishes are based on nitrocellulose lacquers, together with resins and plasticizers like dibutyl phthalate, which prevent cracking of the nitrocellulose film and make it glossier.

Toiletries are manufactured in different ways for different parts of the body. Soap is the most important, but in addition there are products specific to the teeth (toothpaste), hair (shampoo) and face (aftershave, cleansing lotion). Toothpastes contain abrasives, such as calcium carbonate or calcium pyrophosphate, which provide mechanical cleansing action. In addition, they contain a surface active agent to help with the cleaning.

Shampoos are also basically detergents, but they may contain zinc compounds to prevent dandruff, as well as perfumes and polymeric materials which attach to the hair to give it "body". In shaving soaps, there is a higher proportion of free fatty acids and some of the fatty acid salts are in the potassium rather than the sodium form. This gives a greater volume of lather.

Aftershaves are mainly composed of ethyl alcohol, which gives a cool and refreshing feel to the skin as it evaporates. They also contain chemicals which act as astringents, emollients and antiseptics to ameliorate any damage to the skin caused by the mechanical action of the razor.

Underarm deodorants contain ingredients which have multiple functions. The most common is aluminum chlorhydrate. Aluminum salts not only reduce the amount of perspiration, they also have a bacteriostatic effect and are reactive towards the volatile molecules which the bacteria produce from sweat and which are responsible for body odor.

▲ This thermograph shows how the temperature of the skin varies over the surface of the face. Chemicals can help remedy most problems of the complexion.

► Face creams are oil-in-water emulsions that spread easily and provide a base for face powder. Natural oils tend to be preferred, such as almond and lanolin, the grease from wool. It is a complex mixture of esters derived from 37 acids and 33 alcohols. Face powder is mainly talc, a magnesium silicate with a silky feel, plus zinc oxide, chalk, perfumes and dyes. Face powder not only hides blemishes but absorbs and neutralizes oily secretions from the skin. Eye shadow and lipstick are coloring agents held in a hydrocarbon grease. Petroleum jelly with white titanium oxide is the base for eye shadows, with various tints added. Lipstick also uses long-chain hydrocarbons as a base for eosin dyes which are present as their metal salts. The metal enhances the color and prevents its dissolving. Lanolin also plays a part in eye shadow, eyebrow pencils, mascaras and lipsticks.

The advent of detergents

Until the 1950s, personal and laundry cleaning were both based almost entirely on soaps derived from natural triglycerides. A drawback is that the magnesium and calcium salts are insoluble. Thus, in areas where the water contains substantial quantities of these cations ("hard" water), the soaps form insoluble scums. Since the 1950s, detergents based on petrochemicals have taken over the laundry market. A detergent is any synthetic chemical that acts as a soap. The nonpolar part of the molecule is a long-chain hydrocarbon attached to a benzene ring, while the polar part is a sulfonic acid salt.

For many years the most common detergent ingredient was dodecyl benzene sulfonate. This was not a single chemical substance, but a mixture of compounds in which the structure of the dodecyl (12-carbon) group varied. This group was made by condensing four propylene groups to give highly-branched hydrocarbons with a single carbon-carbon double bond. This double bond became the point of attachment of the benzene ring. The highly branched nature of the molecule made it unlike most natural substances, and consequently it resisted degradation by microorganisms. This led to environmental problems. In the 1960s the branched-chain alkylbenzene sulfonates were replaced by linear alkylbenzene sulfonates, and these are still popular in laundry detergents.

At about the same time as the alkylbenzene sulfonates became important, nonionic detergents were also introduced. These are based on hydrocarbons with several oxygen atoms incorporated to provide the polar parts of the molecules. In these linear hydrocarbon alcohols a long-chain alcohol ion reacts with ethylene oxide. The oxide ring splits open and forms an ether linkage with the alcohol group, while forming another alcohol group at the end of the extended chain.

A newer type of nonionic detergent is based on sucrose. The sucrose molecule has a large number of free hydroxyl groups which can be esterified with long-chain fatty acids to make low-foam detergents. These are biodegradable as all the ingredients are biological in origin.

Linear hydrocarbon alcohols are also important in the manufacture of hydrocarbon sulfates such as sodium lauryl sulfate, which is used in many shampoos. More complex sulfates, such as sodium lauryl glyceryl sulfate, are also important in products such as shampoos and toothpastes. These linear alcohol sulfates are high-foaming.

Like the sulfonic ions, sulfates are anionic surface active agents (surfactants). Cationic surfactants are also produced, in which the polar part of the molecule carries a positive charge. Usually, they are long-chain hydrocarbons with a nitrogen group at one end. They do not have detergent action, because most surfaces carry a negative charge and therefore absorb them. They do have commercial applications, however. Some are anti-bacterial and are used in cough sweets. Their surfactant properties also make them useful in concentrating ores by flotation. Anionic surfactants are used in textile treatment as fabric softeners. If added to a washing machine towards the end of the washing cycle, they absorb onto materials to form a monolayer in which the nonpolar part of the molecule faces outwards. This makes the textiles fluffier and reduces static electricity on them.

It is possible to have both negative and positive groups in a molecule. Such molecules are called amphoteric. They behave as either acids or bases according to the acidity of the surrounding medium. They are used in mild shampoos and other cleaning materials. When combined with alkalis they can be used in oven cleaners.

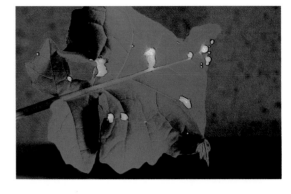

▲ *Detergents improve the wetness of liquids. When added to insecticides, they allow the liquid to spread over the leaf, as seen in UV photographs (normal top; treated below).*

▼ *As a cleaning agent, shampoo is very effective. Not only do its detergent molecules work better than soap, but in addition they do not form a scum with the calcium in hard water.*

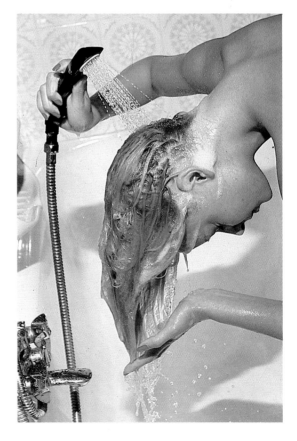

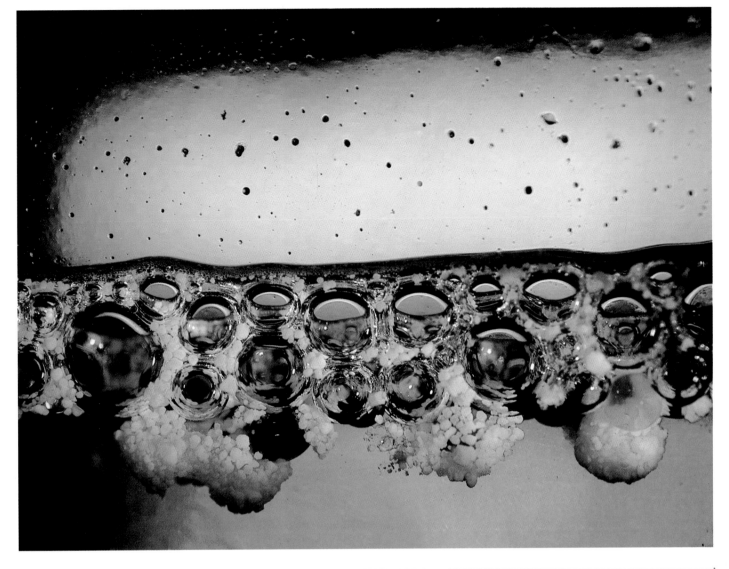

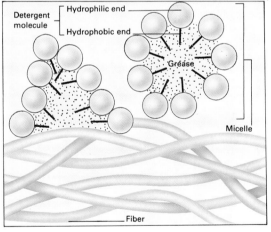

◄▲ From cleaning cars to carpets the principle is the same. Cleaning molecules consist of two parts – a polar end (head) attracted to water, and a nonpolar end (tail) attracted to organic material such as oil and grease. Nonpolar tails of the surfactant molecules are hydrocarbon chains. Polar ends often consist of negatively charged chemical groups, such as sulfate. Such detergents are called anionic surfactants.

▲ Other surfactants carry positively charged ends (cationic), and some carry no charge at all (nonionic). It is likely that surfactant molecules cluster round the particles of dirt and grease with their hydrocarbon tails in the dirt and their polar heads on the outside, these preferring to associate with water. The result is to float the dirt off the article being washed and to keep it suspended in solution until it is rinsed away.

What is there in a packet of detergent?

The first detergent was marketed by the German company Henkel GmbH in 1907. It consisted of a mixture of soap and sodium salts – silicate, percarbonate and carbonate. The silicate helped to keep dirt in suspension after it had been removed from material by the soap, while the percarbonate bleached dirt which remained on the material. This detergent was named Persil for the percarbonate and silicate components, and this name is still in use although for different formulations.

In addition to a cleaning agent, modern detergent powders usually also contain builders, bleaches, optical bleaches, foam stabilizers and soil suspending agents. The builders counteract the activity of calcium and magnesium ions in hard water which interfere with surface activity, although they form soluble salts with modern detergents rather than the insoluble scums which they form with natural soaps. Builders such as sodium tripolyphosphate sequester the ions, so that dirt redeposition is inhibited. They also buffer the solution, or help to maintain its pH at a particular level (in this case, between 9 and 10). This helps to dissolve free fatty acids in the dirt.

Sodium silicate is also used as a builder. Not only does it help to keep dirt from redepositing on materials in the wash, it also makes the detergent particles stronger. As the components of a detergent have widely different densities, if the particles break down, the powder partially separates. Silicate helps to prevent this while making the particles dissolve more easily in use.

Sodium perborate is still the major bleach used in detergent formulations. At temperatures above 50°C it breaks down to generate hydrogen peroxide, which has a more satisfactory effect on textiles than chlorine-based bleaches.

The optical bleaches are not real bleaches but rather brighteners. As textiles age, they yellow. This can be counteracted with colorless dyes which fluoresce blue when absorbed onto fibers.

Foam stabilizers are added mainly because foaminess is considered beneficial by the consumer. The major soil-suspending agent now used is sodium carboxymethyl cellulose.

Some detergents also contain enzymes, or natural catalysts which break down different types of biological molecule. Some of the most difficult soil to remove forms when natural substances in perspiration combine with dirt, for example on collars. Protein and carbohydrate help to bind the dirt to the fabric sufficiently strongly that it may resist the effect of the surfactants in the detergent.

In the early 1960s, enzyme-containing biological soaking powders were introduced as "prewashes". During soaking, they broke down proteins and carbohydrates and thus rendered the dirt more susceptible to the subsequent action of the detergent. Today enzymes are incorporated into the detergent itself, so that the two actions take place almost simultaneously.

► **Oil and water do not mix, and an oil slick can be an environmental disaster (♦ pages 246-247). However oil and water do mix in the presence of detergents, and this principle is also applicable to cleaning up the sea.**

Explosives

Early explosives...Gunpowder and low explosives...High explosives...Designing an explosive...PERSPECTIVE... Fireworks...Alfred Nobel, discoverer of dynamite... Chemistry and war...Uncontroled explosives

▲ *Fireworks were the first, and for a long time the only, application of explosives. Even today they are usually made by hand, requiring careful and intricate packing to ensure even burning and a regular effect.*

The study of explosives combines knowledge of the production of pressure waves with the chemistry of reactions. The explosives used today in commercial and military operations depend on combustion reactions involving oxygen, but this element is not essential to an explosion. If any exothermic reaction (◀ page 94) releases energy much faster than it can dissipate to the surroundings, the temperature rises and the rate of the reaction increases. This self-acceleration may be sufficient to generate a powerful pressure wave, and an explosion is then said to occur.

Explosive chain reactions

For example, mixtures of hydrogen and chlorine explode on ignition. This is a chain reaction. It can be expressed by the overall equation $H_2 + Cl_2 \rightarrow 2HCl$, but it actually occurs by a sequence of steps in which the chlorine molecule divides into two chlorine radicals, each of which combines with a hydrogen molecule to form HCl and creat a hydrogen radical; this encounters another chlorine molecule to create HCl and a further chlorine radical; and so on. Once the initial chlorine molecule dissociates, the reaction is self-sustaining. The chain only stops when two radicals collide and combine.

Some substances explode as the result of a chain reaction in which more radicals are produced at each stage than were present at the start. The combustion of hydrogen in oxygen is an example of such a branched-chain reaction (◀ page 94). Nuclear explosions (▶ page 229) involve branched-chain reactions in which the chain components are neutrons rather than free radicals.

High and low explosives

Chemical explosives are grouped into two main classes: high (detonating) and low (deflagrating) explosives. The high explosives, which are used in blasting operations and for military purposes, decompose rapidly into simple gaseous molecules such as carbon dioxide, nitrogen and water which develop high pressures. The deflagrating explosives burn less rapidly and therefore provide a more controlled release of energy, as in a rifle cartridge.

Under some conditions deflagrating explosives can behave as high explosives, for example if they are tightly contained. Gunpowder is probably the best-known deflagrating explosive, although its use is now almost entirely confined to fireworks and fuses. It has been known in the West since at least the 13th century although its country of origin remains obscure, with claims made for the primacy of the Chinese. Gunpowder is a mixture of potassium nitrate (saltpeter), carbon and sulfur, usually in the proportions 75:15:10. The nitrate group in the saltpeter provides the oxygen required for the combustion of the carbon and sulfur to their gaseous oxides.

Explosives and fireworks

Fireworks operate on the same basic principles as explosives. One major difference is the use in fireworks of a wide range of chemicals to produce particular effects such as colors and sparks. The major colors are supplied by the elements barium (green), copper (blue) and strontium (red).

Gunpowder is used in some fireworks. In this, potassium nitrate is the oxidizing agent. Several other nitrates are also used in fireworks, as are some chlorate and perchlorate salts. The fuel elements of gunpowder, charcoal and sulfur, are supplemented by a range of natural resins such as shellac, as well as other organic materials, such as starch and sawdust. Aluminum, magnesium and titanium all add special effects to the fireworks.

Firework manufacture is complex. Take, for example, a Roman candle. When this is lit, the fuse – usually paper impregnated with potassium nitrate – burns down until it reaches a loosely packed combustible mixture. This burns down until it reaches a star, which is surrounded by gunpowder and has a small packed charge of gunpowder beneath it. The gunpowder around the star ignites it and the charge beneath it explodes, propelling the star into the air. The flame then burns down until it reaches the next star.

The stars themselves are combustible mixtures held together by a resin and containing a salt which will impart color to the star. They must burn fast so that the force of ejection does not extinguish them.

In the mid-19th century single-substance explosives, in which both fuel and oxidizer are combined in the same molecule, were introduced. The German chemist Christian Schönbein (1799-1868) discovered nitrocellulose in 1845-6, by treating cotton with a mixture of sulfuric and nitric acids.

The fully nitrated material is unstable unless very thoroughly washed to remove all traces of unreacted acid. Several disastrous explosions occurred in factories attempting to manufacture nitrocellulose, until Sir Frederick Abel (1827-1902), of the Royal Military Academy in Woolwich, England, discovered this two decades after Schönbein's initial investigation.

Nitroglycerin was discovered at about the same time as nitrocellulose. This too was dangerous to prepare, because it was so easily detonated. It came into widespread use only after the Swedish chemist Alfred Nobel (1833-1896), founder of the Nobel Prizes, discovered in 1867 that a useful explosive, which was also relatively safe to manufacture, could be made by mixing nitroglycerin with the diatomaceous earth kieselguhr. This material he called dynamite. He later improved the formula of dynamite by using combustible materials as the absorbent for the nitroglycerin, rather than the incombustible kieselguhr. He also later produced gelatinous dynamite, a plastic material made by combining nitroglycerin with nitrocellulose.

A number of other organic compounds containing nitro- groups were subsequently introduced as explosives, of which the most famous is probably trinitrotoluene (TNT). This substance has the advantage that it is relatively tolerant to shock and also that it melts at 81°C, so that it can be cast into shells. TNT is still used in military explosives, usually mixed with the more powerful cyclonite (RDX). TNT does not have sufficient oxygen atoms in its molecule for complete destructive oxidation. During World War II, amatol – a mixture of TNT and ammonium nitrate – was used as an explosive.

▲ Vigilance is essential when making explosives. In the early manufacturing plants the one-legged stool meant a worker who dozed off, fell off!

◄ Fired from a submarine this cruise missile, fitted with conventional chemical explosives, is deadly. An explosion comprises a pressure wave, plus the generation of great heat. Here the heat of the explosion has caused the target to burst into flames even before the pressure wave has struck it.

▼ Load a molecule with nitro- (NO₂) or nitrate (NO₃) groups and it becomes dangerously explosive. In the models below the nitrogens are blue, the oxygens red. Nitroglycerin has three nitrate groups (1). This was first made in 1846 from glycerin (glycerol) and nitric acid. Trinitrotoluene (TNT) has three nitro groups on a benzene ring (2). Penterythritol tetranitrate (PETN) has four nitrates (3), and tetryl has four nitros (4). All have been manufactured on a large scale as explosives.

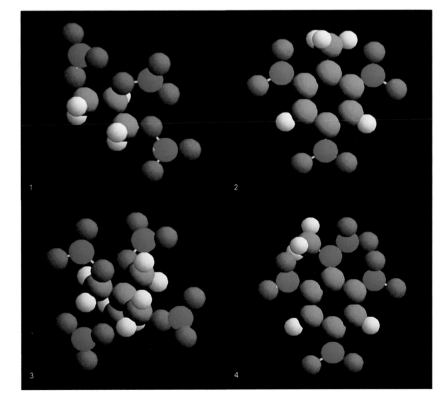

Nobel and dynamite

Alfred Nobel's name is renowned throughout the scientific world for the annual prizes for physics, chemistry, physiology or medicine and peace, which were created by his will. Nobel himself was not trained as a scientist, although his fortune derived from his experimental work on explosives.

Nobel was born in Stockholm in 1833, the son of an inventor and entrepreneur. Alfred Nobel spent his childhood in Russia, but in his late teens, he was sent to the USA to study. When Nobel returned to Russia, his father had become interested in explosives and Alfred also pursued this interest.

Nobel returned to Sweden in 1859, where he continued his interest in explosives, in particular trying to find ways to control the instability of nitroglycerin. After his brother and a number of workmen were killed in a factory explosion in 1864, Nobel transferred his experiments to a barge in the middle of a lake.

In 1867 he patented dynamite, a mixture of nitroglycerin and kieselguhr, which could be handled safely, but when detonated retained the explosive power of nitroglycerin. From this and several other major discoveries in the explosives field, Nobel amassed a fortune. On his death in 1896, a considerable amount of money was set aside to establish the Nobel Prizes. The first of these were awarded in 1901.

Chemistry at War

Explosives at war

Although incendiary arrows date from at least the 9th century BC, the first war chemical was "Greek fire", invented around AD 670. This was natural petroleum thickened with resin. Thrown in pots or squirted from siphons, it was used for centuries by the Eastern Roman Empire against fortresses and ships in the Middle East. Napalm, a modern version of Greek fire, made of oil thickened with rubber latex, has been used for flamethrowers and bombs, particularly in the Vietnam war in the 1960s and 1970s.

Warfare in the West was revolutionized by the development of gunpowder in the 13th century. When harnessed within a tube, its chemical energy of combustion could be turned to impart kinetic energy to a projectile that could kill at long range. But gunpowder had drawbacks as an explosive for use on the battlefield. It was smoky, and had to be ignited by a spark, either from a glowing taper or induced by striking flints or metal. In 1805, the British clergyman Alexander Forsyth (1769-1834) introduced mercury fulminate as an initiator, ignited by percussion rather than spark. Percussion caps led to metal cartridges, breechloaders, and machine guns. Later, gunpowder was replaced in artillery by powerful propellants like guncotton (nitrocellulose). With new smokeless propellants for small arms, such as Nobel's Balistite and the British chemist Frederick Abel's Cordite (extruded mixtures of nitroglycerin dissolved in acetone and absorbed in nitrocellulose), riflemen could remain hidden. Lyddite (trinitrophenol), TNT, amatol and other explosives replaced gunpowder in shell fillings and these high explosive shells became far more destructive.

These advances meant chemical research became essential to modern armies. In 1914-18 British acetone supplies were cut off and the War Office developed a new fermentation process, collecting horse chestnuts to feed acetone-synthesizing microorganisms.

Chemical warfare

Incendiary compositions often included sulfur, and the 19th-century British admiral, Lord Cochrane, advocated the use of sulfur-filled fireships to poison harbor defenders with sulfur dioxide fumes. Chemical advances led to more toxic compounds that were potential weapons.

The era of chemical warfare (CW) opened on 22 April 1915 when German troops released chlorine gas at the Second Battle of Ypres. Henceforth troops were issued with respirators and both sides sought to develop more effective poison gases. Chlorine was followed by phosgene (carbonyl chloride) and then mustard gas – bis(2-chloroethyl) sulfide – an oily liquid that blisters and blinds as well as poisons.

These weapons were introduced by German scientists led by Fritz Haber (1868-1934) and were copied by the Allies. British chemists invented DA, an arsenic smoke; the United States developed Lewisite (nitrogen mustard; 2-chlorovinyl dichloroarsine).

In the 1930s German scientists discovered that low concentrations of organophosphorus compounds are extremely toxic and are absorbed through the skin. These nerve gases were produced by the Nazis but never used. After World War II, Allied scientists learned of the series of nerve gases Tabun, Sarin and Soman. Postwar research culminated in the British discovery of the most toxic known nerve gas, VX. Because of the dangers of leakage, modern nerve munitions are "binary", containing two safer compounds that mix in flight to complete the nerve gas synthesis.

Nerve gases were byproducts of insecticide research. The herbicides 2,4-dichlorophenoxyacetic acid (2,4-D) and 2,4,5-tricholorophenoxyacetic acid (2,4,5-T) were also discovered during World War II and considered for economic warfare by destroying crops. In Malaya and Vietnam they were used as defoliants, robbing guerrillas of jungle cover.

Teargases are non-lethal irritant powders dispersed as aerosols, for instance CS (2-chlorobenzylidine malonitrile). Attempts to find humane incapacitants that would prevent troops fighting have failed. The nearest to success was BZ (3-quinuclidinyl benzilate), related to the "psychedelic" drug LSD (lysergic acid).

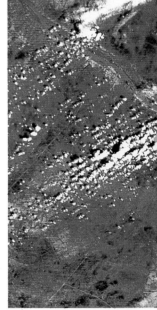

▲ ► Demonstrations seen as threats to authority may be broken up with the aid of chemicals. Tear gas and CS gas are often used. They cause crying or vomiting, which may be unpleasant but not dangerous. They are not true gases but volatile compounds dispersed as dense smoke.

◄ Napalm bombs are designed to cause intense fires and consist of petrol that has a gelling agent dissolved in it. This concentrates the petrol and prevents it burning off too quickly. Indiscriminate use of napalm in an area occupied by civilians is now regarded as unacceptable.

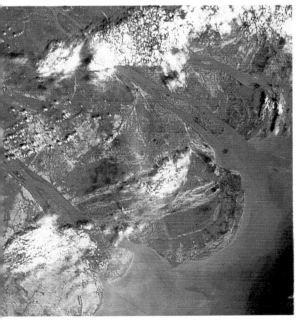

◄ Where the enemy is using forest to hide, his cover can be removed by spraying the trees. Agent Orange was one such herbicide used for this purpose. This consisted of equal parts of 2,4,5-T and 2,4-D. In Vietnam the US airforce sprayed almost 50,000 tonnes of herbicides.

► The German army was the first to use chemicals as effective agents of war, when they released 6,000 cylinders of chlorine gas against British and French troops on 22 April 1915, causing 5,000 deaths and 10,000 wounded. During the next three and a half years both sides used over 125,000 tonnes of poisonous vapors and gases, inflicting over a million casualties. The threat was overcome with the issue of charcoal-filter gas masks and protective clothing.

Explosives for special purposes

Ammonium nitrate is a close relative of saltpeter. Some widely used modern explosives are mixtures of ammonium nitrate with carbonaceous material, thus showing a resemblance to gunpowder. For example, mixtures of ammonium nitrate and coal dust were used as explosives in the early days of strip mining. Then in the 1950s, ammonium nitrate-fuel oil mixtures were introduced.

Another introduction in the late 1950s was that of slurry explosives, initially mixtures of ammonium nitrate, TNT, water and gelling agents. These slurries are now widely used because of the ease with which they can be placed into boreholes in rockfaces prior to blasting.

Uncontrolled explosions

In underground coal mines, only explosives which do not generate very high temperatures are used. This is to prevent ignition of pockets of methane in the mine. For every tonne of coal mined, about $25m^3$ of methane (CH_4) are released. Methane is a major component of natural gas and its controlled combustion provides domestic and industrial heat in parts of the world. However, under some conditions it forms explosive mixtures. If methane explodes in a mine, the pressure wave may stir up a cloud of coal dust, which may explode in its turn, disturbing more coal dust and propagating the explosion. In an accidental explosion in a British coal mine in 1951, flame propagated in this manner traveled a total of 12km in the underground roads.

Any readily oxidizable material may form an explosive mixture with air. This includes not only fossil materials such as coal and methane, but also organic materials such as flour. Where large quantities of organic material which can form dusts are stored, precautions need to be taken to prevent build up of static electricity which could produce a spark and ignite the mixture.

▼ **Certain reactions become dangerous if the energy released cannot be readily dissipated. When such chemical reactions also form a large volume of gases as products, an explosion may result. This flourmill in the United States was destroyed by an explosion of flour in a silo.**

▶ **The controlled use of explosives is an everyday event in mining, tunneling and demolition, as here. Nitroglycerin is an explosive oil, but when absorbed onto the powdery mineral kieselguhr it becomes safe to handle, and requires a detonator to set it off.**

The atom and the nucleus...The discovery of the nucleus...Protons and neutrons...Isotopes...Stable and unstable nuclei...The force holding the nucleus together ...Radioactivity and nuclear decay...Descriptions of the nucleus...PERSPECTIVE...The discovery of radioactivity... Rutherford and the discovery of the nucleus...Units of mass...Exotic radioactive decay

One of the most remarkable discoveries of the 20th century is that the apparently "solid" matter of the everyday world is actually mainly empty space. Matter consists of atoms, and each atom consists of electrons whirling round a nucleus (◀ page 72). The electrons endow the atoms of each element with a unique character, and determine how one atom interacts with others. Yet the electrons represent only a tiny portion – typically around 0·001 percent – of the matter within the atom. Most of the mass of an atom is concentrated in the tiny, dense nucleus; if a pea were as dense as an atomic nucleus, it would have a mass of some 10 million tonnes.

Most nuclei contain two kinds of particle – positively-charged protons and electrically neutral neutrons. (The exception here is hydrogen, the lightest element, which contains a single proton in its nucleus). The positive charge on a proton precisely balances the negative charge on an electron, and the number of protons in an atomic nucleus equals the number electrons in the atom, making atoms electrically neutral overall.

It might be thought that the electrical forces between the densely packed protons would cause them to repel one another and blow the nucleus apart. That this does not occur is due to the strong nuclear force (◀ page 214). This operates only within the confines of the nucleus. The strong force is some 100 times as powerful as the electrical force that would otherwise disrupt the nucleus. The strong force acts equally on protons and neutrons – it does not recognize their differing electric charges. The neutrons therefore assist in keeping the protons bound together within the nucleus.

The discovery of radioactivity

In 1896, the French physicist Henri Becquerel (1852-1908) was investigating the link between fluorescence and the phenomenon of X-rays. He packed together some fluorescent uranium salt, a mask to allow through a specific pattern of radiation and a photographic plate, all of which he intended exposing to sunlight. But for several days the sun did not shine, and eventually, when he removed the package from the drawer, he discovered that an image had formed anyway. This could not be due to fluorescence but had to be due to some new form of radiation emitted by the salt.

Becquerel had discovered radioactivity, the spontaneous transmutation of an atomic nucleus from one state to another, which yields various radiations. At the time, the notion of atoms was far from being fully accepted, and it was to be many years before physicists finally came to a complete understanding of radioactivity. However, Becquerel's discovery opened the door that was to lead to the modern picture of the atom. Here, as other scientists were soon to show, was the first evidence that atoms are not the immutable objects the ancient Greeks had imagined (◀ page 9). Moreover, the radiations produced by uranium and related elements were also to prove powerful tools for probing the atom, and led to the discovery of the nucleus and its contents by the British physicist Ernest Rutherford (1871-1937), and his colleagues (◀ page 188).

◀▼ The first evidence that atoms are not immutable came in 1896 when Henri Becquerel (below) left some uranium salts next to a photographic plate, which became blurred.

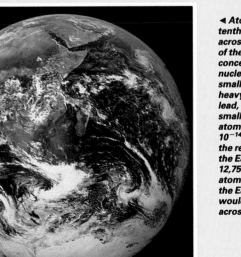

◀ Atoms are typically tenths of nanometers across (10^{-10}m), but most of their mass lies concentrated in a central nucleus which is much smaller. The nucleus of a heavyweight atom, such as lead, is about 10,000 times smaller than the typical atomic dimension – that is, 10^{-14}m across. To imagine the relative sizes, consider the Earth, which is about 12,750km in diameter. If an atom of lead were as big as the Earth, then its nucleus would be only about 1·3km across.

Ernest Rutherford was involved in the discovery of the nucleus, the proton and neutron

Pioneering studies of the nucleus

The founding father of nuclear physics was Ernest Rutherford (1871-1937). He not only discovered the existence of the atomic nucleus and the protons it contains, but also did much to elucidate the nature of radioactivity. Moreover, it was in Rutherford's laboratory at Cambridge University that the neutron was discovered.

Rutherford was born in New Zealand and studied at Canterbury College, Christchurch and Cambridge University. There he worked with the British physicist John Joseph (J.J.) Thomson (1856-1940), who discovered the electron in 1897 (◀ page 72). Rutherford studied radioactivity soon after its discovery in 1896, and he found two types of radiation, which he called alpha and beta.

In 1898, Rutherford moved to Montreal, where he continued his studies of radioactivity, ably assisted by a young British chemist, Frederick Soddy (1877-1956). Together they discovered how radioactivity transmutes one element to another – a natural alchemy. Rutherford later moved back to Britain in 1907, this time to Manchester University. There he worked with the young German physicist Hans Geiger (1882-1945), and showed in 1908 that alpha particles have the same charge and mass as helium atoms. Soon after this, Rutherford encouraged Geiger and Ernest Marsden (1889-1970), a student from Blackburn, England, to study the scattering of alpha particles from gold foil. Marsden made the remarkable discovery that the alpha particles can be turned back in their tracks by the foil, and this led Rutherford to conclude in 1911 that atoms must contain a dense central nucleus.

In further experiments, initiated by Marsden, Rutherford discovered that alpha particles could knock out hydrogen nuclei from a variety of other nuclei. This led him to conclude that hydrogen nuclei are basic components of all atomic nuclei, and he gave them the name of "protons", after the Greek for "first". In 1919, around the time he published his work on protons, Rutherford succeeded Thomson as Cavendish professor at Cambridge. Thereafter he guided the efforts of a remarkable team that grew up around him, making Cambridge the world's most important center for research in experimental nuclear physics in the 1920s and 30s. One of many results that came out of the Cavendish Laboratory in a single year, 1932, was the discovery of the neutron by the British physicist James Chadwick (1891-1974).

Rutherford was a remarkable personality, renowned for his booming voice and distaste for too much theorizing. He received the Nobel Prize for chemistry for his work on radioactivity in 1908, and was made Baron Rutherford of Nelson in 1931. His ashes were placed in Westminster Abbey near the tomb of Isaac Newton (1642-1727).

► *In 1909, Rutherford suggested that Ernest Marsden should investigate whether alpha particles can be deflected through large angles in scattering from a thin foil. Marsden's apparatus consisted of a radioactive source emitting alpha particles, a sheet of gold foil, and a scintillating screen and microscope to view the flashes of light each time an alpha particle struck the scintillator. The screen and microscope could move through 360°. To everyone's amazement, alphas were deflected through all angles, 1 in 20,0000 right back to strike the screen when positioned next to the source.*

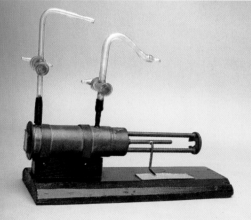

◀▲► *Ernest Rutherford (above right with Geiger) masterminded both the discovery of the atomic nucleus and, within a decade, the protons that the nucleus contains. The apparatus (left) was used in his work. Rutherford proposed that the hydrogen nucleus – the proton – must be a basic constituent in all nuclei. That protons and hydrogen nuclei are identical is shown opposite, where a proton (red) knocks many hydrogen nuclei into motion. The 90° angles between the tracks shows the equivalence of the particles.*

The discovery of the nucleus

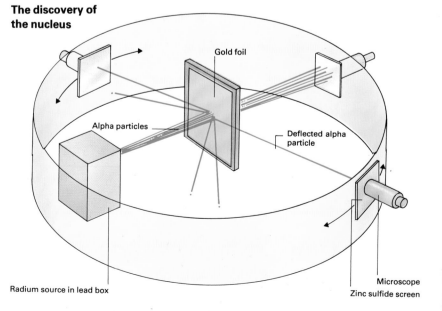

Gold foil

Alpha particles

Deflected alpha particle

Radium source in lead box

Microscope
Zinc sulfide screen

The mass of a nucleus is less than the total mass of the particles it contains

Exotic radioactivity

Naturally occurring radioactive isotopes decay through either beta emission or alpha emission. The same is true of the many radioactive isotopes that have been made artificially, for example by firing energetic neutrons at nuclei in a target material to form new isotopes. Artificial radioactivity was discovered in 1933 by the French physicists Frédéric Joliot (1900-1958) and his French wife Irène Curie (1897-1956).

During the 1980s, however, physicists began to create exotic isotopes with either far too many neutrons or far too few. Isotopes such as the neutron-deficient lutetium-151 (71 protons and only 80 neutrons) are made in the collisions of naturally-occurring heavy nuclei, while the neutron-rich lithium-11 (three protons and eight neutrons) is one of the exotic fragments produced when uranium is bombarded with energetic protons. Such isotopes have revealed new forms of radioactivity which are prohibited to the more usual isotopes on energy grounds. Lithium-11, indeed, exhibits at least six different decay modes. It can undergo a beta decay followed rapidly by the emission of one, two, or three neutrons; or an alpha particle and a neutron; or even a triton (the nucleus of hydrogen-3 which contains one proton and two neutrons). In a similar way, lutetium-151 can emit protons. One of the most exotic forms of radioactivity was found in 1984 in radium, one of the first radioactive elements to be discovered, nearly 100 years previously. Radium-223, which occurs naturally in the decay chain of uranium-235, can emit relatively large clusters of protons and neutrons in the form of carbon-14 nuclei; so too can radium-222 and radium-224. The carbon-14 emission is very rare, occurring once for every 10 billion alpha decays.

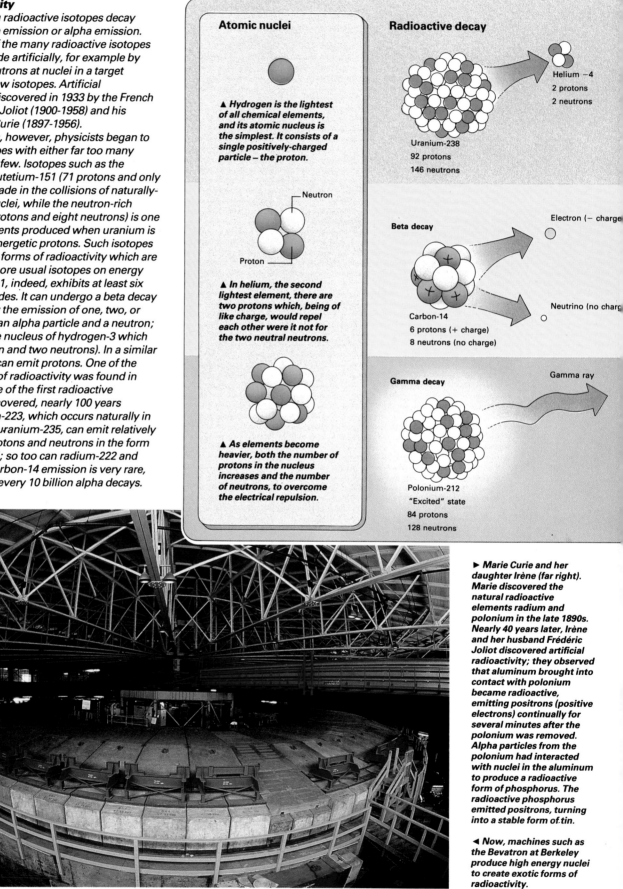

Atomic nuclei

▲ Hydrogen is the lightest of all chemical elements, and its atomic nucleus is the simplest. It consists of a single positively-charged particle – the proton.

Neutron

Proton

▲ In helium, the second lightest element, there are two protons which, being of like charge, would repel each other were it not for the two neutral neutrons.

▲ As elements become heavier, both the number of protons in the nucleus increases and the number of neutrons, to overcome the electrical repulsion.

Radioactive decay

Helium –4
2 protons
2 neutrons

Uranium-238
92 protons
146 neutrons

Beta decay

Electron (– charge

Neutrino (no charge

Carbon-14
6 protons (+ charge)
8 neutrons (no charge)

Gamma decay

Gamma ray

Polonium-212
"Excited" state
84 protons
128 neutrons

► Marie Curie and her daughter Irène (far right). Marie discovered the natural radioactive elements radium and polonium in the late 1890s. Nearly 40 years later, Irène and her husband Frédéric Joliot discovered artificial radioactivity; they observed that aluminum brought into contact with polonium became radioactive, emitting positrons (positive electrons) continually for several minutes after the polonium was removed. Alpha particles from the polonium had interacted with nuclei in the aluminum to produce a radioactive form of phosphorus. The radioactive phosphorus emitted positrons, turning into a stable form of tin.

◄ Now, machines such as the Bevatron at Berkeley produce high energy nuclei to create exotic forms of radioactivity.

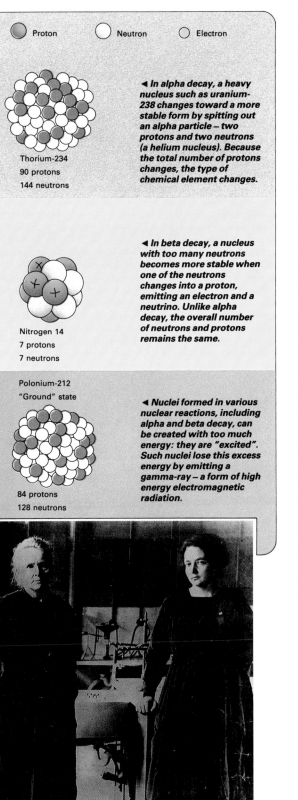

Proton   Neutron   Electron

Thorium-234
90 protons
144 neutrons

◄ In alpha decay, a heavy nucleus such as uranium-238 changes toward a more stable form by spitting out an alpha particle – two protons and two neutrons (a helium nucleus). Because the total number of protons changes, the type of chemical element changes.

◄ In beta decay, a nucleus with too many neutrons becomes more stable when one of the neutrons changes into a proton, emitting an electron and a neutrino. Unlike alpha decay, the overall number of neutrons and protons remains the same.

Nitrogen 14
7 protons
7 neutrons

Polonium-212
"Ground" state

◄ Nuclei formed in various nuclear reactions, including alpha and beta decay, can be created with too much energy: they are "excited". Such nuclei lose this excess energy by emitting a gamma-ray – a form of high energy electromagnetic radiation.

84 protons
128 neutrons

The energy of a nucleus

Certain configurations of protons and neutrons have less energy than others, and the most stable configurations are those with least energy. This is like observing that a ball rolls down a hill to the bottom because its gravitational potential energy (◀ page 22) is less at the bottom of the hill. The "energy" of a nucleus is a subtle concept to grasp; it is in fact the mass of the nucleus. Albert Einstein showed in his special theory of relativity (◀ page 46) that mass is a form of energy. Thus a nucleus with a minimum energy has a minimum mass.

The mass of a stable nucleus is always less than the total mass of the protons and neutrons it contains. For example, the mass of nitrogen-14, the commonest isotope of nitrogen, is less than the total mass of the seven protons and seven neutrons it contains. However, the mass of nitrogen-14 is also less than that of carbon-14 – the radioactive form of carbon used in carbon dating. Carbon-14 also contains 14 nucleons (eight neutrons and six protons), but its mass-energy is 0.0012 percent greater than that of nitrogen-14. Carbon-14 decays to nitrogen-14, shedding its excess mass-energy as it does so.

Radioactive decay

The way in which unstable nuclei rid themselves of their excess energy depends in part on how much energy they have over and above a more stable nucleus. Radioactivity can occur in a variety of ways. The three most common forms are known as alpha, beta and gamma decay. In alpha decay, a nucleus spits out an alpha particle. This is a conglomeration of two protons and two neutrons (in other words, a helium nucleus). To emit an alpha particle, a nucleus must have a larger mass-energy than the total mass of the new nucleus and the alpha particle. This is often true for nuclei heavier than lead, which can lose energy by emitting an alpha particle, but which would need additional energy to give up a single proton or neutron.

In beta decay, the nucleus ejects a different kind of particle – an electron. However, unlike the alpha particle whose protons and neutrons inhabited the initial nucleus, the electron did not exist inside the nucleus before emerging. It is created at the moment the nucleus decays, as a neutron transmutes into a proton. To do this, the nucleus must rid itself of some energy and increase its electric charge by one unit. The electron, with its negative charge, ensures the appropriate change of charge, and it also carries away a small fraction of the excess energy as its own mass. The balance of energy for the whole process, released as the nucleus converts from one kind to another, is, however, shared between the motion of a third product of the decay – a neutral particle called the neutrino (◆ page 210).

Beta decays of this kind occur in nuclei with too many neutrons. This is how carbon-14 decays to nitrogen-14. Nuclei that are too rich in protons can decay by another form of beta radioactivity. In this case, a proton converts into a neutron by emitting a positron along with a neutrino. The positron is in fact a positively charged version of the electron – it is an antielectron (◆ page 209).

The third common form of radioactivity is gamma decay. The nucleus does not emit a particle but a very energetic form of electro-magnetic radiation (◀ page 65). Gamma radiation does not change the number of protons or neutrons in a nucleus, but it does reduce the energy of a nucleus. It occurs when a nucleus has changed through alpha or beta radioactivity to an "excited" nucleus, in which the protons and neutrons jostle each other with more energy than usual.

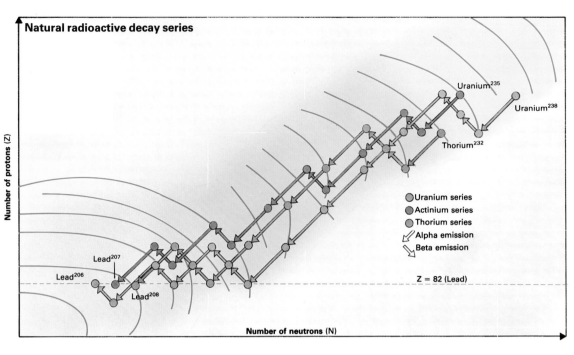

Natural radioactive decay series

Number of protons (Z)

Uranium²³⁵
Uranium²³⁸
Thorium²³²

Lead²⁰⁷
Lead²⁰⁶
Lead²⁰⁸

○ Uranium series
● Actinium series
◔ Thorium series
⇗ Alpha emission
⇘ Beta emission

Z = 82 (Lead)

Number of neutrons (N)

◄ At the top end of the valley of stability are found the heaviest nuclei, all of which are unstable. Through emitting alpha or beta radiation, these nuclei can follow zigzag paths that take them to stable configurations of the element lead. Three such pathways allow the decays of naturally-occurring radioactive nuclei, uranium, actinium and thorium.

The units of atomic mass

Conventional units of mass prove unwieldy when dealing with atomic nuclei. The mass of a proton is 1.673×10^{-27}kg, while that of a neutron is slightly – but significantly – larger at 1.675×10^{-27}kg. Nuclear physicists prefer to work instead with special units. One unit used is the atomic mass unit (amu), which is defined as $\frac{1}{12}$th the mass of an atom of carbon-12. In terms of these units the mass of a proton is 1.0076 amu and that of the neutron is 1.0090 amu.

Because mass and energy are often converted from one to another in nuclear reactions it is also convenient to define masses in terms of energy units. The energy unit used by nuclear physicists is the electronvolt (eV), which is the energy an electron gains when accelerated through an electric potential of 1 volt. It is equal to 0.16×10^{-18} joules. Using Einstein's famous equation $E = mc^2$ (♦ page 46), and setting the velocity of light as one, provides values of masses in units of electronvolts. One atomic mass unit becomes equal to 931 million electron volts (MeV), and in these units the mass of a proton is 938.3MeV, while the mass of the neutron is 939.6MeV.

► A "contour map" of nuclear energies for a wide range of nuclei shows a "valley of stability" where the energies have their minimum values. The nuclei that occupy this valley have particularly stable configurations of protons and neutrons; others attempt to reach it through radioactive decays.

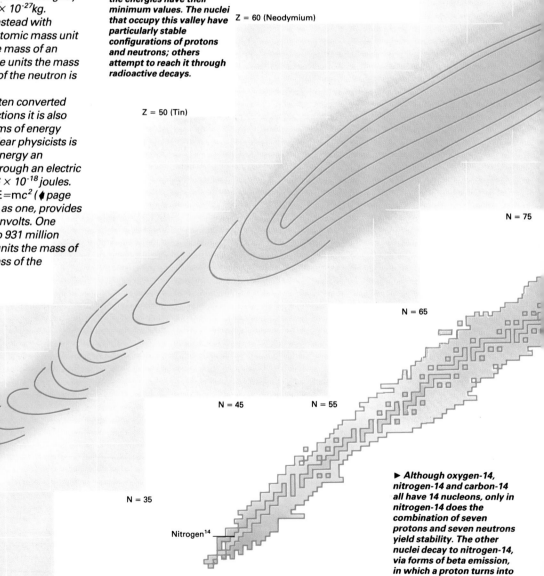

Z = 70 (Ytterbium)

Z = 60 (Neodymium)

Z = 50 (Tin)

Z = 30 (Zinc)

Z = 20 (Calcium)

N = 75

N = 65

N = 55

N = 45

N = 35

N = 25

N = 15

Number of protons (Z)

Number of neutrons (N)

Nitrogen¹⁴

► Although oxygen-14, nitrogen-14 and carbon-14 all have 14 nucleons, only in nitrogen-14 does the combination of seven protons and seven neutrons yield stability. The other nuclei decay to nitrogen-14, via forms of beta emission, in which a proton turns into a neutron or vice versa.

The valley of stability

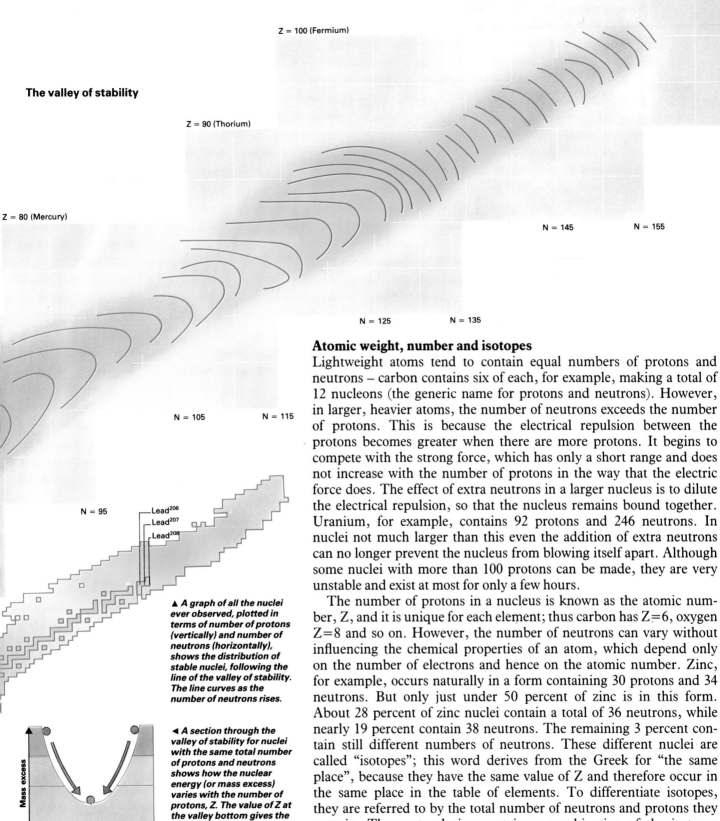

Z = 100 (Fermium)

Z = 90 (Thorium)

Z = 80 (Mercury)

N = 145

N = 155

N = 125

N = 135

N = 105

N = 115

N = 95

Lead²⁰⁶
Lead²⁰⁷
Lead²⁰⁸

▲ *A graph of all the nuclei ever observed, plotted in terms of number of protons (vertically) and number of neutrons (horizontally), shows the distribution of stable nuclei, following the line of the valley of stability. The line curves as the number of neutrons rises.*

◄ *A section through the valley of stability for nuclei with the same total number of protons and neutrons shows how the nuclear energy (or mass excess) varies with the number of protons, Z. The value of Z at the valley bottom gives the preferred stability value.*

Mass excess

Number of protons (Z)

Number of protons (Z)

○ Oxygen¹⁴
○ Nitrogen¹⁴
○ Carbon¹⁴

Z = 7 (Nitrogen)

Number of neutrons (N)

Atomic weight, number and isotopes

Lightweight atoms tend to contain equal numbers of protons and neutrons – carbon contains six of each, for example, making a total of 12 nucleons (the generic name for protons and neutrons). However, in larger, heavier atoms, the number of neutrons exceeds the number of protons. This is because the electrical repulsion between the protons becomes greater when there are more protons. It begins to compete with the strong force, which has only a short range and does not increase with the number of protons in the way that the electric force does. The effect of extra neutrons in a larger nucleus is to dilute the electrical repulsion, so that the nucleus remains bound together. Uranium, for example, contains 92 protons and 246 neutrons. In nuclei not much larger than this even the addition of extra neutrons can no longer prevent the nucleus from blowing itself apart. Although some nuclei with more than 100 protons can be made, they are very unstable and exist at most for only a few hours.

The number of protons in a nucleus is known as the atomic number, Z, and it is unique for each element; thus carbon has $Z=6$, oxygen $Z=8$ and so on. However, the number of neutrons can vary without influencing the chemical properties of an atom, which depend only on the number of electrons and hence on the atomic number. Zinc, for example, occurs naturally in a form containing 30 protons and 34 neutrons. But only just under 50 percent of zinc is in this form. About 28 percent of zinc nuclei contain a total of 36 neutrons, while nearly 19 percent contain 38 neutrons. The remaining 3 percent contain still different numbers of neutrons. These different nuclei are called "isotopes"; this word derives from the Greek for "the same place", because they have the same value of Z and therefore occur in the same place in the table of elements. To differentiate isotopes, they are referred to by the total number of neutrons and protons they contain. Thus natural zinc contains a combination of the isotopes zinc-64, zinc-66 and zinc-68.

All elements have a range of possible isotopes, although often only one isotope dominates, forming 90 percent or more of the substance. This is because all isotopes other than the most common ones are intrinsically unstable, and transform into a different kind of nucleus after a period of time that can vary from fractions of a second to millions of years. These are called radioactive isotopes because they change through one of the several processes that are together known as radioactivity.

Theories of the nucleus

Developing a theoretical model that explains all the observed properties of stable, unstable and excited nuclei has proved a difficult challenge. One of the first models to achieve any degree of success was the "liquid drop model", first developed by two theorists, the Dane Niels Bohr (1885-1962) and the American John Wheeler (b.1911) in 1939. This draws analogies between certain properties of the nucleus and those of a drop of liquid, and it proves particularly useful in discussing the "fission" of heavy nuclei into two fragments (◆ page 229).

However, the liquid drop model cannot deal with one important property of nuclei, namely "spin". Quantum theory shows that protons and neutrons behave as if they possess an intrinsic angular momentum, rather like subatomic spinning tops (◆ page 201). A full theory of the nucleus must take into account the spins of the constituent nucleons, and predict how they add together to give an overall spin angular momentum for the nucleus as a whole. One model based on spin is the "shell model", developed independently in 1948 by physicists Maria Goeppert-Mayer (1906-1972), in the United States, and Hans Jensen (1907-1973), in Germany, both of whom received the Nobel Prize for physics in 1963.

In this model protons and neutrons occupy "shells" analogous to the shells of atomic theory (◆ page 73). One of the driving forces behind such a model is the observation of the so-called magic numbers. Experiments indicate that certain numbers of protons (Z) and neutrons (N) give nuclei that are exceptionally tightly bound. In particular the values 2, 8, 20, 28, 50, 82, 126 are associated with this stability. For example, oxygen-16 (Z=8, N=8), calcium-40 (Z=20, N=20), and lead-208 (Z=82, N=126) are very stable; and tin, with Z=50, has as many as 10 stable isotopes. According to the shell model the magic numbers are associated with closed shells of nucleons.

The collective model of the nucleus

The shell model proves successful in accounting for the observed spins of nuclei, but it runs into difficulty with other nuclear properties. It fails particularly in accounting for non-spherical distributions of electric charge that occur in many nuclei. To deal with problems of this kind, Aage Bohr (b.1922), a Danish theorist and son of Niels Bohr, and Ben Mottleson (b.1926), a theorist in the United States, developed the "collective model" of the nucleus. This had been first proposed in 1950 by another American physicist, James Rainwater (b.1917). The collective model attempts to unite the best features of the shell model and the liquid drop model. It does so by regarding a large nucleus as a closed-shell core surrounded by additional "valence" nucleons. This core can be deformed from its spherical shape by the interactions of the outer nucleons. This model has particular success in describing the vibrations and rotations of atomic nuclei which can be studied experimentally.

A general theoretical description that can be applied to all nuclei remains the goal of nuclear physicists. Data for these studies comes from experiments at machines that can accelerate ions of heavy elements – up to uranium in some cases – and collide them with targets to create "exotic" isotopes with unusual numbers of protons and neutrons, which are sometimes also spinning very rapidly. These isotopes can be greatly deformed, and they provide stringent tests of the theoretical models.

▲ *Niels Bohr (right) and his son Aage both made great contributions to the theory of nuclear structure. Niels, with John Wheeler, put forward the liquid drop model, while Aage has attempted to unify this with the shell model.*

▲ ▼ *Maria Goeppert-Mayer (left) with Hans Jensen. They independently developed a "shell" model for nuclei in the late 1940s, in which protons and neutrons fit into "shells" within a nucleus. The simplest shells can contain two protons and two neutrons – provided the members of each pair spin in opposite directions. In other shells, the nucleons have additional angular momentum and more particles are allowed. Full shells, corresponding to a total of two, eight, or twenty neutrons or protons, for example, yield very stable nuclei. The diagram illustrates the shell allocations for calcium-40, which has closed shells both of protons and neutrons, with 20 particles in each.*

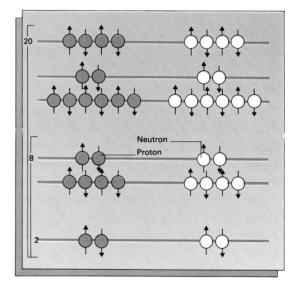

The Quantum World

26

*The concept of quanta...The particle theory of light...
Planck's constant...Bohr's image of the atom...Particles
as waves...Quantum mechanics...PERSPECTIVE...
Theories of light emission...Niels Bohr...The
photoelectric effect...Lasers...Electron microscopy...
Spin and NMR...Philosophical implications of the
quantum theory*

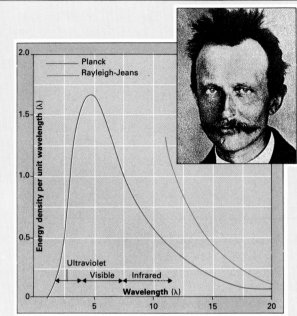

▲ *Max Planck developed his quantum theory of radiation in
an attempt to resolve a difficulty with the existing theory for
the emission of radiation. In the theory of Rayleigh and
Jeans, the amount of energy emitted increased limitlessly at
shorter wavelengths. This was an "ultraviolet catastrophe"
that was clearly at odds with experimental results.*

Many natural physical phenomena are distinguished by their continuity. The Earth smoothly orbits the Sun; the ground slowly warms on a sunny day, only to release its heat in an equally gentle fashion during the night. But in the world within the atom, the story is very different. There, subatomic systems can exist only in certain fixed states. In other words, many subatomic phenomena are "quantized". It is like comparing stairs with a smooth slope – whereas a person can stand at any height on a slope, only fixed "discrete" heights are possible on the staircase.

Quantization is one important feature of subatomic physics. Another is its probabilistic nature. An astronomer can say with certainty where the Earth, Moon and the other planets will be in relation to the Sun on a given date, but an atomic physicist can state only the *probability* that an electron will be in a given state at a given time. In addition, there is a fundamental limit to the precision of statements about the subatomic world. The more precisely the value of one property is known, the less precisely the value of a second related property can be known (◆ page 204). Such features make the subatomic world very unfamiliar from an "everyday" viewpoint.

The particle theory of light

One prime example of quantum physics at work is in the behavior of light. Physicists in the 19th century found good evidence that light is a wave motion (◆ page 48), a view reinforced in the electromagnetic theory of the British physicist James Clerk Maxwell (1831-1879) in 1865. Maxwell's theory predicted the existence of electromagnetic waves traveling through free space at a specific velocity, close to the measured velocity of light. This showed that light is just one variety of a whole range of electromagnetic radiation with varying wavelengths, all of which travel through free space at the same velocity.

The wave theory works well in describing the movement of light through different materials. It explains the phenomena of reflection, refraction, and the diffraction and interference of light (◆ pages 42-43). But the theory breaks down when it comes to explaining the absorption and emission of light.

The first person to come to terms with these difficulties was the German physicist Max Planck (1858-1947). Physicists in the late 1890s were faced with a problem in that they could not explain the way that the intensity of radiation emanating from a "perfect emitter" varies with wavelength. There is a peak in intensity at a wavelength associated with the temperature of the emitting body: the higher the temperature, the shorter is this maximum wavelength. But in the late 19th century, theoretical understanding implied that the intensity should go on rising at ever shorter wavelengths, rather than falling down from a peak. There was clearly something wrong with the theory.

Understanding the emission of light

In the late 19th century Heinrich Hertz demonstrated that light is an electromagnetic wave (◆ page 64). At the same time, observations were made that were difficult to reconcile with Maxwell's theory, such as the emission of light at only a few wavelengths from certain sources. Such sources yield a spectrum of discrete lines when their light is passed through a prism. Even continuous spectra provided a problem, when physicists tried to explain how intensity varies with wavelength.

The British physicists Lord Rayleigh (born John Strutt, 1842-1919) and James Jeans tried to explain the spectrum in the following way. They considered a perfect emitter (or "black body") as a box which can store electromagnetic energy as "standing" waves, like the sound waves in an organ pipe. The spectrum of waves emerging from a hole in the box should, they argued, correspond to the observed emission spectrum from a source like the Sun. They could calculate the number of standing waves within a particular range of wavelengths. And they assumed that the same amount of energy was associated with each wave.

The Rayleigh-Jeans law agreed with observation at high wavelengths, but disagreed at lower wavelengths – indeed, it implied that an infinite amount of energy could be radiated at low wavelengths. Planck discovered the flaw in the theory. The concept of standing waves in a box and the number of waves were not at fault; it was the assumption that the energy is distributed equally between different wavelengths. Planck associated different energies with different wavelengths. He thereby resolved the problem, and gave birth to quantum theory which was eventually to explain the emission lines of other types of source.

Einstein received the Nobel Prize for his work on the photoelectric effect, rather than his work on special relativity

Planck was able to solve the problem of light emission by postulating that the energy in radiation is emitted in fixed amounts, and that the size of each "quantum" of energy depends on the frequency at which the radiation is oscillating. The energy is given by the frequency of the radiation multiplied by a constant: energy equals frequency times a constant. Known as Planck's constant, it is usually symbolized by the letter h. It has a value of $6\cdot6 \times 10^{-34}$ joule seconds. Only in systems where the relevant energies are very small, as in individual atoms, does the quantum concept become important. The equation also reveals how the energy associated with a quantum of radiation varies across the electromagnetic spectrum. Radiation at high frequencies (short wavelengths) consists of high-energy quanta; at low frequencies (long wavelengths) the quanta have low energies.

The German physicist Albert Einstein (1879-1955) used the same idea of quantized radiation to explain the photoelectric effect, in which ultraviolet light causes atoms in the surface of some metals to emit electrons. Energy of the emitted electrons depends not on the *intensity* of the ultraviolet radiation, as might be expected if the radiation is a wave, but on the *frequency*. Also, there is no emission below a certain frequency, characteristic of each metal. Einstein argued that the radiation must arrive at the surface in quantized "wave packets", each with an energy related to the frequency through Planck's equation. Thus, radiation at too low a frequency does not have enough energy to knock out even one electron, but at higher frequencies it knocks out electrons with increasing energies.

Light thus exhibits a dual behavior. In certain circumstances it proves convenient to regard it as a wave, in others as a stream of quanta or "particles" of light. If an electron in an atom absorbs a photon, its energy increases by the amount of energy the photon carried; if the electron emits a photon, its energy decreases. However, the energies of the electron are also quantized. Like a person standing on a staircase, it can absorb or give up energy only in fixed amounts as it moves from one energy level (stair level) to another. Thus an atomic electron cannot emit a photon of any energy.

There is a precise relationship between the situation of an electron within an atom and the energy of the photons it can absorb or emit. The allowed energy levels depend critically upon the positive charge of the nucleus about which the electrons circulate. In other words, each chemical element has its own individual "staircase" of energy levels (◀ page 72). This in turn implies that the photons emitted (or absorbed) by atoms of different elements have different energies (and wavelengths), which correspond to the different gaps in each allowed series of energy levels. Each atom thus has its own unique emission and absorption spectra, which are as good an indicator of an atom's identity as fingerprints are of a human's.

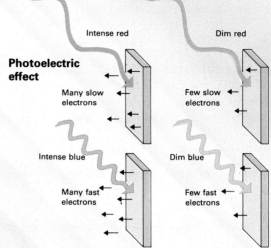

Photoelectric effect

▲ In the photoelectric effect, light knocks electrons out of a material. It can be explained only if light consists of photons, localized wave "packets", as opposed to the continuous waves of classical electromagnetic theory. Changing the intensity of the light (top) changes the number of photons; this alters the number of electrons emitted, but has little effect on their energy. Changing the frequency of the light, however, changes the energy of the photons and therefore the energy of the electrons emitted. In practice, high frequency, ultraviolet light is often necessary.

Energy levels of an atom

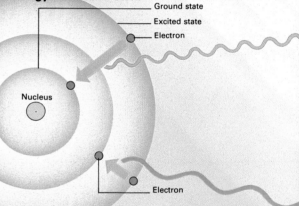

▲ Electrons in atoms emit light when they change energy, often after being excited to higher energy levels than usual, perhaps by heating. With each jump from a higher energy level to a lower one, an electron emits a single photon. The energy (frequency) of the photon depends on the size of the jump: the bigger the jump, the greater the energy of the emitted photon, and the higher its frequency.

▼ The frequencies of light emitted or absorbed by atoms provide their "fingerprints". The emission spectrum of helium (left) shows the light produced as excited electrons jump down from higher energy levels. A spectrum of the light from the Sun (right) shows dark lines where photons of particular frequencies have been absorbed as electrons in atoms jump to higher energy levels.

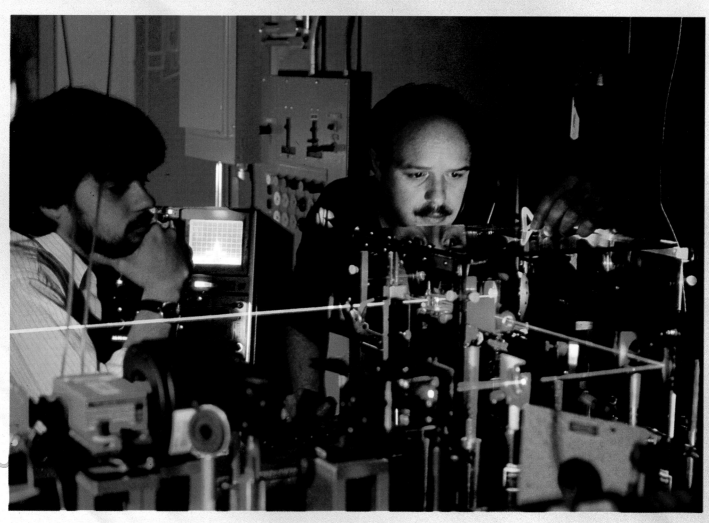

▲ Lasers are key research tools. They provide a source of light that is not only monochromatic (single wavelength) but also coherent – the emitted photons are in perfect step with each other. This greatly enhances the intensity of light.

▶ When an atom absorbs light, electrons are raised to higher energy levels by photons of the appropriate energy (top), they can then emit photons and return to their usual levels spontaneously. However, they can also be stimulated to fall back by a photon of frequency corresponding to the change in energy levels. This is stimulated emission – the emitted photon and the original incident photon move away in step.

Absorption

Incoming photon

Electron knocked into higher orbit

Ground orbit

Outer orbit

Stimulated emission

Incoming photon

Second photon emitted in phase with incoming photon

Electron knocked into lower orbit

Lasers

The acronym laser stands for "light amplification by stimulated emission of radiation". The crucial words here are "stimulated emission". In an emission process, a photon is produced when an electron falls from one atomic energy level to another, lower level. In stimulated emission, this process is encouraged by radiation of the same energy (that is, the same frequency and wavelength) as the emitted radiation. Thus one photon stimulates the emission of a second photon of exactly the same wavelength. Moreover, the two photons are precisely in step such that the peaks and troughs of the resulting radiation match exactly. The radiation produced in this way is said to be "coherent", and it is valuable for carrying information, and for the extremely high energy-density that a beam of such radiation can possess. This is the value of a laser beam.

For stimulated emission to be useful, an atomic system needs more atoms with electrons in the relevant upper energy level than is normally the case. In a laser, this is achieved by "optical pumping". A flashlight, for example, can "excite" many electrons into the upper level. These electrons then begin spontaneously to fall back down to the lower energy level, emitting radiation as they do so.

The principle of a laser is to trap some of this radiation so that it stimulates further emission, and the device emits a coherent beam. This is done by mirrors to ensure that the emission occurs within a "resonant cavity", in which the radiation bounces back and forth, stimulating more emission in the desired way, before it escapes.

The wave nature of the electron is seen in diffraction patterns similar to those of light

Applying the quantum theory to the atom

The Danish physicist Niels Bohr (1885-1962), working from Rutherford's concept of the atom as a solar system in which electrons orbit at random around the tiny nucleus, first put forward the idea of quantized atomic structure in 1913 (◀ page 69). In so doing he took Planck's concept a stage further by quantizing not the energy but the angular momentum (◀ page 16) of the electrons in an atom. Bohr postulated that the angular momentum can take only those values equal to Planck's constant multiplied by an integer (a whole number) and divided by another constant (2 times π, the familiar geometrical constant). The angular momentum of an electron orbiting a central nucleus depends on the radius of the electron's orbit, so this quantization condition implies that only certain orbits are allowed. Bohr used it to calculate the energies of the electrons in these allowed orbits, and so worked out the staircase of energy levels in an atom.

The theory works well with the hydrogen atom, which has only one electron, but it was too simplistic for more complex atoms (◀ page 72). However, it could account for the wavelengths of the observed spectral fingerprints of hydrogen. Moreover, the idea of atomic energy levels allows a relatively simple understanding of physical phenomena that depend on behavior at the subatomic level – for example, the operation of lasers. And it set others on the road to developing mathematical tools for dealing with the quantum world.

Particles as waves

Modern understanding of the quantum nature of the atom includes yet another unusual feature of matter at the subatomic level – the wave nature of particles. In many respects the photon is like a particle of light. It has no mass, but it does have energy, and as the energy of an object is related to its momentum, it is possible to calculate the momentum of a photon. This momentum is equal to Planck's constant divided by the wavelength. In 1924, the French aristocrat and physicist, Louis de Broglie (1892-1987) suggested in his doctoral thesis that this relation could be turned around to define a wavelength for a particle such as an electron. Thus, the wavelength of the particle is equal to Planck's constant divided by its momentum. This proposal revealed for the first time the subtlety of nature that not only makes radiation behave as particles in the subatomic world, but which also makes particles behave as waves. Thus, although the energy levels of the electrons calculated by Bohr are usually depicted as orbits of increasing distance from the nucleus, they can also be seen as waves orbiting the nucleus, their frequency rising with increasing energy.

The wave nature of particles was first shown in 1927 in an experiment by two American physicists Clinton Davisson (1881-1958) and Lester Germer (1896-1971). They directed a beam of electrons at the surface of a crystal of nickel, and detected those which were scattered back from the surface. They found that the pattern of scattered electrons showed a peak at an angle of 50°, and this could be explained only if the regular array of atoms in the surface of the crystal was scattering electron *waves*. Not long afterwards, the British physicist George (G.P.) Thomson (1892-1975) observed a pattern of rings in electrons directed through a thin metal foil: such a pattern is characteristic of diffraction effects, which are well understood in terms of wave motions (◀ page 26). Today the diffraction of neutron beams is also used routinely for the purposes of investigating the structure and properties of materials.

▼ **The wave nature of the electron can be shown by the creation of diffraction patterns similar to those of light (◀ page 46). If a stream of electrons passes through a pinhole, the result is a spray of electrons, rather than a single point hitting the phosphor screen.**

▲ **A scanning electron microscope reveals detail in an image according to the number of electrons that scatter as a fine electron beam scans the specimen. The detail is possible because the wavelength of electrons is much less than that of visible light.**

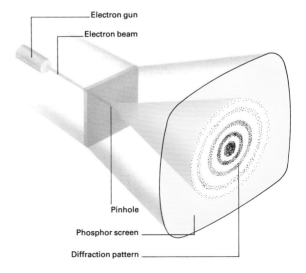

Electron gun
Electron beam
Pinhole
Phosphor screen
Diffraction pattern

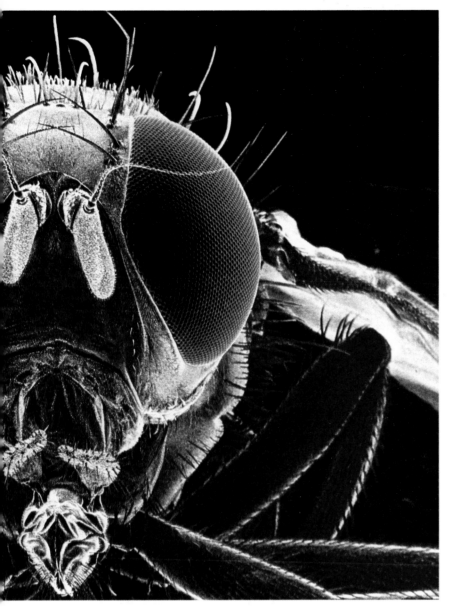

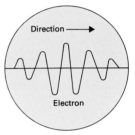

▶ *The particle description of the electron regards it as an object at a point in space and time with a definite velocity. But the concept of an electron as a wave is also valid. In this case the electron is described by a "wave function" which covers an extended region of space. The square of the wave function gives the probability of finding the electron at a given point in space and time. The velocity of this wave "packet" is the same as the velocity in the particle description. But in the wave description the precise position and momentum of the electron cannot be simultaneously determined – this is the uncertainty principle (◊ page 204).*

Bohr and the atom

Niels Bohr had only recently received his doctorate from the University of Copenhagen when he went to work at Cambridge University in 1911. There he encountered the New Zealand physicist Ernest Rutherford (1871-1937) while on a visit to Cambridge, and Bohr was so impressed that he soon moved to Manchester University to work with Rutherford. Rutherford had recently discovered that most of the mass of an atom is concentrated in a central dense nucleus (◊ page 188) and he advanced the concept of the atom as a miniature "solar system" with electrons orbiting the nucleus. One problem with this picture, however, was that according to electromagnetic theory (◊ page 64) the electrons should radiate energy as they whirl round the nucleus, and therefore should spiral inwards towards the center of the atom as they lose energy.

Bohr set about resolving the problem with the aid of the new ideas of quantization. He postulated that an electron does not radiate continuously as "classical" theory demands, but that it emits energy only when it moves from one fixed orbit to another. However, Bohr did not quantize the energy of the electrons in fundamental units. Instead he quantized their angular momentum. In so doing he broadened Planck's concept, which had applied to the energy of radiation, and showed that in the subatomic world other basic quantities can be quantized. He was awarded the Nobel Prize for physics in 1922 for his work on quantum theory, and became one of the most influential physicists of the century, developing the philosophical understanding of quantum theory.

◀ *Davisson and Germer were the first to show the diffraction of electrons and hence their wavelike nature. Davisson is holding the electron tube they used.*

▼ *Classical physics found it hard to explain how electrons flow easily along a metal wire when conducting electricity (◊ page 56). The waveform of an electron moving through the lattice of ions (◊ page 78) in a metal crystal is modified close to the ions. The net effect is to allow the electron to move easily, without colliding many times.*

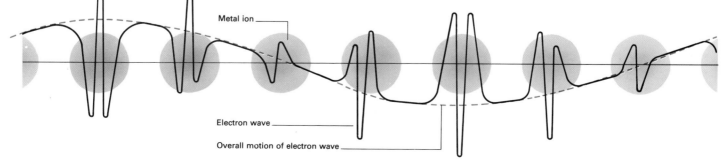

Metal ion

Electron wave

Overall motion of electron wave

Wave mechanics

The discovery of the wave nature of particles provided a deeper understanding of the meaning of Bohr's momentum quantization. The quantization condition allows only those orbits into which a whole number of electron wavelengths fit. With this realization, and the representation of electrons as wave "packets", it is possible to develop a theory of "wave mechanics" that describes not only the structure of atoms, but many aspects of subatomic behavior.

The key feature of wave mechanics is that a particle like an electron is represented mathematically by a "wave function", which is a measure of the probability that the particle is in a given state. In the atom, Bohr's fixed orbits are replaced by "orbitals" – regions of space in which the probability wave of an electron can fit (◀ page 73). Thus the orbital does not give a precise location for an electron; rather it gives the probability that an electron is more likely to be remote from the atomic nucleus, for example, than close to it. This modern quantum theory of the atom may sound very "woolly", yet it provides a surprisingly cast-iron framework for much of what scientists today know about physics, chemistry and even aspects of biology. Quantum theory is here to stay – at least until anyone finds anything better.

Nuclear magnetic resonance imaging

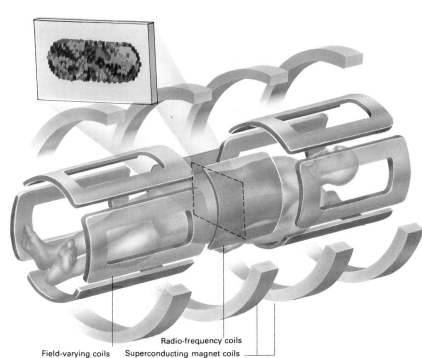

Field-varying coils   Radio-frequency coils   Superconducting magnet coils

◀▲ *To produce an NMR image, such as this image of an adult head, the patient lies within the coils of a big superconducting magnet. Points within the body are selected by varying the strength of the magnetic field in three dimensions using coils above, below and along the magnet's axis. The radiofrequency field supplied by other coils make hydrogen protons resonate.*

◀◀ *This electrostatic potential map of a water molecule is derived using quantum mechanics. It shows the strong negative potential below the oxygen atom, and positive potential around the hydrogens.*

▼ *Protons in hydrogen nuclei behave like tiny magnets, as a result of their spin (1). In a magnetic field, the proton magnets tend to align with the field, but continue to wobble (precess) about the field at a frequency that depends on the strength of the field (2). Pulsing the protons with a radiofrequency field that is oscillating at the precession frequency flips them into the opposite orientation (3). The protons soon flip back to their original orientations and radiate radio signals as they do so. The key to NMR imaging is that the frequency of these signals depends on location, because the field varies.*

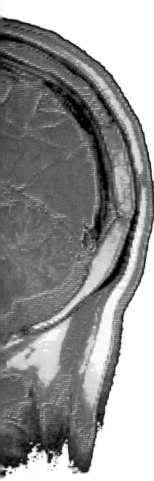

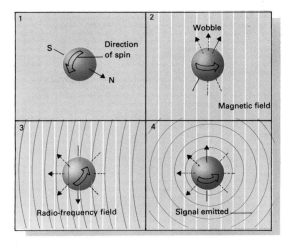

Electron spin and wave mechanics

Planck quantized energy and Bohr quantized angular momentum, and each involved the use of an integer – or "quantum number" – to specify how many basic quanta of energy or angular momentum a system is allowed to possess. Wave mechanics gave meaning to these quantum numbers by showing that they are associated with solutions to the basic wave equation. Indeed, a solution describing a given state of electrons in an atom is associated with several quantum numbers. Thus each energy level corresponding to each line in an atomic spectrum can be defined in terms of a unique set of quantum numbers.

In 1925, two Dutch physicists, Samuel Goudsmit (1902-1978) and George Uhlenbeck (b.1900), postulated another member for this set – a quantum number associated with an intrinsic angular momentum of the electron, in other words a "spin" quantum number. Goudsmit and Uhlenbeck were attempting to explain the fine structure of hydrogen spectra, in which single emission lines are seen on close inspection to consist of two closely-spaced lines. Their solution was to introduce the spin of the electron, which they asserted can exist in only two states, specified by quantum numbers + ½ and -½. This is like saying that a spinning top can spin about the vertical in one of only two ways, either clockwise or anticlockwise. As a shorthand, the electron is said to have a spin of ½. In 1928 the British theorist Paul Dirac (1902-1984) successfully united quantum mechanics with special relativity (♦ page 46) and showed that in the resulting theory a spin of ½ is fundamental to the electron.

The electron is not unique in possessing spin; the protons and neutrons in the nucleus also each have a spin of ½. This means that the nucleus itself can have a net spin angular momentum, depending on how the individual contributions from the protons and neutrons add together. The effect of the quantization of nuclear spin can be detected in atomic spectra as hyperfine structure, or splitting on scales a thousand times smaller than the fine structure splitting due to the electron's spin. (This is related to the large difference between the mass of an electron and the mass of a nucleus.)

The fact that some nuclei, and in particular the proton, spin like children's tops has important applications. A proton (or any nucleus) has a positive charge, so a spinning proton is a moving electric charge and therefore generates a magnetic field (♦ page 61). In other words, protons and any nuclei with net spin are like subatomic magnets. This property is put to good use in the effect known as nuclear magnetic resonance (NMR). In a magnetic field, a proton wobbles about the direction of the field at a specific frequency. The art of NMR is to apply a field and measure this frequency. The technique is important to chemists because the precise frequency depends on the location of the proton – that is, a hydrogen atom – within a particular chemical. NMR is also valuable in medical imaging, because it shows the distribution of hydrogen across the body and can thereby reveal different tissues in a way that is potentially less hazardous than X-ray imaging.

Near absolute zero, superfluid helium defies gravity to escape from a container

Quantum effects at low temperatures

At high temperatures, the average energy of the atoms and molecules is many multiples of frequency times Planck's constant, so the restrictions imposed by quantum theory on the vibrations of an atom about its lattice site in a solid pass unnoticed. At low temperatures the effect is dramatic. It is as if we could use only whole numbers of dollars to make purchases: the effect on the price of an automobile would be minor, but the cost of a newspaper would change drastically.

In 1907, Albert Einstein first applied the ideas of quantization to a theory of the heat capacity of solids. In 1912, the Dutch scientist Peter Debye (1884-1966) developed a simple theory that fitted well experimental data on the variation with temperature of a solid's heat capacity. At high temperatures, the spread in energy of the different atoms means that it is highly probable to find atoms with one or two units of energy more or less than the average value. When the temperature is increased a little, all that happens is that more atoms are to be found with higher energy than before. At very low temperatures the situation is different. When the temperature is raised, most of the atoms remain with the minimum allowable energy – their ground state – and only a few atoms are promoted to higher energy levels. Thus the solid can accept only a small fraction of the heat energy that it would accept at high temperatures.

Long before the introduction of quantization, many of the properties of metals were successfully described by the "free electron theory", in which the conduction electrons are not bound to a particular atom but are free to move about from atom to atom within the metal. However, the theory did not describe the magnetic and thermal properties very well. It was as if the electrons did not contribute to such processes. The major advance in understanding metals came only with the use of the "exclusion principle", due to the Austrian physicist, Wolfgang Pauli (1900-1958). The energies of electrons in a metal are also quantized, but unlike the atoms (or ions) which vibrate independently about fixed lattice sites, the electrons are all part of one system. According to Pauli's exclusion principle, no two electrons can have precisely the same energy. Once the lower energy states are filled up, the electrons must have higher energy to occupy the next available levels.

When the metal is heated, only those electrons in the highest energy level can be excited to the unoccupied higher energy levels; the electrons in the lower energy levels are unaffected because far too much energy would be needed to excite them. It is like an apartment block where the first family lives in the bottom apartment, the second family immediately above, family five on the fifth floor, and so on, with unoccupied appartments being at the very top of the building. If family two or seven wants to change apartment they cannot move one floor; they must go right to the top of the building. Only families already right at the top of the building near to unoccupied levels can move easily. At room temperature only 1 percent of electrons contribute to the thermal properties of metals; no wonder that a problem existed before quantum theory.

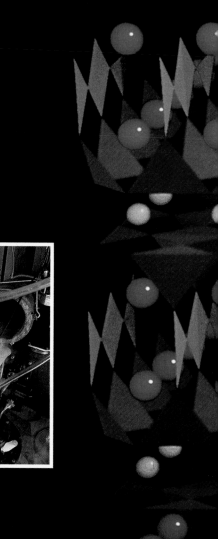

► The structure of a "high temperature" superconductor, $YBa_2Cu_3O_{(7-x)}$, displays its properties at temperatures that can be reached with liquid nitrogen.

▼ In the mid-1980s, research into the high-temperature superconductors found several such materials, offering great hopes for the applications of superconductivity.

▼ A bar magnet lowered onto superconducting tin creates a magnetic image in the dish, and is therefore repelled by the force between like poles, so that it remains levitated above the dish. The field of the magnet induces persistent currents in the tin, which form the magnetic image.

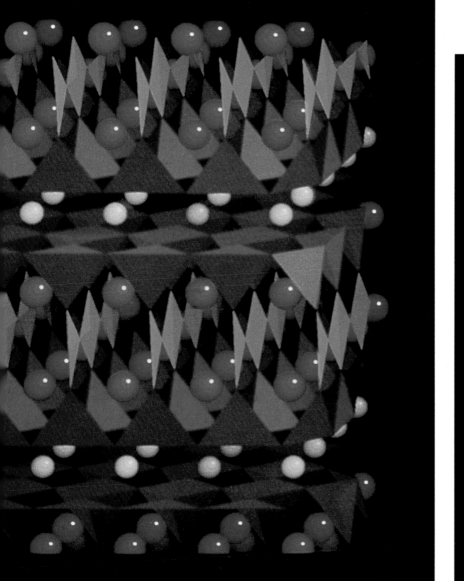

Superconductivity

Following the liquefaction of helium in 1908, the Dutch physicist Heike Kamerlingh Onnes (1853-1926) and other workers began a systematic investigation of the mechanical, electrical and thermal properties of many substances at temperatures a few degrees above absolute zero. Onnes discovered that at temperatures below about 4.25K, the electrical resistance of mercury abruptly disappears. Lead and tin also appear to lose their resistance or become "superconducting".

Onnes was able to show that the resistance falls to zero, not only to some very small amount. He set up a current in a coil of lead in a bath of liquid helium, and then isolated the coil from the battery. The current in the coil produced a magnetic field, which Onnes could monitor with a compass needle. The needle pointed steadily in the same direction, both before and after the battery was disconnected, indicating that the current encountered no resistance at all. In such circumstances, an electric current circulating in a loop will flow forever, so that no energy is required to maintain the current.

How does the resistance fall to zero? It is not

possible to remove the obstacles that normally hinder the drift of electrons. Rather, in a superconductor the obstacles become ineffective; they no longer obstruct. In superconductivity, the drifting electrons become linked together in pairs via the lattice of atoms they travel through. These pairs are in the lowest possible energy state, and cannot lose energy when they meet obstacles. The interaction that creates these pairs is feeble: they are usually destroyed by an increase in temperature or by a high magnetic field. The material then becomes a normal conductor.

Until 1986, the highest temperature at which superconductivity had been observed was at 23.3K, in an alloy of niobium and germanium – although for practical purposes alloys of niobium-titanium, which become superconducting around 9K, have proved most useful. Then, in 1986, two physicists working in Switzerland, Alex Müller and Georg Bednorz, discovered a new class of materials – ceramic metal oxides – some of which are superconducting at much higher temperatures, around 100K. Such temperatures are reached with a refrigeration system that works with nitrogen (boiling point 77K) rather than helium (4.2K).

▲ **Another spectacle from the strange world of low temperatures is the fountain effect – a result of one of the peculiar properties of superfluid helium. In its superfluid state, the liquid acts rather like a superconductor of heat, and flows toward a heat source as if attempting to eliminate it. To produce this effect, a heater in the glass vessel "attracts" superfluid helium which rushes into the vessel below the heater, and forces the jet of liquid through a small hole at the top.**

▲ *The motion of a racing car at full speed is frozen by the camera, but at a price. In panning the camera to track the car, the background has become blurred. Similarly, but more fundamentally, at the quantum level the better motion is known, the more uncertain position becomes.*

▼ *The Austrian physicist Erwin Schrödinger formulated the theory of wave mechanics in the mid-1920s. He was greatly influenced by the work of the French prince, Louis de Broglie (1892-1987), who first proposed that particles such as electrons could be described as waves.*

Heisenberg and uncertainty

In the early 1920s a German, Werner Heisenberg (1901-1976) was wary of taking the idea of electrons in fixed orbits too literally, and began to work on a theory that depended on observable quantities only. The result, published in 1925, was the first theory of quantum mechanics – a mechanics to describe the subatomic world. The theory was based on mathematical structures called matrices. It did not at first reach such wide recognition as Schrödinger's theory of quantum wave mechanics, published the following year. This was more easily understood, but in fact the two approaches are mathematically equivalent and lead to the same answers. Heisenberg was rewarded with the 1932 Nobel Prize for physics; the prize for the following year went to Schrödinger.

Two years after discovering quantum mechanics, Heisenberg went on to postulate his famous "uncertainty principle", in 1927. This states that certain pairs of physical quantities, in particular position/momentum and energy/time, can never both be known exactly. Indeed, the better one is measured, the more uncertain the other becomes. Imagine, for example, trying to measure the location of an electron with a powerful microscope that uses a beam of very short-wavelength gamma rays. According to quantum theory the beam consists of a stream of photons, which can reveal the electron by bouncing off it up the microscope. But each photon that bounces off the electron will change its momentum, just as in subatomic billiards. Indeed, the more accurately we try to pinpoint the electron, using gamma rays of shorter and shorter wavelength, the more momentum the individual photons give to the electron because their energy increases with decreasing wavelength.

Schrödinger and his cat

The Austrian physicist, Erwin Schrödinger (1887-1961) never liked the "dual" picture of matter that behaves both as waves and particles, and he tried to construct a theory based on waves only. He encapsulated some of his concerns about quantum theory in his "cat paradox". Schrödinger imagined a cat locked in a box with a phial of poison and a radioactive nucleus, and he supposed that the phial would be broken by the decay of the nucleus, in which case the cat would be killed. The question is, at any given time can anyone say what is the state of the cat – whether it is dead or alive – without opening the lid? The quantum physicist can only calculate the probability that the nucleus has decayed (◆ page 190). Only when the lid is opened and an observation made, is the answer known.

So what is the state of the cat (representing a quantum system) between observations? Does it really exist in this interim period? And, to take an extreme view, does it exist only in the mind's eye of the physicist, even when being observed? Such questions tax the minds of philosophers and physicists alike. One view is that the present theory of quantum mechanics is not complete; that there are "hidden variables" of which physicists are as yet unaware but which describe the quantum system in a fully deterministic way. Another view is that all possible states between observations really do exist. In this "many universes" interpretation of quantum theory, the Universe splits each time there are different possible outcomes at a quantum level. Thus the real Universe consists of a host of separate universes, but each of us is conscious of only one at a time. Each time we observe we proceed at random to the universe corresponding to only one of the possible outcomes.

Building blocks of the nucleons...Quarks and antiquarks...Leptons...Neutrinos...How elementary particles are made...PERSPECTIVE...Naming the particles...Identifying particles...Particle accelerators and detectors...Antimatter...Cosmic rays...Predicting new types of particle

Since the beginning of the 20th century the picture of the fundamental nature of matter has changed dramatically. In the closing years of the previous century there were already signs that atoms could not be the indivisible, immutable objects that had once been imagined. In 1897 the British physicist Joseph John (J.J.) Thomson (1856-1940) discovered the electron, a tiny fragment of atomic matter about one two-thousandth the mass of an atom of hydrogen, the lightest element. Then with the discovery of radioactivity, physicists found that atoms could spit out other fragments – namely alpha particles (that is, helium nuclei) and so transform from one element into another. In 1911, Thomson's former student Ernest Rutherford (1871-1937) and his colleagues at Manchester University in Britain discovered that most of the mass of an atom is concentrated in a tiny central nucleus, which by the 1930s was known to consist of protons and neutrons.

Today the electron is still regarded as one of a small number of truly fundamental particles. As far as experiments can show, it behaves like a "point" of matter, with no internal structure, even down to distances as small as 10^{-19}m, one ten-thousandth the size of a proton. With protons and neutrons, however, the story is very different. Experiments in which energetic beams of electrons (and protons) are scattered by thin "targets" of various materials show that protons and neutrons are extended in space, having a definite diameter of some 10^{-15}m. More detailed experiments, using electron beams of a high enough energy to probe within the protons and neutrons themselves, show that they contain an internal structure of pointlike objects. These "points" are particles called quarks and, like electrons, they appear to be truly fundamental.

Elementary particles

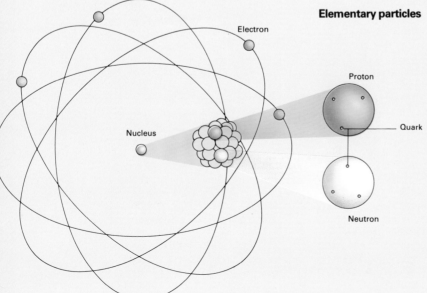

Naming particles
The profusion of subatomic particles discovered since the late 1940s seems often to have been named with scant regard for logic. Yet beneath the bizarre terminology lies some sense. The first subatomic particle to be discovered – the electron – was given the name the British physicist George Stoney (1826-1911) had proposed several years previously for the "unit" of electric charge lost or gained by atoms during electrolysis. The proton is named for the Greek for "first", as it was the first component of the atomic nucleus to be identified. "Neutron", for the proton's neutral partner, then seemed a logical addition to the list.

With the discovery of a number of unpredicted particles by the early 1950s, Greek letters were introduced to label the different varieties. In the case of those initially labeled pi (π) and mu (μ), the particles have become known as the pion and mucn.

In 1962 two physicists, Murray Gell-Mann (b.1929) and George Zweig (b.1937) independently proposed that many of the growing number of subatomic particles, including the proton and neutron, must be composed of more fundamental objects. The protons and neutrons would each contain three of these objects, and Gell-Mann chose the name "quark", which occurs in the phrase, "Three quarks for Mister Mark", in the work "Finnegan's Wake" by James Joyce. Zweig chose the name "aces", but it was Gell-Mann's choice that stood the test of time.

Probing matter

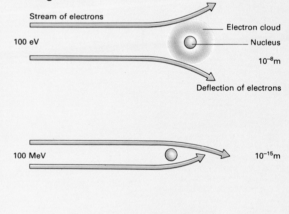

▲ *Beams of electrons of increasing energy can be used to probe matter on smaller and smaller scales. The electrons, themselves electrically charged, are deviated, or scattered, by charged structures. At energies of 100 electronvolts (eV) or more, the electrons are scattered by the cloud of electrons around an atom. At energies a million times greater than this (100 MeV or more), the electrons penetrate this cloud but come under the influence of the positive nucleus at the heart of the atom. At energies of billions of electronvolts (GeV), the electrons scatter from the tiny charged quarks within the larger protons and neutrons.*

Seeing Inside the Atom

Accelerators

To see inside the atom requires techniques that can probe distances of 10^{-15}m and smaller. This is far beyond the capabilities of the most sophisticated microscopes, but the principles used are similar. An optical microscope uses light scattered from an object, but its resolution (or ability to reveal detail) is limited by the wavelength of visible light, around 10^{-7}m. An electron microscope forms images from energetic electrons scattered from an object. The higher the energy of the electrons, the shorter the wavelength associated with them, according to quantum theory.

Studying matter on scales of 10^{-15}m or less requires very high energy beams of electrons, or other kinds of particle, produced at machines known as particle accelerators. Again, information about structure at these scales comes from observing the ways in which objects – in this case, subatomic particles within the nucleus – scatter the high energy beam. Although physicists do not reconstruct "images", the patterns of the scattered particles can be detected in specialized equipment. Moreover, the high energies can often reveal new, heavier (or more energetic) states of matter which cannot be observed at low energies.

Most particle accelerators have a format in which the beam travels in a circle through an evacuated pipe. Magnetic fields keep the particles on the correct path as they orbit the machine thousands of times. On each circuit, the particles pick up small amounts of energy from radio waves set up in sections of the machine known as the accelerating cavities. As the particles increase in energy, so do the magnetic fields guiding them until eventually the particles reach the maximum energy that the magnetic ring can contain. The beam is then directed out of the accelerator towards awaiting experiments, or to targets where it may be used to produce secondary beams of particles such as pions, muons or neutrinos.

► **In this photograph of particle tracks in a bubble chamber colors have been added. Negative particles (kaons) cross the picture from below. Their tracks curve to the right showing that the chamber's magnetic field bends negative particles (purple, pale blue and green) to the right, positive ones (orange and red) to the left. The two spirals must be due to electrons, the only particles light enough to curl so much in the magnetic field. The tiny spiral is from an electron knocked from an atom in the liquid; the large spiral comes from the decay of the particle that made the pale blue track – this must have been a muon. The gap between the two "V"-prongs shows that a neutral particle, which could not leave a track, has been produced at the first "V" and decayed at the second "V".**

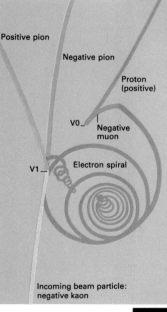

Positive pion

Negative pion

Proton (positive)

V0

Negative muon

V1

Electron spiral

Incoming beam particle: negative kaon

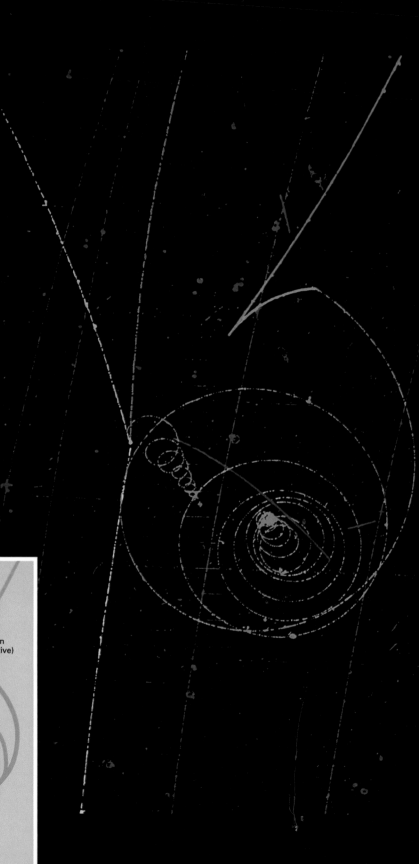

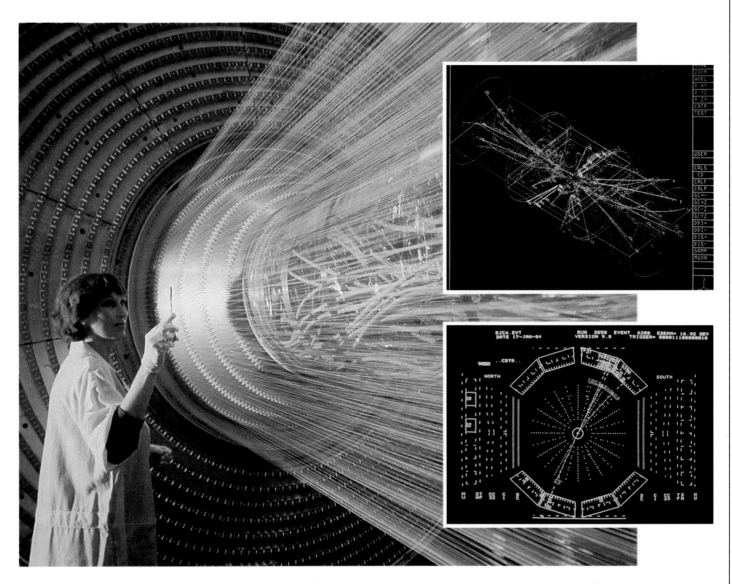

Bubble chambers

To work out what happens when a high energy particle interacts with a proton or a neutron within a nucleus, experimenters must record the tracks of the scattered particles. These often include many new particles created in the energy of the collision. Electrically-charged particles are relatively easy to detect. These leave trails of ionized atoms as they move through matter, and these trails can be revealed in a variety of ways. Moreover, charged particles are influenced by magnetic fields and the curvature of their paths through magnetic fields gives clues to their charge and momentum.

One device for tracking particles is the bubble chamber. This is a vessel containing a liquid under pressure, close to its boiling point. If the pressure is reduced, the boiling point should also drop. But if the pressure change is rapid enough, the liquid becomes superheated (remains liquid above its boiling point). If the pressure change occurs just after a beam of particles has passed through, however, the liquid is unstable enough to begin to boil along the path of the ionized particle tracks. Tiny bubbles form, revealing the paths of the charged particles, which can be photographed.

Wire chambers

Physicists using a bubble chamber must measure photographs of the tracks and feed the relevant numbers to a computer in order to calculate back to what happened in the original collision. Devices called wire chambers can eliminate one stage in this process by producing electrical signals that can be recorded directly in a form appropriate to a computer. In a wire chamber, charged particles leave ionized trails through a volume of gas, splitting the atoms into electrons and positive ions. Wires spread uniformly throughout the chamber are made electrically positive so that the negatively charged electrons are attracted towards the nearest wire. As the electrons approach the wire, they knock additional electrons out of the gas, until eventually an avalanche of electrons arrives at the wire, inducing a sizeable electrical signal. By piecing together the pattern of signals from a whole array of wires, a computer can reconstruct the paths of the original charged particles that passed through the wire chamber. If the chamber is within the coils of a large electromagnet the curvature of the tracks allows experimenters to measure the charge and momentum of the particles.

▲ In wire chambers, thousands of wires cross the space traversed by the particles, as in this drift chamber (left) seen during construction. When complete, the wires in such a device are contained in a gas-filled vessel. Charged particles ionize the gas, releasing electrons which travel toward the wires under the influence of an electric field, and produce signals in circuits attached to the wires. Computers then reconstruct the track of each particle and display them in images. The top picture shows a three-dimensional view of tracks in a cylindrical drift chamber. Below is a cross-sectional display from a different experiment; this not only shows tracks in a drift chamber (red lines) but also "hits" in other types of detector which help in identifying the different particles.

Antimatter is created along with matter in equal amounts, but on meeting again matter and antimatter annihilate

The neutrons and protons in stable nuclei can live forever (or at least, for an extremely long time; ◗ page 216). However, the energetic collisions of protons and nuclei reveal other subatomic particles. These are similar in many ways to the proton and neutron, but they are short-lived and cannot form part of the atoms and matter of the familiar world. These particles are also built from quarks, and analysis of the relationship between them has shown that there are probably at least six types of quark in all. The six are called "up", "down", "charm", "strange", "top" and "bottom" – although evidence for the top quark is still scant.

Particles such as the neutron and proton are built from three quarks, and are known collectively as "baryons". The "up" quarks and "down" quarks are the lightest varieties, and so together they form the lightest known baryons – the proton and neutron. The other four kinds of quark – strange, charm, bottom and top – are successively heavier and they form heavier particles. The neutral particle called the lambda contains an up quark, a down quark and a strange quark. It is like a neutron but with a heavier strange quark replacing one of the down quarks. This makes the lambda about seven percent heavier than the neutron. It is also unstable. After a life of only 2×10^{-10} seconds it decays, usually to a proton and a particle called the pion, or to a neutron and a pion. In each case, the strange quark transmutes into a lighter quark and the difference in mass is taken away by the pion that has been produced at the same time. This transmutation of quarks is the same basic process that underlies the decay of a neutron to a proton in beta-radioactivity (◗ page 191). In beta decays, a down quark in the neutron transmutes into an up quark, thus forming a proton. Such changes between quarks bring all the heavier baryons down to the level of the proton, the lightest baryon of all.

There are also six antiquarks which form antibaryons such as the antiproton and antineutron. The lightest antiquarks are readily created in the collisions of protons and nuclei, in cosmic rays as well as in particle accelerators. However, the creation of antibaryons is quite costly in terms of energy, so the antiquarks pair up more easily with quarks created at the same time out of the available energy. The resulting particles, built from a quark and an antiquark, are called mesons, meaning "between", since the first known examples had masses between that of an electron and that of a proton.

In charged mesons the quark and antiquark are of different types. For example, the positively charged version of the pi-meson, or pion, is built from an up quark and a down antiquark. Its antiparticle, which has negative charge, is built from a down quark and an up antiquark. Neutral mesons, on the other hand, can be built from a quark and the antiquark of the same variety. In this case the quark and antiquark temporarily orbit around each other rather like an electron and a proton in a hydrogen atom. However, such neutral mesons are shorter-lived than their charged counterparts. The neutral pion is a mixture (at the quantum level) of up-antiup and down-antidown. The mesons are all short-lived and ultimately decay to lighter particles. The quarks are held together within baryons and mesons by the strong nuclear force – the strongest of nature's four fundamental forces (◗ page 214). It is due to the nature of the strong force that the quarks can form only clusters of three quarks (baryons) and quark-antiquark pairs (mesons). Other combinations are forbidden, including the possibility of single quarks. In practice, therefore, the quarks and antiquarks appear always to be confined within baryons and mesons.

Antimatter

One of the main discoveries of modern physics concerns the existence of antiparticles. These are particles that have the same mass as the familiar electron, proton and neutron and the other more exotic species, yet which have exactly opposite properties such as electric charge. Thus the antielectron, or positron, has positive charge, while the antiproton has negative electric charge.

Antiparticles were postulated in a theory due to the British physicist Paul Dirac (1902-1984). In 1928 he combined quantum theory (◗ page 195) with the special theory of relativity (◗ page 46), but found that his equation describing the electron had two possible solutions. One described the familiar electron, while the other turned out to describe the antielectron, or positron. This interpretation did not become clear until 1932 when physicists studying cosmic radiation discovered tracks due to positively charged electronlike particles in detectors called cloud chambers. These particles occurred in "showers" with equal numbers of electrons. The conclusion was that electron-positron pairs were materializing from gamma rays generated as the cosmic rays passed through sheets of metal.

Dirac's theory also revealed the fate suffered by electrons and positrons that come too close to each other. They mutually self-destruct in a reversal of the process that creates electron-positron pairs, releasing an amount of energy equivalent to their total mass. This process is known as annihilation.

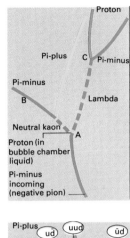

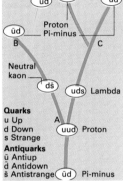

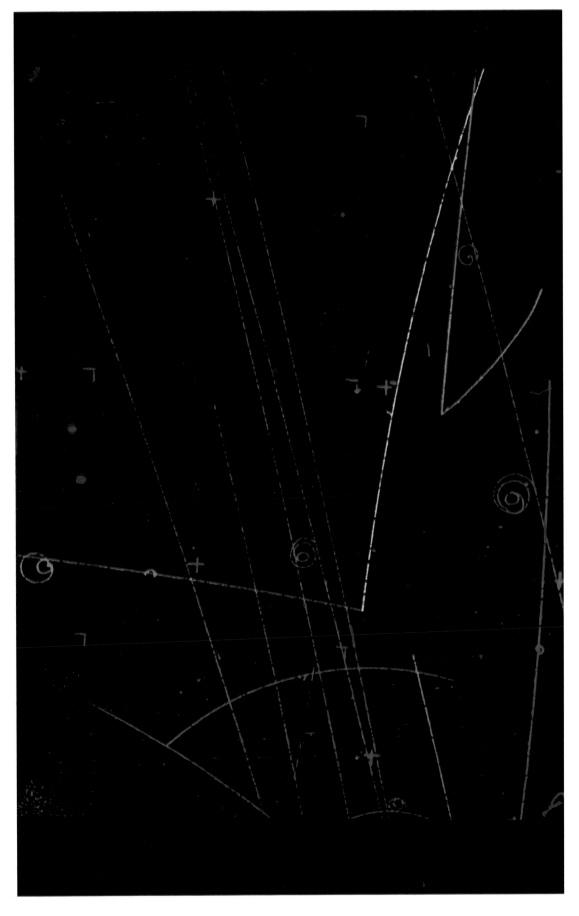

▲ ▶ *A bubble chamber photograph shows the "associated production" of strange quarks and antiquarks. The long tracks are due to pions crossing the chamber; one, colored green, interacts with a proton and produces two neutral particles which leave no tracks (A). One contains a strange quark, the other a strange antiquark. The overall "strangeness" is thus zero both before and after the reaction. This is a property of reactions that occur via the strong nuclear force. The two particles betray themselves when they decay (B, C) via the weak nuclear force, and in this case strangeness can change. In the decay of the lambda particle, for example, the heavier strange quark gives birth to an up quark, a down quark and an up antiquark (C).*

◀ *An antiproton leaves a track in a bubble chamber (pale blue) before eventually annihilating with a proton. Their properties exactly cancel, creating a burst of pure energy. This rematerializes in the form of entirely different particles and antiparticles: four positive pions (red) and four negative pions (green).*

Only two quarks, one charged lepton and one neutral lepton are required to explain the everyday world

Elementary particles

Baryons		Quarks	Mass (MeV)	Lifetime (s)
Proton	p	uud	938.3	$>10^{32}y$
Neutron	n	udd	939.6	898
Lambda	Λ	uds	1115.6	2.6×10^{-10}
Sigma	Σ^+	uus	1189.4	0.8×10^{-10}
	Σ^0	uds	1192.5	5.8×10^{-20}
	Σ^-	dds	1197.3	1.5×10^{-10}
Xi	Ξ^0	uss	1314.9	2.9×10^{-10}
(cascade)	Ξ^-	dss	1321.3	1.6×10^{-10}
	Λ_c^+	udc	2282.0	2.3×10^{-13}

Quarks		Charge	Mass (MeV)
Up	u	$+\frac{2}{3}$	5
Down	d	$-\frac{1}{3}$	7
Charm	c	$+\frac{2}{3}$	1400
Strange	s	$-\frac{1}{3}$	150
Top	t	$+\frac{2}{3}$	?
Bottom	b	$-\frac{1}{3}$	1,800

Mesons			Mass (MeV)	
Pion	$\pi^\pm$	$u\bar{d},\bar{u}d$	139.6	2.6×10^{-8}
	π^0	$(u\bar{u},d\bar{d})$	135.0	0.8×10^{-16}
Kaon	$K^\pm$	$u\bar{s},s\bar{u}$	493.7	1.2×10^{-8}
	K_S^0	$(d\bar{s},s\bar{d})$	497.7	0.9×10^{-10}
	K_L^0			5.2×10^{-8}
	$D^\pm$	$c\bar{d},\bar{c}d$	1869.4	9.2×10^{-13}
	D^0	$c\bar{u}$	1864.7	4.4×10^{-13}
	$F^\pm$	$c\bar{s},\bar{c}s$	1971	1.9×10^{-13}
	$B^\pm$	$u\bar{b},\bar{u}b$	5270.8	14×10^{-13}
	B^0	$b\bar{d}$	5274.2	

$\bar{d}, \bar{u}, \bar{s}, \bar{c}, \bar{b}, -$ Antiquarks

Leptons		Charge	Mass (MeV)	Lifetimes (s)
Electron	e	−1	0.51	stable
Electron-neutrino	ν_e	0	$<0.46\times10^{-4}$	stable
Muon	μ	−1	105.6	2.197×10^{-6}
Muon-neutrino	ν_μ	0	<5.0	stable
Tau	τ	−1	1,784	3.4×10^{-13}
Tau-neutrino	ν_τ	0	<164	stable

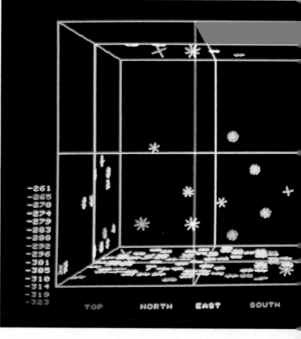

The quarks are only one type of fundamental particle. The electron is clearly not built from quarks. Instead it belongs to a second family of fundamental particles – the leptons. This name comes from the Greek for slight, for at one time all leptons seemed to be lightweight. Like the family of quarks, the family of leptons contains six members. There are three negatively-charged members: the electron, the muon and the tau. And there are three neutral members which are called neutrinos, and which seem to have little or no mass. However, neutrinos are clearly of distinct types. In particle interactions one sort will produce only electrons, one sort will produce only muons, and the third variety should give rise only to taus, although this has still to be proved. They are referred to as the electron-neutrino, the muon-neutrino and the tau-neutrino. As with the quarks, the six leptons all have corresponding antiparticles, which are positively charged in the case of the antielectron (positron), antimuon and antitau.

The electron and the three types of neutrino seem to be stable, but the muon and tau can both decay to lighter particles. The muon, which has only about 11 percent the mass of the proton, can decay only to an electron, because no other charged particles lighter than the muon exist. When the muon decays, it also gives rise to a muon-neutrino and an electron-antineutrino. The tau's mass is nearly double

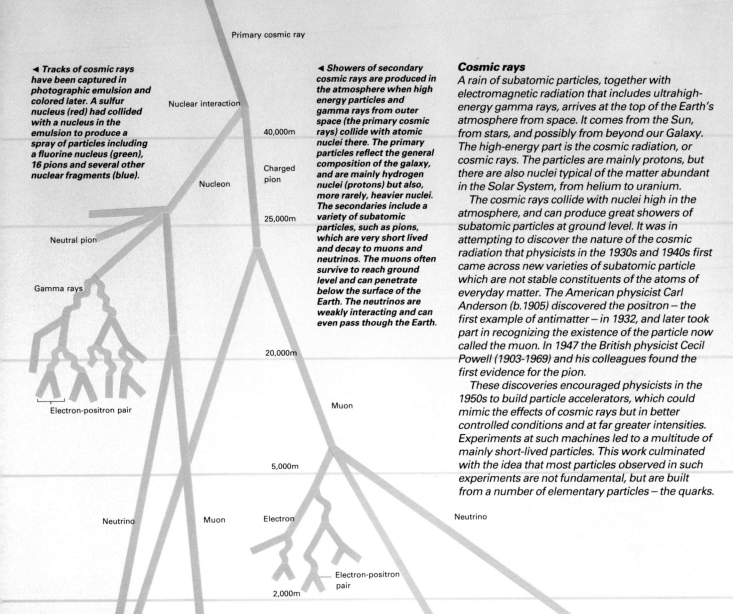

◄ Tracks of cosmic rays have been captured in photographic emulsion and colored later. A sulfur nucleus (red) had collided with a nucleus in the emulsion to produce a spray of particles including a fluorine nucleus (green), 16 pions and several other nuclear fragments (blue).

Primary cosmic ray

Nuclear interaction

40,000m

Charged pion

25,000m

Nucleon

Neutral pion

Gamma rays

20,000m

Electron-positron pair

Muon

5,000m

Neutrino   Muon   Electron

Neutrino

2,000m

Electron-positron pair

Ground level

◄ Showers of secondary cosmic rays are produced in the atmosphere when high energy particles and gamma rays from outer space (the primary cosmic rays) collide with atomic nuclei there. The primary particles reflect the general composition of the galaxy, and are mainly hydrogen nuclei (protons) but also, more rarely, heavier nuclei. The secondaries include a variety of subatomic particles, such as pions, which are very short lived and decay to muons and neutrinos. The muons often survive to reach ground level and can penetrate below the surface of the Earth. The neutrinos are weakly interacting and can even pass though the Earth.

Cosmic rays

A rain of subatomic particles, together with electromagnetic radiation that includes ultrahigh-energy gamma rays, arrives at the top of the Earth's atmosphere from space. It comes from the Sun, from stars, and possibly from beyond our Galaxy. The high-energy part is the cosmic radiation, or cosmic rays. The particles are mainly protons, but there are also nuclei typical of the matter abundant in the Solar System, from helium to uranium.

The cosmic rays collide with nuclei high in the atmosphere, and can produce great showers of subatomic particles at ground level. It was in attempting to discover the nature of the cosmic radiation that physicists in the 1930s and 1940s first came across new varieties of subatomic particle which are not stable constituents of the atoms of everyday matter. The American physicist Carl Anderson (b.1905) discovered the positron – the first example of antimatter – in 1932, and later took part in recognizing the existence of the particle now called the muon. In 1947 the British physicist Cecil Powell (1903-1969) and his colleagues found the first evidence for the pion.

These discoveries encouraged physicists in the 1950s to build particle accelerators, which could mimic the effects of cosmic rays but in better controlled conditions and at far greater intensities. Experiments at such machines led to a multitude of mainly short-lived particles. This work culminated with the idea that most particles observed in such experiments are not fundamental, but are built from a number of elementary particles – the quarks.

◄ This display records the collision of a cosmic ray neutrino in a detector consisting of 10,000 tonnes of ultrapure water. The tank, in a mine 600m below ground, has walls lined with light-sensitive phototubes. This view shows the phototubes that fired with the light from particles created when the neutrino collided in the water.

that of the proton, so it has access to many different decay routes. It can even decay to particles containing quarks, provided that they together contain equal numbers of quarks and antiquarks.

There seems to be a natural symmetry between the two families of fundamental particles. Indeed, the resemblance runs even deeper because each family can be divided into three "generations". The matter of the everyday world is built from up quarks and down quarks (in atomic nuclei) and electrons. In addition, electron-type neutrinos and antineutrinos are emitted in the radioactive decays of some unstable nuclei. Thus only two quarks, one charged lepton and one neutral lepton are required to explain the everyday world.

However, studies of cosmic rays and at accelerators reveal muons and their neutrinos, and strange and charmed quarks. Once again this group of four particles contains two quarks, a charged lepton and a neutral lepton. Indeed, this "second generation" of quarks and leptons seems to mirror the first almost precisely, but in a higher-energy world. The quarks and the charged lepton of the second generation are all heavier than their counterparts in everyday matter. Lastly, since the mid-1970s, experiments at higher energies have suggested a third generation of still heavier quarks and leptons. This should contain the bottom and top quarks, the tau and its neutrino.

Free quarks and monopoles

The quarks have electric charges that are fractions of the size of the electron's charge. The up quark has a positive charge of ⅔ the size of the electron's (negative) charge, while the down quark has a negative charge of ⅓ the electron's charge. If single quarks exist, this "fractional" charge should reveal them as other particles have zero or integer charge.

There have been no convincing examples of particles with fractional charge. One intriguing result came from United States physicist William Fairbank (b.1917), who found evidence for charges of ⅓ located on tiny balls of niobium. He levitated the balls magnetically, and then made them oscillate in an alternating electric field. By measuring the amount of oscillation he found that on some occasions some balls carry a charge of ⅓. Many physicists remain skeptical of these results.

In 1982 Blas Cabrera (b.1946) claimed evidence for the monopole. Monopoles are particles that carry a single magnetic charge or pole. Though predicted by Dirac's theory of 1931, there is no experimental evidence that they exist. Theory suggests that monopoles were formed in great numbers during the Big Bang. Some of these "relics" may still be around. As time passes and new experiments fail to find any new evidence, the existence of these particles remains in doubt.

▼ **William Fairbank at Stanford University has observed indications of fractional charge when he levitates tiny spheres of niobium between two plates like the one he holds.**

► **Blas Cabrera, in his basement laboratory at Stanford University, together with an improved, three-coil version of his monopole detector. He has recorded one event in 1982.**

Antimatter such as positrons occurs naturally in the showers of cosmic rays that cascade through the Earth's atmosphere. Antiprotons and antineutrinos can also be created. But the everyday world is manifestly built from electrons, protons and neutrons, and it seems that the nearby region of the Universe must all be built from matter rather than antimatter, or there would be evidence of large releases of energy every time pieces of matter and antimatter encountered each other. However, in the collisions of cosmic rays and in experiments, particles and antiparticles are always created in equal numbers. So why is there no evidence of large amounts of antimatter in the Universe?

Some theories of the forces at work in the very early Universe suggest that matter and antimatter were indeed created in equal amounts in the hot Big Bang with which the Universe began. However, subtle asymmetries in the behavior of the fundamental forces between particles (◆ page 213) tipped the balance in favor of matter. While most of the matter and antimatter annihilated, a small excess of matter remained to form the Universe that exists today.

Why should matter have replicated the fundamental particles at least three times, revealing only one set in the physical reactions of life on Earth? The answer to this basic question may be tied up in an understanding of the evolution of the Universe. In the Big Bang, the heavier generations would have existed on equal terms with the familiar quarks and leptons of the first generation. The total number of generations may have had an influence on how the Universe developed. One of the challenges to particle physicists is to search for evidence of additional generations of the quarks and leptons.

Fundamental Forces

28

The forces of the Universe: Strong nuclear force, weak nuclear force, electromagnetic force and gravitational force...Bosons, particles carrying the forces... PERSPECTIVE...Einstein's theory of general relativity ...Gravitational lenses...Grand Unified Theories and the Theory of Everything...Proton decay experiments

Quantum electrodynamics

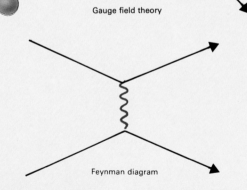

Electric field around a point charge

Electric field between two similarly charged points

Electron

Boson

Gauge field theory

Feynman diagram

One surprising feature of the physical world is that much of its great diversity can be described in terms of only two fundamental forces – the gravitational force and the electromagnetic force. Gravity controls the motions of the stars and planets, and will even determine the fate of the Universe as it expands from its initial Big Bang. The electromagnetic force keeps electrons tied to the nuclei of atoms, and governs the motion of electrons in all kinds of systems, ranging from the "silicon chip" to the human nervous system. However, there are two additional fundamental forces, which are less apparent because they operate only within the confines of the atomic nucleus. These are the weak nuclear force and the strong nuclear force.

Modern theories of these four forces are all generally based on the mathematical framework known as a "gauge field theory". Each force is regarded in terms of a field describing the strength and direction of the force throughout space and time. (The term "gauge" is related to the concept that measurements at different locations in space and time should always give the same results). An important feature of these theories is that there should exist particles which in a sense "carry" the force. These are called gauge bosons. The term "boson" refers to the fact that these particles have integer values of the intrinsic angular momentum, or spin ($\blacklozenge$ page 201). This makes the gauge particles fundamentally different from the particles of matter – the quarks and leptons – all of which have a spin value of $\frac{1}{2}$.

The best developed gauge theory, called quantum electrodynamics or QED, deals with the electromagnetic force. It accounts for the behavior of charged particles even down to the level of the fundamental quarks and leptons ($\blacklozenge$ page 206). In QED the source of the electromagnetic force is electric charge; its gauge bosons are the familiar photons of light. Charged particles interact by exchanging photons between themselves, rather as in a game of quantum football.

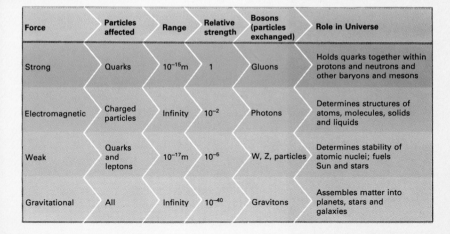

Force	Particles affected	Range	Relative strength	Bosons (particles exchanged)	Role in Universe
Strong	Quarks	10^{-15}m	1	Gluons	Holds quarks together within protons and neutrons and other baryons and mesons
Electromagnetic	Charged particles	Infinity	10^{-2}	Photons	Determines structures of atoms, molecules, solids and liquids
Weak	Quarks and leptons	10^{-17}m	10^{-5}	W, Z, particles	Determines stability of atomic nuclei; fuels Sun and stars
Gravitational	All	Infinity	10^{-40}	Gravitons	Assembles matter into planets, stars and galaxies

◀ *Physicists recognize four basic forces that appear to underlie all phenomena, from atoms to the workings of the Universe. Each has its own exchange particle.*

▲ *At the quantum level, the electric field between charged objects is mediated by an exchange particle – the photon – and depicted in Feynman diagrams.*

The pion was predicted to exist in theory 12 years before it was actually detected

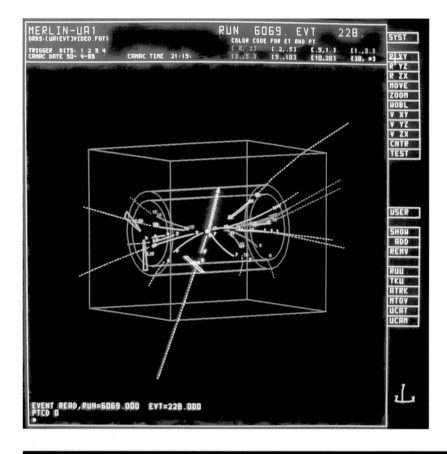

Pions and the strong force

The first viable theory of the strong nuclear force was put forward in 1935 by the Japanese theorist Hideki Yukawa (1907-1981). The electromagnetic force was already understood in terms of the exchange of photons at the quantum level and Yukawa considered a similar mechanism for the strong force. He argued that, as the strong force acts only over very small distances, it must be carried by a heavy particle. (The electromagnetic force, with infinite range, is by contrast carried by a particle of zero mass – the photon). Yukawa proposed that the carrier of the strong force must be a particle with a wavelength equal to about 10^{-15}m, and that the mass of such a particle is about 15 percent of the mass of the proton. He proposed that such particles are responsible for binding protons and neutrons together as they are constantly exchanged.

Yukawa's particle – now known as the pion – was discovered in 1947. His picture of pion exchange is still useful; however, the strong force is now known to originate in the quarks within the protons and neutrons. The binding of the protons and neutrons in a nucleus is due to a leakage of the strong interquark force.

◀ **A proton and an antiproton have collided at the center of an electronic detector, producing some low energy particles (yellow and red tracks) and a W particle. The W has decayed to an electron (long blue track) and a neutrino (arrowed).**

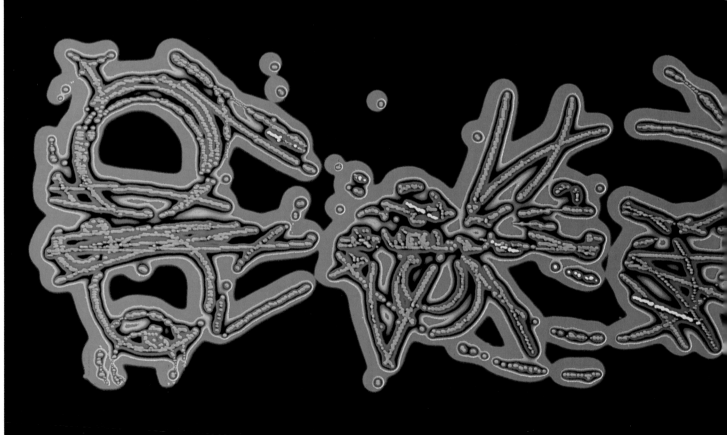

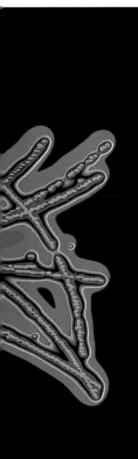

▲ *The lightweight particles called pions are produced in abundance in energetic nuclear collisons. Here a high energy proton (yellow) collides with one at rest in the liquid in a bubble chamber. The collision produces as many as seven negative pions (blue) and nine positive particles (red) which include seven positive pions.*

◀ *This photographically colored image shows the decay of a Z particle – the neutral carrier of the weak nuclear force and partner of the W particle. Again, a proton and an antiproton have collided at the center of the image, producing a melée of tracks. Two of the tracks (yellow) are particularly straight. They belong to an energetic electron and positron from the Z's decay.*

The weak and strong nuclear forces

During the 1960s QED was incorporated into a larger theory in an attempt to describe the electromagnetic force and the weak nuclear force at one and the same time. The successful outcome is now known as the electroweak theory. The weak nuclear force is responsible for the decay of particles such as the neutron and hence underlies beta-radioactivity (◀ page 190) and was recognized as distinct from the strong force that binds the nucleus in the early 1930s. It is carried by three gauge bosons – a neutral particle called Z, a positive particle called W^+ and a negative particle called W^-.

Unlike the photon, which has zero mass, these particles are all heavy, being 90-100 times as massive as the proton. It is the large mass of these particles that makes the weak force weak. In exchanging a gauge boson, matter particles (quarks and leptons) must temporarily violate the principle of energy conservation (◀ page 22). This can occur at the quantum level through the uncertainty principle (◀ page 204), but it is allowed for only a short time – so short that the weak force acts only across distances the size of the nucleus or less, and with a strength some 100,000 times weaker than the electromagnetic force within ordinary matter.

"Colored" particles

The strong force acts within the same domain as the weak force, but it is very different in its behavior. It is the strongest of nature's fundamental forces, and within the nucleus it dominates the electromagnetic force by a factor of more than 100. The gauge theory of the strong force is called quantum chromodynamics, or QCD. According to this theory, the source of the strong force is a property called "color" (this has, however, nothing to do with the color of light), and to carry the strong force there are eight gauge bosons called "gluons", which act between "colored" particles. Quarks carry the property of color, but leptons do not. This explains why baryons and mesons (which contain quarks) interact strongly with each other, while leptons (which do not contain quarks), do not interact through the strong force with baryons, for example.

Color can be thought of as similar to electric charge, but it differs in that it occurs in three different varieties – called red, blue and green after the primary colors of light – where electric charge occurs only as positive and negative. An important rule of QCD is that quarks can form only those clusters that have a total "color" of zero – or white in the analogy with real colors. Thus only clusters of three quarks (baryons) and quark-antiquark pairs (mesons) exist in the world at large. A baryon, for instance, must contain a red quark, a blue quark and a green quark; it cannot, say, contain two red quarks and a green quark. Moreover, single quarks, which are by definition colored objects, cannot exist alone – and indeed there is no convincing evidence for single quarks, despite many searches (◀ page 212).

The rule about color is tied in with the nature of the gluons, the gauge bosons of QCD. A gluon, like a photon, has no mass, but they differ in a crucial way. The photon has no electric charge: it is not a source of the electromagnetic force that it conveys. Gluons, on the other hand, are colored, and as a result they can interact among themselves via the strong force. This has the profound effect of making the strong force stronger at greater distances within a particle, and also of confining the colored quarks within the dimensions of the subatomic particles that they form.

Gravity and the general theory of relativity

Newton's theory of gravity can be used to calculate motions above the Earth, from the path of a ball to the orbit of a satellite. It also reveals how the gravity of the Moon and the Sun produces tides on Earth. However, despite its many successes, Newton's law of gravitation has failed in some small but crucial areas.

Mercury is the closest planet to the Sun. Its orbit is an ellipse, and the closest point of approach to the Sun on this orbit is called the "perihelion". In successive orbits, this point advances slightly around the orbit. The rate of advance is very slow – about 0.159° per century. The gravitational forces from the other planets cause about 0.147° of this advance, but Newton's laws cannot explain the remaining 0.012°.

The explanation for the discrepancy did not come until more than 200 years after Newton, with the work of the German physicist Albert Einstein (1879-1955). Einstein is famed for his special theory of relativity, one of the basic assumptions of which is that no information can be transmitted faster than the speed of light (◀ page 46). According to Newton, however, gravitational forces act instantaneously throughout all space. The two theories disagree.

These problems stimulated Einstein to look again at the theory of gravity. In 1907 the Russian mathematician Hermann Minkowski (1864-1909) combined the three dimensions of space with the dimension of time, and proposed that all events in the Universe occur in a four-dimensional "space-time".

Einstein adopted Minkowski's idea of space-time, and proposed that space-time is curved in the presence of massive bodies. All particles, including light, travel on the shortest route (a geodesic), and where space-time is curved these routes are curved. According to Einstein, mass produces a curved space-time which makes particles move as if attracted. This has been summarized as, "Matter tells space how to curve and space tells matter how to move".

Einstein's theory is known as the general theory of relativity. The German theorist Karl Schwarzschild (1873-1916) solved Einstein's equations to discover the shape of space-time in the presence of a massive spherical object. His solution can be used to reveal the planetary orbits around the Sun. For most practical purposes Einstein's orbits are identical to those of Newton, but Einstein's theory predicts the extra 0.012° advance in the perihelion of Mercury, which Newton's theory was unable to explain.

Another prediction of Einstein's theory proved a crucial test. Light always travels by the quickest path. This means that, in the curved space-time near the Sun, light should be deflected by 0.0049°. Newtonian gravitation predicts a deflection of half this. A way to measure the deflection is to observe a star whose light passes close to the Sun. This is possible only when there is a total solar eclipse. In 1919, the British astronomer Arthur Eddington (1882-1944) organized an expedition to Principe, an island in the Gulf of Guinea, and to Brazil, to measure the deflection of light during the total eclipse of the Sun that year. The results, first announced at the Royal Society in London, were in complete support of Einstein's theory.

Three hundred years after Newton, gravity is still not properly understood. Physicists in many countries continue to work on gravity to understand what causes it, and how it is transmitted. Some physicists are trying to merge gravity with the quantum theories of the subatomic world, in an attempt to show that all Nature's forces are only different manifestations of a single underlying force.

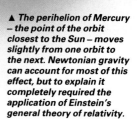

▲ The perihelion of Mercury – the point of the orbit closest to the Sun – moves slightly from one orbit to the next. Newtonian gravity can account for most of this effect, but to explain it completely required the application of Einstein's general theory of relativity.

Sun

Orbit

Mercury

Perihelion

Quasar

Light path

Gravitational lens

Galaxy

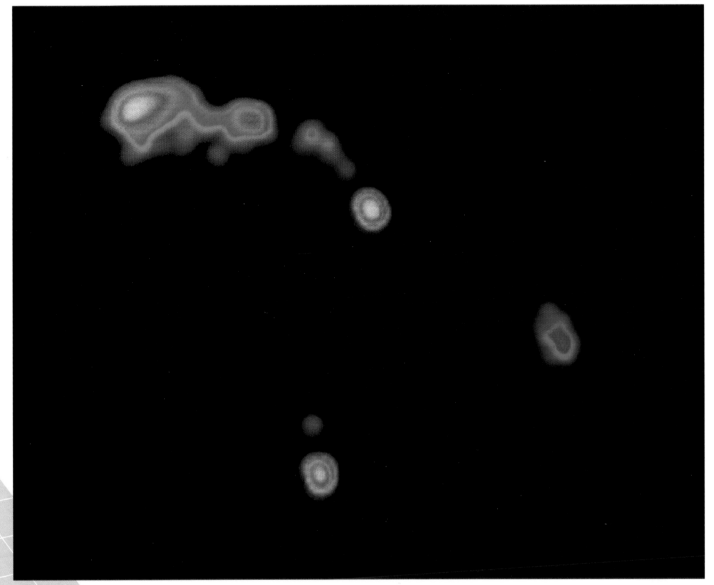

◀▲ A large mass warps the space around it so that light passing the object does not travel in a straight line, but follows a bent path rather as when traveling through a lens. The effect of such a "gravitational lens" is believed to have been observed in the "double quasar", 0957+561. In the radio telescope image above, the two bright white blobs correspond to objects that both appear to be quasars at identical distances, 10,000 million light years away; it seems likely therefore that they are two images of the same object. The upper image is thought to be the direct view of the quasar, while the lower is presumed due to light bent around an intervening galaxy.

Earth

Superstrings

The dream of many theorists is to find a "theory of everything" (TOE) that incorporates gravity along with the strong and electroweak forces. But so far gravity has defied a complete description at the quantum level. There is no equivalent of QED for gravity. It is possible, however, to use general relativity to hypothesize the properties of a gauge boson for gravity. This is the graviton, and it differs from the photons, gluons, and W and Z particles in that it has spin 2; the others have spin 1. However, while there is good evidence for the other gauge bosons, there is none at all for the graviton.

Most attempts to build a quantum theory of gravity have proved inconsistent in a fundamental way, but in 1984 a British theorist, Michael Green, and his United States colleague John Schwarz, discovered a class of theories that appear to avoid these inconsistencies. The theories are types of "superstring" theory, in which the fundamental particles – or quanta – are represented as tiny string-like objects. This contrasts with the "point-like" representations of conventional quantum field theories. The strings are about 10^{-35}m long, a dimension that is far smaller than can be studied with the most powerful particle accelerators. But on the larger scale at which physicists can make observations, the strings appear like points, just as in the conventional theories.

Superstring theories contain a symmetry called supersymmetry – hence the name superstrings. Supersymmetry directly relates bosons (force-carrying particles with spins 0,1,2,...) to fermions ("matter" particles such as quarks and leptons, with spins ½, ³⁄₂,...). This symmetry has an elegance that is attractive to theorists because it unites these apparently disparate sets of particles together; it also seems to be a crucial ingredient of superstring theories.

Superstring theories appear at first to have the drawback that they work only in 10 dimensions – that is, they describe a universe with six dimensions over and above the familiar three dimensions of space and one of time. But this proves not to be a problem, for the idea of extra dimensions is not new. Theorists know how to make the unwanted dimensions disappear by "curling up" so that they are not apparent in the everyday world.

The remarkable discovery of Green and Schwarz in 1984 was that two – and only two – specific types of superstring theory appear to be properly consistent. The inclusion of gravity in these theories does not have the devastating effect it has in other theories. Whether superstrings prove to be the key to the ultimate theory of everything remains to be seen, but they offer one of the most exciting advances in theoretical physics for many years.

▲ ▶ *One test of the GUT is that it should be possible to detect the occasional decay of protons. An experiment to do this involves a cave 600m below Lake Erie, Ohio, filled with water and edged by strings of phototubes. If a proton in the water decayed, it would register on these tubes.*

▼ *The four fundamental forces eventually combine to a single force at ever higher energy levels.*

Grand unified theories

The electroweak theory "unifies" two of nature's four fundamental forces. It has inspired theorists to incorporate it with the gauge theory of the strong force – QCD – to form a single "grand unified theory" (GUT) that would embrace the strong, weak and electromagnetic forces. Such theories generally link quarks and leptons in a way that implies that quarks can decay into leptons, albeit very rarely in the everyday world, at which energies are generally low. One important implication is that protons should not be stable, but have an average lifetime of some 10^{32} years or more. As this is an average it should be possible to monitor a large enough collection of protons and observe a few decays in the space of a year or so. Such experiments, involving hundreds of tonnes of water or some other material, are underway in many parts of the world. However, none of them has yet found any convincing evidence for the predicted decays of protons.

Another important feature of unified theories is that they contain an inherent symmetry between the forces they unite. Electroweak theory predicts that, at energies higher than those typical in atomic nuclei, the weak force becomes stronger until it eventually equals the electromagnetic force in strength. At these higher energies the W and Z particles can be exchanged as readily as photons, and the two forces appear as symmetric facets of a single electroweak force. These ideas have been verified in high-energy experiments at particle accelerators, where W and Z particles have been produced and observed. In a similar way, grand unified theories predict that at a very high energy – far beyond the reach of present accelerators – the strong and electroweak forces become symmetric.

Ideally, theorists would like to include gravity in a single "theory of everything" (TOE). In such a theory, all four forces would be symmetric facets of a single underlying force. Only in the extreme heat of the Big Bang at the start of the Universe would energies have been high enough for the four forces to appear the same. As the Universe expanded and cooled, each of the forces would have separated out, first gravity, then the strong force, and finally – at energies now reached in particle accelerators – the weak and electromagnetic forces parted company. The original symmetry of the Universe would have become hidden in the diversity apparent in its cool, low energy state of today.

The Grand Unified Theory

Weak

Electromagnetic

Strong

Gravitational

Electroweak

Grand unification

Unity

| Energy (GeV) | 10^2 | 10^4 | 10^6 | 10^8 | 10^{10} | 10^{12} | 10^{14} | 10^{16} | 10^{18} | 10^{20} |

Radiation and Radioactivity

29

Alpha, beta and gamma radiation...Stopping radiation...Natural sources of radiation...Effects of radiation on the body...Using radiation to see inside the body...PERSPECTIVE...Pierre and Marie Curie...The discovery of X-rays...Units of radiation and dosage... Predicting the effects of low-level exposure

Radiation is all around. It rains down from outer space, and leaks out of the ground and from the walls of buildings. It is almost totally natural. Only 13 percent of the radiation dose that people receive in Britain, for example, comes from artificial sources, and most of that is in medical uses. Much less comes from the burning of coal, the testing of nuclear weapons, and only 0·1 percent of the total comes from radioactive wastes and effluents. But what is radiation?

The term radiation can refer to electromagnetic waves (from gamma rays to radio waves: ◀ pages 64-65) or to energetic subatomic particles. However, in common use and especially in the context of radiation protection, the term refers more specifically to ionizing radiations – in other words, radiations that lose energy by ionizing atoms (knocking electrons out of them) or sometimes even by disrupting atomic nuclei. Such radiation can be electromagnetic or it can consist of particles. The particle component includes the alpha rays (helium nuclei: two protons and two neutrons) and beta rays (electrons) emitted in radioactive decays (◀ page 191), as well as the showers of cosmic-ray particles generated in the atmosphere by "primary" cosmic rays from outer space. Strictly speaking only charged particles can ionize, because the basic reaction is due to the electrical force between the charged particle and the atomic electrons. But some neutral particles, in particular neutrons, and the X-rays and gamma rays at the high-energy end of the electromagnetic spectrum, are indirectly ionizing, because they produce charged particles when they interact with matter.

Radiation

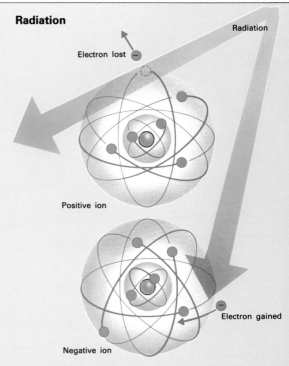

Electron lost

Radiation

Positive ion

Electron gained

Negative ion

▲ *Radiation causes damage when it knocks electrons out of atoms (left), creating positive ions. The electrons can then attach to other neutral atoms and create negative ions.*

The Curie family
The French physicist Henri Becquerel (1852-1908) discovered radioactivity in 1896, when he found that radiation from a uranium salt would darken a photographic plate kept well away from light. He was soon able to show that uranium was the element responsible for the emissions. However, it fell to two other scientists working in Paris, Marie Curie (1867-1934) and her husband Pierre Curie (1859-1906), to show that other elements could produce similar rays. In 1898 Marie found that thorium emitted the same kind of radiation, and she also discovered that uranium ore (pitchblende) emitted more radiation than pure uranium. Together the Curies laboriously purified kilogram after kilogram of pitchblende, and in 1899 they announced the discovery of two new elements, which they called polonium, after Marie's native Poland, and radium. They also gave the effect the name still in use today – radioactivity.

Pierre Curie was tragically killed in a road accident in 1906, but Marie continued to do much important work on radioactivity, identifying beta-radiation as a stream of negatively charged particles. In 1903, Marie and Pierre shared the Nobel Prize for physics with Henri Becquerel for their work on radioactivity. Marie later became one of the few people to receive two Nobel Prizes, when she was awarded the prize for chemistry in 1911, for her discovery of radium and polonium.

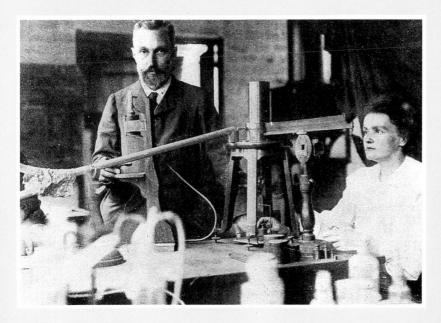

◀ *Pierre and Marie Curie in their laboratory in 1898 during their early work on radium. The instrument here is a quartz electrometer, which they used to measure the amount of ionization in air caused by various radioactive samples.*

The value of X-rays in imaging the inside of the body was recognized almost as soon as X-rays were discovered

▶ Alpha particles are heavily ionizing – their tracks in a cloud chamber (a type of particle detector) are thick and short. Beta rays (center), which consist of energetic electrons, are poorly ionizing and leave thin tracks. Gamma rays (right) are non-ionizing and leave no tracks, but give rise to pairs of electrons and positrons (◀ page 208), which are ionizing.

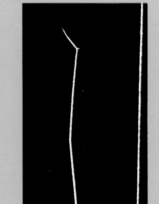
Alpha

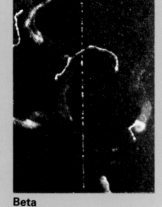
Beta

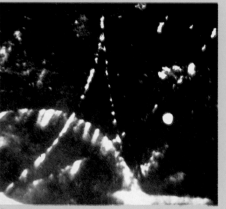
Gamma

▲▶ Röntgen's discovery of X-rays in 1895 captured the world's imagination. Their ability to "see" inside things was appreciated as a valuable medical aid, but the harmful effects of even quite low-level exposure was less clear.

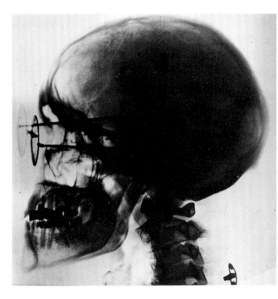

Ionizing radiation

The effects of ionizing radiation on materials vary greatly with the type of radiation. The higher the charge of a particle, the more rapidly it ionizes. Alpha particles with the same velocity as protons lose energy four times as rapidly as the protons because they have double the charge of the protons. And the lower a charged particle's velocity, the greater its rate of ionization. As a charged particle loses energy it slows down and the energy it loses through ionization increases, until suddenly the particle no longer has enough energy to ionize an atom. The particle becomes neutralized and comes to a halt; it is said to have come to the end of its range.

Alpha particles from radium-226, for example, have enough energy to travel about 20 micrometers in aluminum. Beta particles (electrons), on the other hand, are far more penetrating. This is because for the same energy an electron, which has only 0·01 percent the mass of an alpha particle, has a much higher *velocity*, and energy losses are inversely proportional to the square of the velocity. An electron from the beta decay of carbon-14 travels about 100 micrometers through aluminum. If it had the same energy as the alphas from the radium-226 it would travel as far as one centimeter.

X-rays and gamma rays can also penetrate materials. Indeed, X-rays

Röntgen and X-rays

On 8 November 1895, Wilhelm Röntgen (1845-1923), professor of physics at the University of Würzburg, was studying the passage of electricity through a gas at low pressure in a cathode-ray tube. This was a glass tube with a positive electrode (a metal wire) at one end and a negative electrode at the other. As the gas was pumped out of the tube, the residual gas would glow, and eventually the wall of the tube opposite the cathode (the negative electrode) would glow green. Scientists had been investigating these phenomena for several years, but on this particular evening Röntgen noticed for the first time that a nearby screen was fluorescing, even though the tube was covered in black card. Invisible rays emanating from the cathode-ray tube were causing salts of barium platinocyanide to fluoresce. Over the next few weeks Röntgen discovered that the rays would ionize the air, would darken photographic plates, and, most remarkably, would pass through a variety of materials that are opaque to light. He announced his discovery of X-rays, as he called them, at the beginning of 1896, and immediately the rays captured the imaginations of scientists and the public alike.

The X-rays in Röntgen's experiment originated at the place where the "cathode-rays" within the tube struck the glass wall opposite the cathode, making it glow green. Two years later, in 1897, the British physicist J.J. Thomson (1856-1940) showed the cathode rays to be streams of electrons, emitted from the cathode (◀ page 72). What Röntgen was observing was the emission of high-energy electromagnetic radiation – X-rays – from atomic electrons in the glass wall, which had been raised to higher energy levels by the cathode rays and were now returning to their usual energies.

Röntgen received the first Nobel Prize for physics in 1901. By this time X-rays were well known, and tubes were made commercially to produce the rays for medical use, where they were especially useful for observing fractured bones and metal objects lodged inside the body. Today, medical X-ray imaging can be a highly sophisticated technique. X-rays are also used in radiotherapy at far greater intensities to damage tissue deliberately in the treatment of cancerous growths.

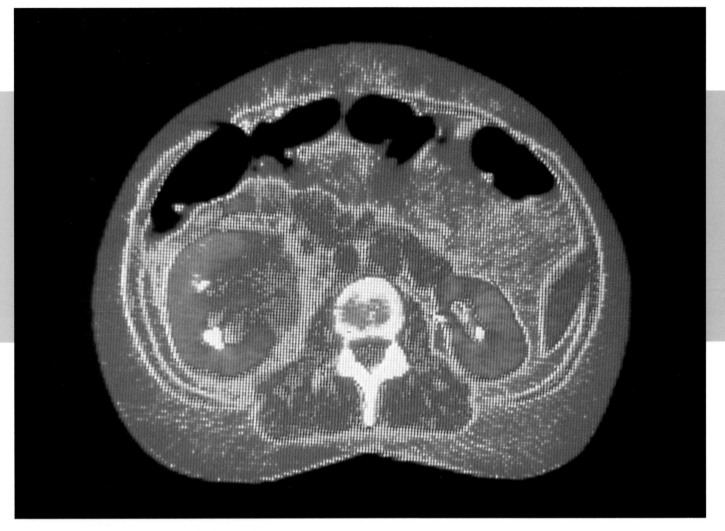

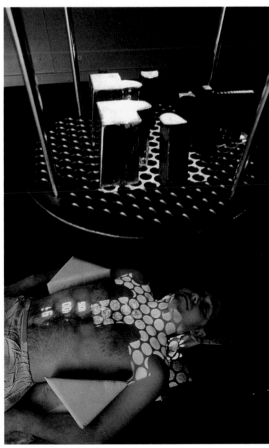

◄ X-rays damage body tissues by knocking energetic electrons from atoms. This effect is put to good use in directing X-rays to kill tumor cells. Here a patient lies beneath lead blocks that define the area that X-rays will reach.

▲ Simple X-ray images, like the one shown on the opposite page, record the net effect of absorption by all the organs and tissues between the source of X-rays and the photographic plate. The more refined technique of computer assisted tomography (CAT) allows images to be formed of "slices" through the body, as in this example showing a section through the trunk – the spinal cord appears as white near the bottom of the image. CAT scanners work by rotating the X-ray tube around the body to define the "slice".

are well known for this ability. The energies of X-rays are generally typical of the energies between electron shells in atoms (◀ page 75), and an X-ray is absorbed when its energy is sufficient to eject an electron from its shell. This electron may receive enough to leave the atom entirely and to ionize other atoms in the vicinity. In this way, the X-ray has the same general ionizing effect as a charged particle. The probability that an X-ray is absorbed varies approximately with the fourth power of a material's atomic number. Thus, lightweight elements, such as make up skin and muscle, transmit X-rays more easily than the heavier elements, as are found in bone or metals.

Gamma rays are of too high an energy to be absorbed easily by atomic electrons. They do, however, lose energy in collisions with the electrons. The discovery of this effect by the United States physicist Arthur Compton (1892-1962) played a part in establishing the particle-like nature of electromagnetic radiation (◀ page 196). These collisions can give the electrons enough energy to ionize. The collisions also reduce the gamma ray's energy, so that eventually the gamma ray becomes absorbed like an X-ray, but only after penetrating much farther than an X-ray can. Gamma rays from radioactive decays can pass through 10 centimeters of aluminum, a far greater distance than the alpha particles or even the electrons from radioactivity.

Most of the radioactive substances contained in the early Earth have long since decayed away completely

The uranium and thorium that occur in the ground and in the materials used for buildings provide the greatest natural exposure to ionizing radiations, along with potassium-40. This isotope has a half-life of 1·3 billion years, but forms only 0·1 percent of all potassium. However, it emits gamma rays of relatively high energy, and it is the gamma radiation from radioactive materials that has the greatest effect on humans. Alpha particles, for example, cannot penetrate the dead outer layers of human skin whereas gamma rays can penetrate much farther. The uranium and thorium decay chains and potassium-40 are also responsible for most of the radioactivity taken internally in food and drink. Within the body, of course, alpha particles and beta particles produce greater effects. Here the alpha particles from polonium-214, one of the members of the uranium-238 decay chain, have a dominant effect, along with the gammas of potassium-40. However, the biggest effect of building materials lies in the

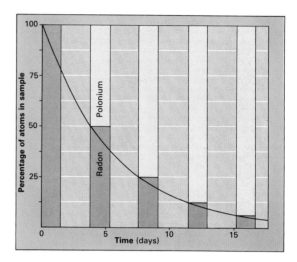

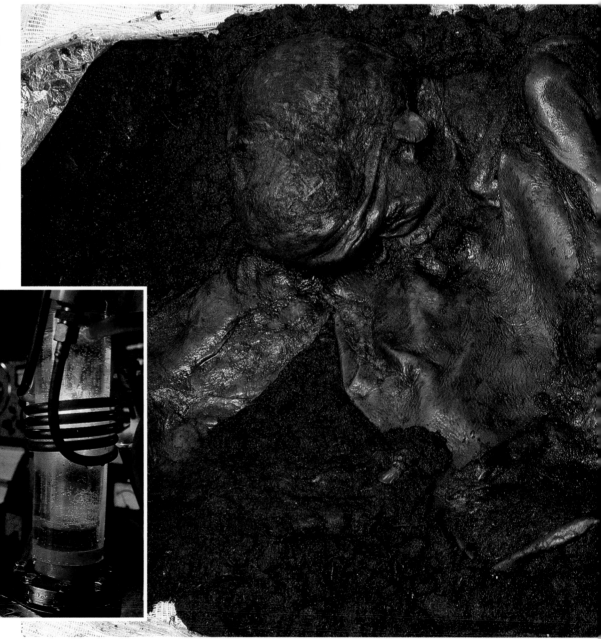

▶ This body, well-preserved in the bogs of central England, has been radiocarbon dated to about AD 200.

▼ Radiocarbon dating is used to determine the age of an organic substance by measuring the proportion of C-14 in a sample. The sample is cleaned and then oxidized in a sealed chamber, so that the C-14 is converted to radioactive CO_2. This gas is then placed in a counting chamber, comprising a mass spectrometer (◆ page 71), for about 20 hours. During this time a sample 5,000 years old – the oldest that can be dated by this method – might register a count of 22,000 C-14 atoms.

◄ *Radioactive isotopes are characterized by their halflife – the time for half the nuclei in a sample to decay. The graph here shows how a sample of the isotope radon-222 decreases as it decays to polonium-218, through the emission of alpha particles. The halflife of radon-222 is 3·8 days, so after 3·8 days half the original amount is left; 7·6 days (two halflives) later, a quarter of the radon remains; after 11·4 days (three halflives), an eighth of the sample is left, and so on. The halflives of radioactive isotopes vary from fractions of a second to billions of years.*

► *The Geiger counter measures levels of radioactivity – here it is being used to check the activity of produce at a market after the explosion of the nuclear reactor at Chernobyl. Invented in 1928 by the German physicist Hans Geiger and his colleague Walther Müller, the counter consists of a gas-filled tube with a wire along its axis. When ionizing radiation passes through the gas, it liberates electrons which, under the influence of a high electric field in the tube, set off avalanches of electrons that are picked up by the central wire, and produce a signal.*

radioactive gases they emit, for these can be inhaled, and inhalation like ingestion makes radiation more effective in ionizing critical parts of the human body. The main gas is radon, and its major isotope is radon-222, which again occurs in the decay chain of uranium-238. People living in areas where granite, for example, is common, are exposed to higher amounts of radiation from radon-222 in their houses than people living in other areas.

Natural radiation comes both from cosmic rays and from natural radioactivity. The best-known naturally-occurring radioactive element is uranium. Some 99 percent of uranium is in the form of uranium-238, the isotope (◀ page 70) that combines 146 neutrons with the standard 92 protons. It emits alpha particles, and it was in an experiment involving uranium salts that radioactivity was first discovered (◀ page 187).

Radioactivity transmutes the atomic nuclei of an isotope of one element to those of an isotope of another element. Once all the nuclei of the original isotope have "decayed" in this way, that isotope no longer exists. The time this process takes varies from one isotope to another, because it depends on subtle balances of the changes in mass and energy involved in the transition. The process is also fundamentally random. The nuclei of a radioactive isotope do not all decay simultaneously, nor is it possible to say which nucleus is going to decay next. One nucleus may decay quickly, another may take a long time to decay. The rate of decay of a particular isotope is therefore characterized by a statistical quantity called the halflife. This is the time it takes for half the nuclei in a given sample to decay. The halflives of the elements vary from fractions of a second for some isotopes, to millions of years for others and this provides a clue as to which isotopes occur naturally. At the time the Earth was formed some 4.5 billion years ago, it could have contained many radioactive isotopes. Those with halflives that are short compared with the age of the Earth will have long since decayed away completely. Only a few remain to this day, including uranium-238, with its halflife of 4·51 billion years. However, the continuing decays of uranium-238 give rise to nuclei that are themselves radioactive on shorter timescales. Indeed, uranium-238 is the start of a chain of decays by both alpha and beta emission. The chain passes through a dozen or so isotopes of different elements and stops eventually at a stable form of lead – lead-206. This chain includes radium-226, a far more powerful emitter of alphas than uranium.

The harmful effects of radiation are put to good use in the treatment of cancers

The units of radiation and radioactivity

The basic measure of radioactivity is the rate at which a substance decays. Historically the unit of radioactivity was defined in relation to the activity of one gram of radium, the radioactive element that Pierre and Marie Curie discovered. For many years the basic unit was the curie (Ci), defined as 3.7×10^{10} decays per second, approximately the rate from 1g of radium. Now the "unit of activity" is the becquerel (Bq), named for the discoverer of radioactivity, and defined as one decay per second.

The number of decays says something about the amount that there is of a particular radioactive substance, such as 1g of radium, or the relative activity of different substances, but it says nothing about the ionizing effects of the radiations produced. For gamma rays, and also for X-rays (which are not produced by radioactivity but which are akin to gamma rays), a unit is defined to quantify the ionizing effect of the radiation in air. This is called the "unit of exposure" and it is a measure of the amount of X- or gamma radiation that produces enough ionization to create a total electric charge of one sign (negative or positive) of 1 coulomb in 1kg of air. Units called röntgens are often used to express exposure. One röntgen is equal to 2.58×10^{-4} coulombs per kilogram. These units reveal in a sense the amount of X-rays or gamma rays to which something or someone is exposed. A different unit is used to quantify the amount of radiation actually absorbed by a material. The "unit of absorbed dose" is called a gray (Gy), and it is defined as a deposited energy of 1 joule per kilogram. This is the energy the radiation gives up in ionizing the material. (The related unit, the rad, is equal to 0.01Gy.)

For the purposes of calculating the amount of radiation that is safe for a human to receive, none of these units is satisfactory. This is because the different kinds of radiation have different effects on human tissues, and the importance of the effect depends critically on the part of the body that receives it. Radiobiologists work in "units of dose equivalent", called sieverts (Sv). These units are grays multiplied by a quality factor which takes into account the type of radiation and the long-term risks it has. Gamma rays and electrons have a quality factor of 1; for protons it is in the range 1 to 2, depending on the energy of the particles; for neutrons it ranges up to 10 or so for the most energetic particles; and for alpha particles and heavier ions, assuming that they are inhaled or ingested (because otherwise skin and clothing prevent any damage), the quality factor is as high as 20. Typical annual doses from natural radioactivity total in the region of 1 millisievert (one thousandth of a sievert). A lethal dose, which would kill within a month, is over 100,000 times greater.

▲ ▶ **High levels of radiation may not kill straight away, but "early effects" (occurring within a few weeks) include vomiting, secondary infections, and blistering or reddening of the skin (erythema). Many victims of the nuclear bomb dropped on Hiroshima during World War II showed such effects. "Late effects" refer mainly to an increased incidence of cancers appearing up to 30 years or more after exposure to radiation. Certain organs are more susceptible than others to the radioactive isotopes present in nuclear fallout.**

▼ **Different types of radiation have different potential for damaging tissue. Alpha particles are heavily ionizing and a few particles can deposit a significant amount of energy, as measured by the absorbed dose in grays (J/kg). A gamma ray, on the other hand, does not directly ionize atoms, but releases a few electrons which are only weakly ionizing.**

Harmful isotopes

Iodine[131]
The body easily absorbs iodine either through the digestive system or through the lungs. The iodine quickly travels to the thyroid gland and can stay there several months. Radioactive iodine[131] can be blocked from the thyroid by taking stable iodine pills.

Cesium[134/137]
Muscles are the tissues that accumulate cesium and retain it for many months. Radioactive cesium[137] is particularly problematic, with a halflife of 30 years. Both cesium[137] and cesium[134] enter the food chain via vegetation grown on contaminated soil.

Strontium[90]
Bone cancer is one of the likely results of the absorption of too much strontium[90]. The strontium at first replaces calcium atoms near the surface of bones, but eventually it also leads to damage to the bone marrow itself, causing leukemia.

Carbon[14]
Carbon[14], used for radiocarbon dating, is easily taken into the food chain in the form of carbon dioxide gas. Its halflife is very long – 5,730 years – but fortunately it is quickly incorporated into carbon dioxide in the body, and is exhaled.

Effects of ionizing radiation

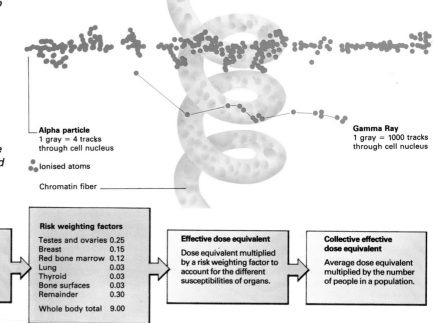

Alpha particle
1 gray = 4 tracks
through cell nucleus

Ionised atoms

Chromatin fiber

Gamma Ray
1 gray = 1000 tracks
through cell nucleus

Calculating radioactive dosage

Absorbed dose	Dose equivalent	Risk weighting factors	Effective dose equivalent	Collective effective dose equivalent
A measure of the energy that radiation deposits in a material in the process of ionizing the atoms.	Absorbed dose multiplied by a quality factor to show variations in damage by different kinds of radiation.	Testes and ovaries 0.25 Breast 0.15 Red bone marrow 0.12 Lung 0.03 Thyroid 0.03 Bone surfaces 0.03 Remainder 0.30 Whole body total 9.00	Dose equivalent multiplied by a risk weighting factor to account for the different susceptibilities of organs.	Average dose equivalent multiplied by the number of people in a population.

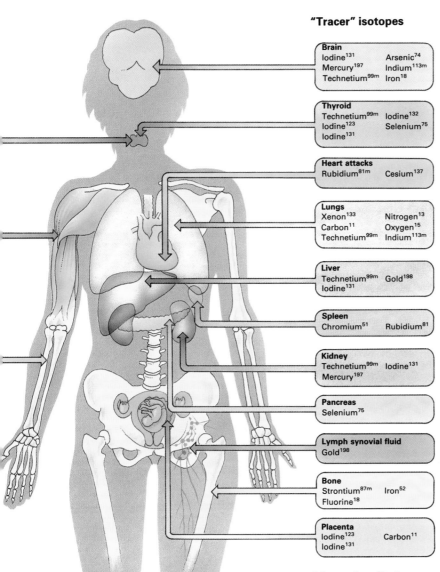

"Tracer" isotopes

Brain
Iodine[131]   Arsenic[74]
Mercury[197]   Indium[113m]
Technetium[99m]   Iron[18]

Thyroid
Technetium[99m]   Iodine[132]
Iodine[123]   Selenium[75]
Iodine[131]

Heart attacks
Rubidium[81m]   Cesium[137]

Lungs
Xenon[133]   Nitrogen[13]
Carbon[11]   Oxygen[15]
Technetium[99m]   Indium[113m]

Liver
Technetium[99m]   Gold[198]
Iodine[131]

Spleen
Chromium[51]   Rubidium[81]

Kidney
Technetium[99m]   Iodine[131]
Mercury[197]

Pancreas
Selenium[75]

Lymph synovial fluid
Gold[198]

Bone
Strontium[87m]   Iron[52]
Fluorine[18]

Placenta
Iodine[123]   Carbon[11]
Iodine[131]

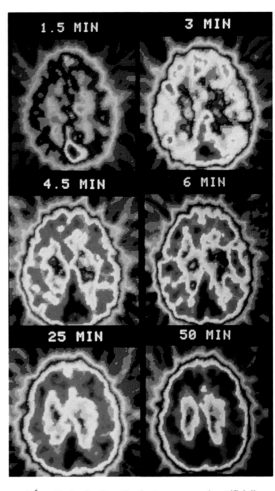

◀▲ *Many kinds of radioactive isotope are made artificially for use in medicine. The various substances home in on different tissues and their radiations can be used either to produce images of organs or to damage tumor cells, in radiotherapy. The scans show views of the human brain revealed by the absorption of radioactively-traced glucose.*

Biological effects

Ionizing radiations can be harmful to biological tissues; the more radiation a tissue is exposed to, the greater the likelihood of damage and the greater its severity. The damage to cells is done when the DNA (the molecule that contains the instructions for the cell's operation and reproduction) is broken. This occurs either directly, when the radiation itself breaks the molecular chain, or indirectly, by the attachment of free radicals (highly reactive groups of atoms; ◀ page 94) that the radiation has released elsewhere in the cell. Direct damage is usually caused by heavy charged particles, such as protons, alphas and heavier ions. Neutrons also give rise to direct damage, because they collide with nuclei in a cell, and produce nuclear fragments, including protons and alpha particles, which then break the DNA. Indirect damage generally occurs with electrons and high-energy protons, which ionize atoms in the cell and thereby release free radicals. X-rays and gamma rays also damage the DNA indirectly, because they energize atomic electrons within the cells, which then ionize other atoms. These harmful effects are put to good use in treating cancers.

Natural radiation accounts for far fewer deaths than, for example, air crashes or accidents in the home. Artificial radiations, from nuclear power plants, say, are equally low risk, provided adequate care is taken to keep the levels of radiation under control. Exposure to too much radiation causes cancers by damaging cells to the extent that they can still reproduce but no longer function effectively; still larger amounts of radiation produce severe effects, including burning, and can result in a swift death. Certain radioactive isotopes, such as strontium-90, iodine-131 and cesium-137, are particularly dangerous if they collect in specific parts of the body such as the bone marrow.

Artificial isotopes can also be extremely useful, especially in medicine where they are used as radioactive "labels" that reveal the way a particular substance is taken up in the body. Minute amounts of iodine-131 – which is dangerous only in large amounts – once injected into the body, will travel to the thyroid gland. By measuring the (temporary) radioactivity from this gland, doctors can discover if it is overactive or underactive, both of which are unpleasant conditions. Radioactive "tracers", such as the positron-emitters carbon-11 and oxygen-15, are used to image blood flow and oxygen metabolism in the brain. Positrons annihilate with electrons, producing gamma rays which are detected by a "gamma-camera" surrounding the head.

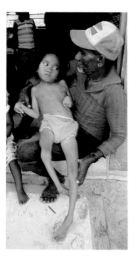

▲ *Carol, aged eight, lives on Rongelap, an atoll of the Marshall Islands in the Pacific. She was deformed at birth, possibly as a result of US nuclear tests on the neighboring Bikini atoll in the 1950s. Data on longterm effects of radiation come only from such unfortunate cases.*

▼ *US servicemen cover their ears as they watch an atomic bomb explode in the Nevada desert in 1957, displaying the ignorance at that time about radiation.*

The risks of low level radiation

One issue that worries many people is the risk from exposure to radiation in excess of the natural background level, either at work or in occupations that involve radiation, or accidently as in the case of the explosion at the nuclear reactor at Chernobyl in the Soviet Union in 1986. Particularly disconcerting for the nonspecialist are the wide variations in the risks that experts in different countries or different organizations claim to be safe.

After the accident at Chernobyl, the Soviet authorities suggested that an additional 42,500 deaths due to cancer would result from the fallout from the explosion; twelve months later they claimed that this figure was too high by 10 or 20 times. In May 1987, experts in the United Kingdom concluded that there would be about 1,000 extra deaths in the European Economic Community (EEC) attributable to the accident, while a study in the United States put the figure ten or twenty times higher. All the figures are small in comparison with the total number of deaths from cancer, which will be about 30 million in the EEC alone in the 50 years following the accident. Yet the discrepancies can still be worrying.

Calculating the longterm risks from a radiation release such as this is a complex matter. It involves, for example, a knowledge of the composition of the fallout produced, an understanding of how the global atmospheric circulation might distribute the fallout, an understanding of how different kinds of radiation produce cancer, and so on. Each parameter in this calculation has its uncertainties, but one major problem and a major reason for

scientific debate lies in estimating the risks due to the relatively low levels of radiation received by people away from the main region of the accident.

Very high levels kill, while somewhat lower levels cause cancers of various kinds. But it is hard to extrapolate from the known effects of high levels of radiation to calculate the number of deaths that might result from much lower levels. Much of the data scientists have at their disposal comes from specific groups of people – the survivors of the bombs dropped on Japan in 1945, people who lived in the Marshall Islands in the Pacific during bomb tests in the late 1940s and early 1950s, workers in uranium mines, workers using luminous paints (which contain radium) who licked their brushes to obtain fine points, people deliberately exposed to radiation in the course of therapy for a variety of ailments.

The basic and most common technique used to deal with this information is to develop a model for calculating the risk from radiation received by these specific groups, which in some cases applies to particular organs or tissues. The models should predict reasonably correctly the number of cancers observed in each group. The information from these models is then combined together to give an overall risk factor to the whole body. The arguments arise in extrapolating the results to lower levels of radiation that the subjects in the various groups actually received. Some scientists question whether a simple linear extrapolation, where the risk remains directly proportional to the dose even at low levels, is valid. Monitoring people around Chernobyl will help resolve this question.

The energy stored in the nucleus...Splitting the uranium atom...Nuclear fission in peace and war...Energy from fusion...The force that fuels the Sun...PERSPECTIVE...The race for fission...The Manhattan Project...A natural nuclear reactor...The missing solar neutrinos

Atoms are tiny storehouses of energy. Chemical reactions can release some of this energy by rearranging the way in which atoms combine in different substances. This happens, for example, when coal burns and in the spectacular explosions of gunpowder and fireworks (◀ page 181). Reactions such as these involve changes in the energy of the electrons within atoms and molecules (◀ page 91). But far more energy is released when an atomic nucleus rearranges itself. The energies binding the nucleons (protons and neutrons) together in a nucleus (◀ page 190) are typically a million times greater than those that bind the electrons in atoms and molecules (◀ page 61). Thus the energy required for and released by nuclear reactions is a million times greater than the energies involved in chemical reactions. In other words, roughly the same energy can be produced from nuclear reactions in one gram of uranium as from the burning of a million grams (1 tonne) of coal.

The best known methods for a nucleus to release energy are fission (splitting into two smaller fragments) and fusion (joing two nuclei together). Another way that a nucleus can release energy is through radioactivity (◀ page 190). This occurs in nuclei that can change to a configuration of nucleons with a lower total energy, for example by emitting a cluster of two protons and two neutrons (alpha decay), or by converting a neutron into a proton (beta decay). For any total number of nucleons – represented by the so-called mass number, A – the arrangement with the least energy is the most stable, and gives the most common form of nucleus for a particular element.

The discovery of fission

Nuclear fission was discovered in 1938 by two German physicists, Otto Hahn (1879-1962) and Fritz Strassmann (b.1902). They were bombarding uranium with neutrons when they identified barium nuclei among the fragments from the collisions. Barium, with 56 protons, is much smaller than uranium, which has 92 protons.

Two of the researchers' colleagues, who had fled from the Nazis in Germany, Lise Meitner (1878-1968) and her nephew Otto Frisch (1904-1979) who was working in Copenhagen in Denmark, suggested that a neutron could cleave a uranium nucleus in two – a process they named fission. Meitner and Frisch tested their idea experimentally and proved that the uranium nucleus does break into two more or less equal parts, such as barium and krypton (with 47 protons).

The Danish theorist Niels Bohr (1885-1962) leapt upon this idea, and together with the American John Wheeler (b.1911), he developed the liquid drop model of the nucleus, which describes several properties of atomic nuclei (◀ page 194). One feature of neutron-induced fission soon became apparent. It released neutrons which could produce more fissions and so establish a chain reaction.

In September 1939 Germany invaded Poland and World War II began. The possibility of harnessing the energy released in a fission chain reaction to make a powerful bomb was not lost on scientists outside Germany, especially on those who had fled the Nazis. In the United States, physicists urged the government to support research in nuclear physics as a matter of urgency. But it was not until the end of 1941 that a concerted effort on establishing the possibility of a chain reaction really began.

The task fell to Italian physicist Enrico Fermi (1901-1954). In December 1941 he was summoned to Chicago, where American physicist Arthur Compton (1892-1962) was head of the "uranium project". A year later Fermi's team had built the first "atomic pile" and on 2 December 1942 it achieved the first artificial chain reaction.

► The Italian physicist Enrico Fermi, who had emigrated to the United States and was awarded the Nobel Prize for physics in 1938 for his work on radioactivity, built the world's first atomic pile underneath the squash courts at the University of Chicago. On 2 December 1942 his team produced the first sustained chain reaction on it.

The world's first nuclear reactor occurred naturally almost 2,000 million years ago

One way to compare the stability of the nuclei of different elements is to calculate the difference between the mass of each nucleus and the total mass of the individual nucleons it contains (◀ page 200). In general, the commonest (most stable) nuclei of each element have a mass that is less than the total mass of the constituents. The effect is greatest for medium-sized nuclei, from about strontium (A=90) to cerium (A=140). It is smallest for the lightest nuclei (A less than 20) and the heaviest nuclei (A greater than 210).

The total mass of two medium-sized nuclei is *less* than the mass of a nucleus of double the size. Thus when a large nucleus is split in two, a net reduction in the total mass occurs, and the amount by which the mass changes appears as energy (◀ page 47). This process is known as nuclear fission, and the way in which it occurs can be understood by regarding the original heavy nucleus as a drop of liquid.

If a drop of liquid becomes sufficiently distorted from its natural spherical shape, it splits into two drops, because the two smaller spherical drops require less energy than the distorted large drop. A similar process occurs if a large nucleus becomes distorted. This can happen spontaneously in some nuclei because an effect of quantum wave mechanics (◀ page 210) allows the nucleus to distort. But such "spontaneous fission" is very rare. For example, in the common form of uranium, uranium-238, the halflife (◀ page 223) due to spontaneous fission is 8×10^{15} years. However, nuclear fission can be made to occur by bombarding appropriate nuclei with neutrons. A nucleus of uranium-238 will capture an energetic neutron and form an excited (energized) nucleus of uranium-239. This new nucleus distorts far more easily, and within a fraction of a second – about 10^{-20} seconds – it splits into two smaller fragments.

The two nuclei produced in fission are usually accompanied by a few much smaller pieces, including neutrons. A large nucleus contains a greater proportion of neutrons than a smaller nucleus, in order to dilute the repulsion between the many electrically-charged protons. Thus the two smaller nuclei produced in fission contain too many neutrons, and these are shed almost immediately after the fission occurs. In a sample of fissionable material these neutrons can induce further fissions, thereby producing a self-sustaining chain reaction with a release of energy at each step in the chain. If controlled, such a chain reaction yields a steady flow of energy, the principle behind the operation of nuclear reactors. If the chain reaction is uncontrolled, it can lead quickly to a catastrophically huge release of energy, and the result is an "atom" bomb.

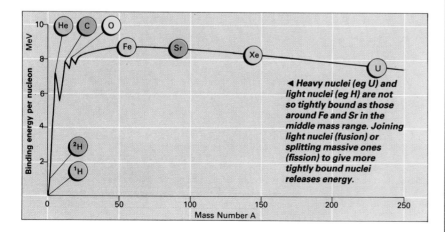

◀ Heavy nuclei (eg U) and light nuclei (eg H) are not so tightly bound as those around Fe and Sr in the middle mass range. Joining light nuclei (fusion) or splitting massive ones (fission) to give more tightly bound nuclei releases energy.

▲ ◄ *Robert Oppenheimer, (left) and General Leslie Groves survey the remains of the tower on which the world's first nuclear explosion had occurred on 16 July 1945 – the "Trinity" test. The explosion in the desert at Jomada del Muerto, near Alamogordo in New Mexico, had crystallized the earth at the base of the tower. Groves, who was in charge of the project to build the bomb, had recognized the need for a special laboratory and had appointed Oppenheimer, one of the world's most brilliant theoretical physicists, as its director.*

The father of the atomic bomb

In November, 1942, General Leslie R. Groves of the US Army visited Los Alamos, a remote area of New Mexico. By the end of World War II, the site had become home to some 6,000 people, all working on the Manhattan Project – America's effort to build an atom bomb. The director of the Los Alamos laboratory was J. Robert Oppenheimer (1904-1967), a theoretical physicist who has become one of the most fascinating figures of 20th-century science.

Oppenheimer headed the laboratory at Los Alamos with great flair, collecting about him a team of scientists whose brilliance and dedication remains unparalleled. Their work came to fruition on 16 July 1945, with the "Trinity Test", the explosion of the world's first atomic bomb, in the desert in New Mexico. At this stage, Oppenheimer and his fellow scientists were jubilant with success; later, after the horrors of the bombs dropped on Hiroshima and Nagasaki became apparent, many of the same scientists became concerned about the spread of these weapons. Oppenheimer himself opposed the development of the hydrogen bomb, and in 1954 he was declared a security risk and banned from government work. Only in 1963 did the government make amends by awarding Oppenheimer the Fermi medal for his work.

A natural nuclear reactor

In the 1970s, scientists measuring samples of uranium ore from a quarry in Gabon, West Africa, found evidence of a chain reaction that occurred 1,780 million years previously. The Oklo quarry is evidently the site of a natural nuclear reactor that was active in the Precambrian geological era.

Today, uranium-235 normally forms 0·72 percent of naturally-occurring uranium, a percentage that is fixed by the relative halflives of the two isotopes. The Okla sample, however, contained unusually low proportions of uranium-235, especially where the ore was rich in uranium. Traces were found of other rare elements in proportions typical of artificial nuclear reactors. Presumably, when the proportion of uranium-235 was as high as 3 percent, there was enough fissile uranium in the rock for the fission process to go critical and a chain reaction to begin. Water trapped in the surrounding sandstone acted as a moderator, and the chain reaction continued for almost a million years.

▲ *In nuclear fission, a heavy nucleus, such as uranium-238, absorbs an extra neutron. This destabilizes the nucleus so that it distorts and splits into two smaller nuclei, at the same time releasing a few neutrons. These neutrons can be used to induce more fissions in a chain reaction.*

Nuclear fusion

An alternative way to release the energy contained in atomic nuclei involves light nuclei such as hydrogen and helium. In this case the total mass of two light nuclei is *greater* than the mass of a nucleus of double the size. Thus combining two light nuclei to form a heavier nucleus gives a net reduction in mass, and is accompanied by a release in energy to keep the total balance of mass-energy constant. This type of reaction is known as nuclear fusion, and it is the process that fuels the Sun and other stars.

Physicists have been trying for several decades to produce energy from hydrogen fusion here on Earth. They have been successful in making "hydrogen" bombs, which release fusion energy in a sudden, violent manner, but they have still to achieve the production of energy in controlled conditions. The main problem in inducing nuclear fusion lies in overcoming the natural electrical repulsion of the like-charged protons. This happens in the center of the Sun because matter there is a hundred times more dense than water, and at temperatures of 15 to 20 million degrees the particles in the core collide very energetically. Even so, fusion reactions in the Sun are relatively infrequent. A proton will exist in the core on average for 10 billion years before it joins with another to release deuterium; such is the weakness of the weak nuclear force responsible for the basic reaction. Physicists need to create high temperatures, like those in the Sun, and very high densities if they are to achieve practical rates in fusion reactions here on Earth. They also intend to "improve" upon the Sun by using reactions that occur millions of times faster than the Sun's basic proton-proton reaction (page 232). The fusion of deuterium nuclei, either with each other or with tritium nuclei, offers the best opportunities. These materials are heavy isotopes of hydrogen – that is, they are forms with the same number of protons (one) but with an additional one or two neutrons in deuterium and tritium respectively.

To achieve suitable temperatures and densities in a fusion reactor the "fuel" of deuterium and tritium must be in the form of a plasma. This is a state of matter in which electrons and nuclei are completely separated: it is a totally ionized gas. A major problem lies in containing such a plasma, which must not be allowed to touch conventional walls. The favored option is to use magnetic fields because they can exert a force on the charged particles of the plasma.

Fusion offers the possibility of an inexpensive alternative to fission for releasing nuclear power, because naturally-occurring deuterium could be extracted from sea-water. It also promises to be inherently safer than fission power, which is based on a self-sustaining chain reaction that can in principle surge out of control. However, as physicists struggle to understand the behavior of plasmas, the practical application of fusion power still seems to lie far in the future.

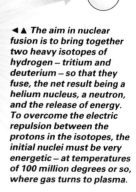

◄▲ *The aim in nuclear fusion is to bring together two heavy isotopes of hydrogen – tritium and deuterium – so that they fuse, the net result being a helium nucleus, a neutron, and the release of energy. To overcome the electric repulsion between the protons in the isotopes, the initial nuclei must be very energetic – at temperatures of 100 million degrees or so, where gas turns to plasma.*

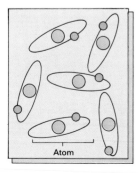

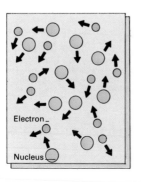

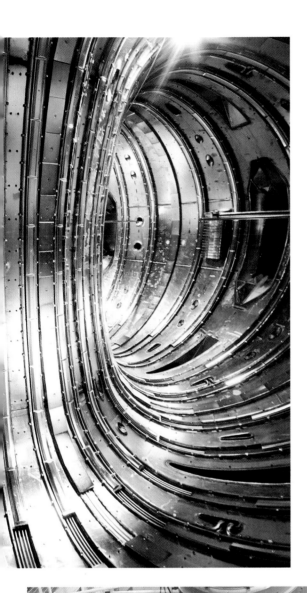

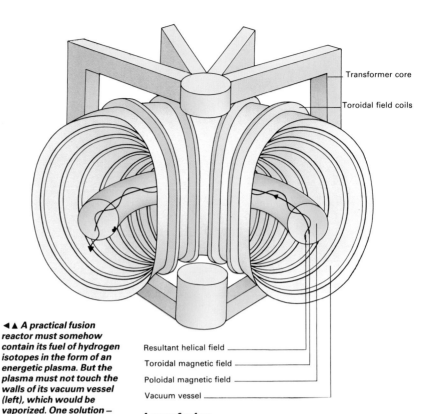

Transformer core

Toroidal field coils

Resultant helical field
Toroidal magnetic field
Poloidal magnetic field
Vacuum vessel

◀▲ **A practical fusion reactor must somehow contain its fuel of hydrogen isotopes in the form of an energetic plasma. But the plasma must not touch the walls of its vacuum vessel (left), which would be vaporized. One solution – the tokamak – is to contain the plasma within a magnetic field shaped like a torus. The magnetic field is set up by D-shaped coils around the vacuum vessel.**

▼▶ **An experiment in laser fusion at the Lawrence Livermore laboratory in California has a complex of massive lasers to deliver thousands of joules of energy for tiny fractions of a second to a deuterium-tritium pellet target, smaller than a grain of sand.**

Laser fusion

One alternative to the magnetic confinement of fusion fuel is the approach known as inertial confinement. This relies on the inertia of the fuel to keep it together once it has been compressed to a very high density. The aim is to use many beams of particles or laser light to bombard a glass-walled pellet of fuel so as to generate the conditions to "burn" all the fuel. The beams would evaporate away the outer layers of the pellet, inducing the remainder to implode. The implosion would compress the fuel and could lead to fusion.

Most research on inertial confinement fusion has so far been done with powerful laser beams directed at pellets a few tenths of a millimeter across. For a useful power station, pellets would have to be several millimeters across and the lasers used would have to deliver many megajoules of energy over periods of several hundredths of a microsecond.

▲▶ *Heavy elements, particularly those heavier than lead, are probably produced in processes that occur in a supernova – a stellar explosion. Supernova SN1978a (right), observed in 1987, was the first nearby supernova to be seen for nearly 400 years. The image above shows the same region of sky before the explosion.*

The missing solar neutrinos

Several steps in the chain of reactions that fuel the Sun emit particles called neutrinos. These particles have little or no mass (certainly less than one ten-thousandth the mass of the electron), no electric charge, and they interact only via the weak nuclear force. This means that, although it has been estimated that every cubic centimeter of space contains between 100 and 1,000 neutrinos, they interact only very rarely with matter. Indeed, neutrinos from the Sun can pass right though the Earth.

However, it is still possible to detect the very rare interactions. All that is needed is a large detector built from a suitable material, and a reasonable amount of patience. One such detector exists 1,500m below ground in the Homestake gold mine in South Dakota. Here United States physicist Ray Davis, of the Brookhaven National Laboratory, has been picking up neutrinos from the Sun since the mid 1960s. His detector consists of 400,000 liters of perchlorethylene (dry-cleaning fluid). The neutrinos interact with chlorine in the liquid to produce a radioactive form of argon-37. Davis flushes the argon out of the tank periodically and measures the amount of radioactivity. He can then work out how many solar neutrinos his detector has intercepted.

The surprise is that the experiment consistently detects only one third the number of neutrinos that are expected according to theory. No one can find fault with the experiment, nor with the theory of the Sun's interior. Possibly the neutrinos have a small mass. Another intriguing possibility concerns the nature of the neutrinos themselves. Do some of them change character from being so-called electron-neutrinos, which the Sun emits, to become one of the other types that particle physicists know exist (◆ page 210)? Davis has not been able to detect other kinds of neutrino. The answer to this question is one of the challenges facing particle physicists, and has given rise to a number of new kinds of solar neutrino detector with which physicists hope to cast new light on the problem during the 1990s.

Fusion in the heart of the Sun

The basic process that fires the Sun's furnace is the fusion of four hydrogen nuclei (protons) into a nucleus of helium-4 (which consists of two protons and two neutrons). For this to occur, two of the four protons must somehow convert into neutrons. This happens each time two protons combine to form a nucleus of deuterium, which contains a proton and a neutron. This reaction is in a sense the opposite to the beta decay of a neutron to a proton. Like beta decay it occurs via the weak nuclear force, and is accompanied by the emission of a positron (positively-charged antielectron) and a neutral particle known as the neutrino. The deuterium produced in this way combines with another proton to form a nucleus of helium-3, and this in turn reacts with another nucleus of helium-3 to form helium-4 and two protons. The net result of this "proton-proton chain" – named after the initiating reaction – is the formation of helium-4 from hydrogen, accompanied by a release of energy which is all-important for life on Earth.

Once there is sufficient helium in a star it can form heavier nuclei through additional fusion reactions, leading quickly to the production of carbon-12. Carbon-12 contains six protons and six neutrons and is therefore directly equivalent to three nuclei of helium-4. The simplest way that it forms in stars is for two helium-4 to fuse, making beryllium-8 which then fuses with a third helium-4 nucleus to make carbon-12. The carbon-12 reacts with hydrogen initiating a sequence of events that has the same net result as the proton-proton chain – namely, the formation of helium-4 from four protons, with the attendant release of energy. During this chain of reactions the carbon-12 is converted into nitrogen and then to oxygen, which then reverts back to nitrogen and finally to carbon-12. Thus the carbon is not used, but acts as a catalyst for the hydrogen "burning". This carbon-nitrogen cycle is important in stars that are hotter than the Sun.

31

Transportation of chemicals and wastes...Safety procedures in a large plant...Waste disposal...Dumping unwanted products...PERSPECTIVE...Predicting the harmful effects of chemicals...Beyond the safety limits...Case studies...Bhopal, Flixborough and Seveso...Transportation accidents

The manufacture, storage, transportation and use of chemicals can all be hazardous. Often, the extent of these hazards is connected with population density. Many older chemical plants were built in areas which were sparsely populated but have subsequently become increasingly urbanized.

Good design and careful physical monitoring can help to cut down the number of accidents within chemical plants. Nevertheless, chemicals still have to be transported widely to their points of use. Road tankers and rail freight are the major means of transport. Physical techniques are being developed to ensure that vehicles carrying hazardous cargos are monitored continuously for defects.

Despite all precautions, chemical accidents will inevitably continue to occur. It is essential that, when they do, emergency services are prepared to cope effectively. A key element in this is a knowledge of what processes are being carried out in chemical plants or what materials are being transported. Many countries now have legislative systems to keep records of hazardous sites and to have standard symbols to indicate particular hazards. Many local authorities also have procedures for dealing with a wide range of major emergencies, whether they are chemical disasters, hurricanes or terrorist attacks.

Predicting the harmful effects of chemicals
The effects of a release of chemicals to the environment may not be known in detail until after a disaster actually occurs. One such example is dioxin, which caused widespread alarm at Times Beach, Missouri, and Seveso, Italy (♦ *page 237).*

The effect of dioxin has been studied on animals, and here it has been shown to be extremely toxic, carcinogenic and teratogenic (causing birth defects). On the strength of these animal tests dioxin was labeled as extraordinarily toxic.

At Seveso, the main result of dioxin exposure was the rise in skin ailments, plus a temporary rise in the spontaneous abortion rate, digestive upsets and effects on the nervous system. There has been no rise in the incidence of cancer, although some forms may take 20 years or more to develop.

▼ *Floods can move chemically contaminated soil and this happened at Times Beach, Missouri, in the early 1980s. Environmental Protection Agency (EPA) technicians tested debris before its removal to hazardous waste tips. Local residents had protested about the dumping of dioxin in their town, and in 1983 the entire town was purchased for $33 million by the EPA because of the scale of the contamination.*

The accident at the Chernobyl nuclear reactor was the result of unauthorized experiments testing the plant's safety system

▲ *An engineer installs a radiation detector on the outside of a reactor vessel. Two such detectors can be used to measure the flow of chemicals. A sharp pulse of radioactive tracer is injected into the process line, and this mixes with the process stream by the time it reaches the first detector. The time taken for the tracer to pass between the two detectors gives an accurate indication of the flow rate.*

Safety monitoring

A range of physical techniques has been developed to provide online monitoring of a chemical plant, to provide warning of failure before it becomes catastrophic. Frequently, where online monitoring has been introduced for safety reasons, it has also led to more efficient operation of the plant.

Radioactive isotopes (◀ page 225) are important in monitoring plant safety. They can be used in two ways, either as open or as sealed sources. If a leak, whether of liquid or gas, is suspected in a chemical plant, it can be detected by addition of a radioisotope. For example, in the case of a gas stream, a radioactive gas which does not react with the chemicals in the stream can be added. This gas can then be searched for throughout the plant using detectors such as geiger counters, and the source of the leak traced.

A similar technique can also provide important information about flow behavior inside different parts of the plant. This may reveal blockages or other disruptions to the flow of material, which could have hazardous consequences.

Some sealed sources emit gamma radiation, which is absorbed in different ways by different materials (◀ page 222). Consequently, a gamma-ray souce placed on a distillation column, with a detector on the other side of the column, can show whether the column is operating efficiently, or whether there are problems such as liquid hold-up, which could lead to hazardous operation.

Neutron sources can also be used. For example, in the petrochemical industry, an important part of most plants is the flarestack which burns off residual hydrocarbons. Combustion produces carbon dioxide and water. In cold weather, the water may turn to ice and block the exit from the stack. A neutron detector can spot this and warn the operators before the blockage becomes dangerous.

Another important tool in safety monitoring is the laser. Laser scanning of a chemical plant can detect any unwanted emission of chemicals over a wide area. Although it cannot pinpoint the leak in the way that radiotracer techniques can, it may provide early warning of a previously unsuspected hazard.

The reactor vessels are particularly liable to degradation. Many reactions mixtures are corrosive and if a vessel develops a flaw, it is essential to spot it at an early stage. Many reactors are now fitted with acoustic detectors because metals "cry out" when they are in trouble.

◀ *An engineer prepares to inject the radioactive tracer into the process stream. As well as measuring flow rates, this technique can be used to detect leaks quickly and categorically. Isotopes with halflives of only a few minutes are used, to avoid contaminating the process stream.*

▶ *X-ray or gamma ray transmission techniques permit the testing of components such as these heat exchanger tubes. The quality of pipework, welds and corrosion can all be seen. A sealed source is placed on the equipment; an image of the interior can be made up from the radiation absorbed.*

Pushing beyond the limits of safety

In April 1986 the world's worst nuclear accident occurred at the Soviet graphite-moderated reactor at Chernobyl, near Kiev. The reactor was destroyed by a fire that took several days to put out, by which time a cloud of radioactive gases had escaped from the plant, causing thousands of people in nearby towns to be evacuated and giving rise to fears about contaminated water and foodstuffs throughout Europe in the following months.

Although the Chernobyl reactor was of a type unknown in the West, and subsequently denounced as less safe than other designs, the Chernobyl plant had an excellent safety record. In the enquiry that followed the accident, it transpired that the accident had been the result of unauthorized experiments being carried out by operators without supervision.

The safety systems that should automatically have come into play to prevent overheating were intentionally circumvented in an attempt to learn more about the operation of the plant. The aim of the experiment was to discover whether there was sufficient energy in the generator turbine during shutdown to provide a temporary emergency electricity supply in the event of a blackout.

The operators put the reactor onto low power and withdrew the safety rods which absorb neutrons from the reaction and control the reaction speed by being raised or lowered into the reactor core. Some of the important safety systems were switched off, in order to make the experiment possible. However, the reactor became unstable and power output quickly rose. It proved impossible to replace the safety rods, either automatically or manually, and an explosion resulted, causing the reactor lid to be blown off, radioactive steam to be released and the fire to begin. More than 30 workers lost their lives in the accident and in attempting to put the fire out.

◄▼ The nuclear power station at Chernobyl in the Soviet Union exploded in 1986 in the world's most serious nuclear accident. The plant had a graphite moderator, and more than 1,600 pressure tubes filled with water. After the explosion, the plant was buried under tonnes of sand and concrete (left), and the topsoil for miles around was removed to clear away residual radiation.

Chernobyl reactor layout

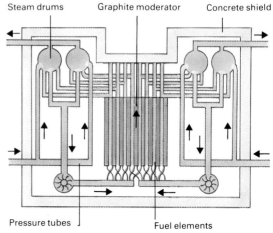

Steam drums   Graphite moderator   Concrete shield

Pressure tubes   Fuel elements

Chemical disasters – case studies

Bhopal – the world's worst chemical disaster
At about midnight on 3 December 1984, the city of Bhopal in central India was engulfed in a cloud of poisonous gas. This cloud, consisting primarily of methyl isocyanate (MIC), was released by an explosion at the Union Carbide chemical plant, which was making carbamate insecticides.

This disaster killed about 2,000 people and probably caused long-term damage to the health of many more. Tests on rodents indicate that exposure to a large dose of MIC leads to persistent obstructive lesions in the lungs. The airways become partially blocked and oxygen exchange between air and blood is hindered.

Methyl isocyanate is volatile and highly reactive. It is one of three organic isocyanates in commercial use. The other two, toluene di-isocyanate and 4,4,diphenylmethane di-isocyanate, are both used in the manufacture of urethane polymers. They are much less volatile than MIC.

The isocyanate group $(-N=C=O)$ undergoes addition reactions but because of the two double bonds in the isocyanate group it is very reactive. MIC reacts exothermically (releasing heat) both with water and with itself.

At Bhopal, a tank containing MIC appears to have become contaminated with water. This caused a reaction. The water itself may have been contaminated with metal ions, or these may have been produced by corrosion of the vessel walls during the reaction with water. However they arose, metal ions probably catalyzed the self-reaction of MIC.

MIC itself boils at 39·1°C. Pressure in the vessel rose rapidly, from the reactions taking place inside, and it eventually ruptured a pressure release valve, sending a cloud of toxic gases over Bhopal. MIC vapor is twice as dense as air and therefore tends to stay close to the ground, rather than dispersing rapidly through the atmosphere. At one stage, the cloud covered $40km^2$, affecting one quarter of Bhopal's population.

A major element of the safety system at Bhopal was refrigeration to keep the MIC from vaporizing. The refrigeration system had been switched off, however. Once the pressure release valve on the MIC tank had blown, the vapor should have passed through a gas "scrubber" and been neutralized by sodium hydroxide. The scrubber failed to work. Once through the scrubber the gas should still have been destroyed by burning in a flare tower. This also failed to work.

Only a few months after the Bhopal incident, another Union Carbide plant had a leak. At this one, in West Virginia, United States, the material which leaked was aldicarb oxime, a pesticide intermediate. This substance is much less toxic and affected only just over 100 people. Nevertheless, it led to an investigation by the US Occupational Safety and Health Administration (OSHA).

In 1986, Union Carbide in the United States was fined more than $1 million by OSHA for violations of safety regulations. The OSHA investigation pinpointed design weaknesses in the plant and also found a number of "wilful" violations – transgressions in which the plant's management knew that hazardous conditions existed.

◄ This boy from Seveso in Italy was exposed to the cloud of dioxin released by an explosion at the chemical works there. This toxic chemical produces a disfiguring condition known as chloracne.

▼ The tragedy of Bhopal killed 2,000 people when a cloud of methyl isocyanate rolled over the town. The chemical escaped after safety devices at a chemical plant failed to function.

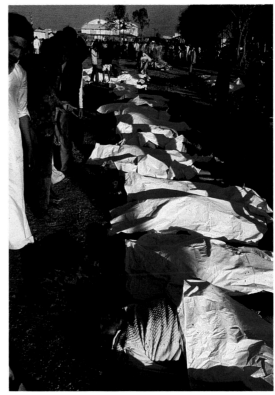

▲ The Flixborough disaster, 1974. One spark was all it took to ignite a ruptured pipeline carrying the highly flammable chemical cyclohexane used in the manufacture of nylon. The factory comprised eight reaction vessels, each of which converted a small amount of cyclohexane to cyclohexanone; this reaction was carried out under 8 atmospheres pressure and at 150°C.

◄ Relief workers clearing up after the dioxin release at Seveso in Italy. Rising temperature and pressure in the reaction vessel caused a pressure disk to rupture, and vapor was released into the atmosphere for about 20 minutes.

Seveso and the dioxin controversy

Shortly before 1pm on 10 July 1976, a pressure disk ruptured at the Icmesa plant at Seveso, near Milan in Italy. The overheated contents of the reaction vessel was propelled 50m into the air, contaminating about 1800ha. Sixteen days later, the Italian authorities evacuated 179 people. Eventually a total of 733 people were evacuated.

A program of therapeutic abortions for pregnant women was undertaken. It was predicted that many of the people exposed to the fallout would develop cancer. The reason for these claims was the discovery that the material emitted from the plant had contained 2,3,7,8-tetrachlorodibenzo-dioxin (TCDD or dioxin). Dioxin has been called the most toxic synthetic chemical known; it has been found in many parts of the world, deriving from the burning of PVC and other industrial wastes.

Yet despite the panic at Seveso, no human death has been unequivocally attributed to TCDD, as a result of that incident or exposure elsewhere.

Flixborough — an unconfined vapor cloud

On 1 June 1974, an explosion destroyed the Nypro Works at Flixborough in northern England. Of the 70 workers on site, 28 were killed.

The Nypro works made caprolactam, the starting material for the synthetic fiber Nylon 6 (◀ page 135). Part of this process was normally carried out in linked series of eight reaction vessels. A few months before the explosion, one of the reactors had to be taken out of service, and a metal pipe was used to join together the reactors on either side. This pipe had a smaller diameter than the outlet and inlet of the reactors. Consequently, it was connected at each end via a bellows.

The reactors were on different levels, so the pipe connecting them was not straight. This put unexpected stresses on the bellows, which consequently ruptured. A large amount of cyclohexane vaporized and formed a cloud over the Nypro works. It encountered a flame and exploded with a force equivalent to 15-45 tonnes of TNT.

Chemical wastes are increasingly able to be reprocessed to minimize the dangers of pollution

Waste disposal

Many industries produce chemical wastes which can harm people or their environment. In recent years, most countries have enacted laws of growing stringency to control the disposal of such wastes.

In the past, the major methods of disposal have been dumping at special sites and incineration. The latter process destroys the waste, but may produce harmful vapors in the process. It is only effective, also, where the waste is primarily organic and therefore combustible.

Much waste has been dumped in special tips since the 1950s. This has a number of drawbacks, notably the possibility of toxic materials leaking from the tip site into water courses. Some wastes, including low-level radioactive waste, have been stored in disused salt mines. For a salt deposit to exist, it must be protected by layers of impermeable rock, otherwise the salt would have been washed away. However, the number of disused salt mines is limited.

New approaches are now being adopted for waste disposal which either destroy the waste completely or immobilize it. One possibility for destruction is the use of a plasma torch, which operates at a much higher temperature than an incinerator, to atomize organic molecules to relatively harmless mixtures of gases. Another possibility is the use of microorganisms, either naturally occurring or genetically engineered, to detoxify wastes.

A major problem in the past has been the complex mixtures which develop in tip sites. Now attempts are being made to keep different types of waste separate, so that at least some of it can be reprocessed.

Chemical waste treatment

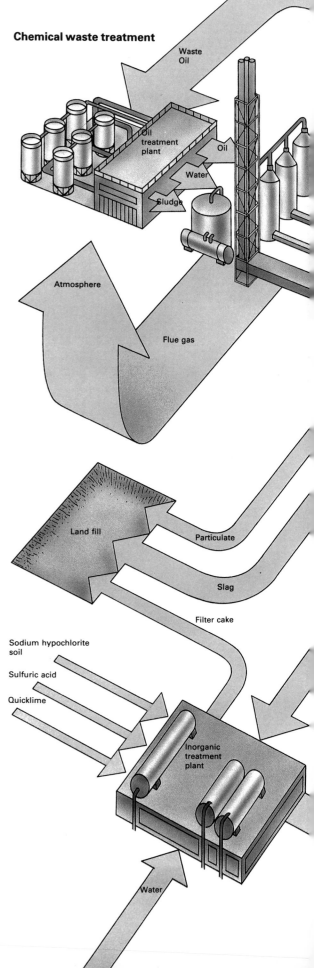

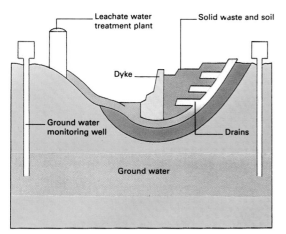

◄ *A properly designed tip, for the disposing of solid chemical wastes, should prevent water that leaches from it contaminating the natural water table. A clay layer is laid down at the bottom to prevent seepage, and the water that collects can be treated if necessary.*

▼ *Sometimes chemical tips cause serious problems, such as the one at Love Canal near Niagara Falls, New York State. Dioxin leached from contaminated waste was affecting the community. The remedy was to remove sediment.*

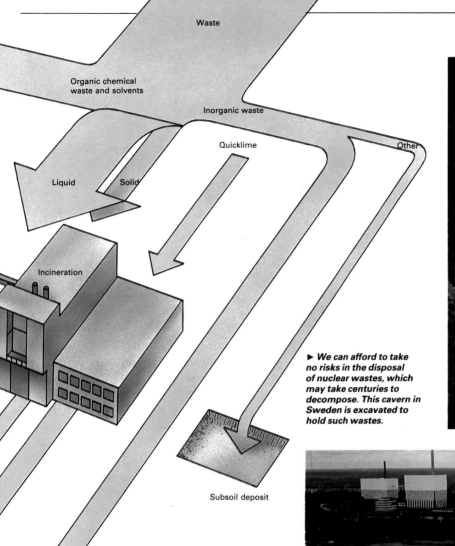

Waste

Organic chemical
waste and solvents

Inorganic waste

Quicklime

Other

Liquid

Solid

Incineration

▶ We can afford to take
no risks in the disposal
of nuclear wastes, which
may take centuries to
decompose. This cavern in
Sweden is excavated to
hold such wastes.

Subsoil deposit

Sewer

▶ The elaborate Swedish
program for nuclear waste
disposal includes the
transportation of all wastes
by sea on specially
designed ships, such as
this one.

◀ The chemical waste
disposal plant at
Kommunikemi, Denmark, is
the world's most advanced
plant for reprocessing.
It treats the chemical
wastes, industrial,
domestic and institutional,
of the whole nation. The
incinerator is the largest
element in the plant.

Transporting chemicals

Chemicals are transported in large quantities by
boat, train and truck, and to a lesser extent by air.
Conditions for sea and air transport are agreed
internationally; road and rail transport are usually
controlled by national governments.

In each case, the purpose of the regulations is to
minimize the chances of an accident, either to the
environment or to people. In the case of ships
which carry chemicals in bulk, the more hazardous
the cargo, the greater the degree of containment
which is required in the ship's design.

Ships for bulk chemical transport have to be able
to remain afloat after a collision and to have
sufficient reserve buoyancy to be refloated if they
run aground. The ships also have to have special
venting arrangements so that crew will not be
harmed by any vapors emitted from the cargo.

Bulk chemicals transported by sea often require
purpose-built cargo handling facilities to ensure
safe transfer of cargo from shore to ship and vice
versa. However, substantial amounts of chemicals
are shipped as general cargo, taken on passenger
ferries, or transported on container vessels.

These containers can often be transferred directly
to rail or road vehicles, to minimize the chance of
cargo handlers coming into direct contact with
harmful substances.

Where chemicals are transported by road and
rail, they are often identified by number and letter
codes, which are attached to the outside of the
vehicle. These codes enable the authorities to
identify very quickly the nature of the chemical
being transported if there is an accidental spillage
following a collision or derailment. They should
also help to prevent mixed rail cargos from
combining chemicals which should be kept well
separated because of their reactivity.

"Gas! Gas! Quick, boys!"

The harmful effects of chlorine gas are well documented. It was first used as a chemical warfare agent on the Western Front in World War I by the German army at Ypres in April 1915. In that initial attack, there were 15,000 casualties, of whom one-third died. Survivors of World War I gas attacks often had their lives shortened by the long-term effects of the gas on the respiratory system. Yet chlorine is also an important element in chemical manufacture and large quantities of it are made in and transported around industrial countries.

Shortly before midnight on 10 November 1979, a 106-car freight train derailed at Mississauga, Ontario. The freight cars contained a variety of different substances, including one with a 90 tonne load of chlorine. In all, 24 freight cars derailed. Eleven contained propane (C_3H_8). Within a few minutes, fires broke out and one propane car exploded. Another behaved like a rocket and was propelled 750m from the site.

The chlorine-containing car was found to have been ruptured. Within 24 hours, nearly a quarter of a million people had been evacuated from an area covering nearly 120km^2. Some of them were not able to return home until five days later.

The evacuation, although costing millions of dollars in terms of lost work, was carried out peaceably and effectively. This was because the area had well-developed emergency plans, providing a clear system for dealing with a crisis. It was operational within six minutes of the derailment.

The Chlorine Institute, which represents chlorine manufacturers in North America, also has emergency procedures for dealing with accidental chlorine releases. This has existed since 1972 and there are chlorine emergency teams throughout the continent. The one nearest to Mississauga was 270km away, which meant that it could be at the site within a few hours of the disaster.

The chlorine team was not as successful in dealing with the emergency as the evacuation team, however. The chlorine tank had a hole in it about 75cm in diameter. It was estimated that 70 tonnes of the chlorine had escaped, which meant that the derailed car still contained 20 tonnes of toxic element. An attempt to patch the hole failed. Eventually, on the ninth day after the disaster, the residual chlorine was removed from the tank.

It may seem that the disaster was caused primarily by mixing toxic and explosive materials in the same train load. However, the inquiry into the disaster found that the chlorine which had escaped had probably been forced rapidly into the upper atmosphere by the updraught from the burning propane. If the derailment had holed the chlorine tank, but there had been no fire, the gas might have spread along the ground to inhabited areas.

Since the accident, considerable efforts have ben made to ensure that train crews receive prompt warning of component failure on freight cars.

▶ **Moving dangerous chemicals in bulk is risky, as shown by this photograph of the train derailment at Mississauga, 1979. Some 70 tonnes of liquid chlorine gas were sent skyward by burning propane from other freight cars.**

Pollution

32

Natural and artificial pollution...Atmospheric pollution from hydrocarbons...Industrial effluents into rivers... Pollution of the soil...PERSPECTIVE...Acid rain... Photochemical smog

Pollution of the environment occurs when substances reach abnormal concentrations which adversely affect living systems. It may be physical or chemical, local or global, natural or manmade. A volcanic eruption, for example, is a natural event which affects a limited area adversely, largely because of the particulate matter emitted. Some eruptions, such as that of Tambora in 1815 and Krakatoa in 1883, may emit sufficient particles and gases to produce global physical effects. Over such natural pollution, humans have little if any control. Consequently, most discussions of pollution focus on that which is artificial and thus preventable.

The types of environmental pollution

Pollution is usually considered according to where in the environment it occurs: air, water or land. Because of the mobility of air and water, pollution of these is more likely to have widespread effects than land pollution. The different types are, however, not mutually exclusive. Pollutants may be leached from land by rain and thence find their way into underground aquifers, rivers or the sea, while atmospheric pollutants may dissolve in water vapor and be carried to earth with the rain.

▲ *The cloud of gas and dust escaping from an erupting volcano can affect the climate. All we can do is wait for the dust to settle, and this may take over a year, and affect a huge area.*

▼ *When the Italian volcano Vesuvius erupted in AD 79 it buried the nearby towns of Pompeii and Herculaneum under meters of ash. Some of the victims can be seen, preserved under the ash.*

◄ *When Lake Nyos in Cameroon released an enormous bubble of carbon dioxide gas in August 1986 it engulfed the surrounding area, asphyxiating sleeping villagers and their cattle. Over 1,700 people died. The lake sits over an old volcanic vent and escaping gas had saturated the lake so that it had become like a bottle of fizzy drink. Suddenly the gas spurted out and 600 tonnes of carbon dioxide escaped.*

Most of the gases that pollute the atmosphere are produced naturally as well as in industrial activity

Air pollution

In tonnage terms, the main source of air pollution is combustion of fossil fuels. Burning of coal increased rapidly during the 19th century. In addition to the physical pollution caused by the release of particulate matter in smoke, this added large quantities of carbon oxides to the atmosphere. As much coal contains sulfurous impurities, gaseous sulfur oxides were also produced.

During the 20th century, with the spread of automobiles and airplanes, consumption of fossil fuels has switched largely from coal to hydrocarbons. Combustion of these produces water vapor in addition to the carbon and sulfur oxides, while the conditions in automobile cylinders are severe enough to convert some of the atmospheric nitrogen and oxygen present to nitrogen oxides. A further important source of atmospheric carbon dioxide is the destruction of forests by "slash and burn" agricultural development.

All of these gases occur naturally and cycle through the atmosphere. A streak of lightning, for example, produces nitrogen oxides as it descends through the air. It may also cause a forest fire, thus releasing carbon oxides. The atmosphere can be seen as a giant reaction vessel, where many chemical reactions are proceeding all the time.

Carbon dioxide and the greenhouse effect

Human activities have greatly increased the turnover of the different cycles, and concern has been expressed that the natural systems may be overloaded so that a buildup in concentration of some of the gases may occur. Adding more of some of the reactants to this giant reaction vessel may produce harmful consequences.

There has been a noticeable increase in concentration of atmospheric carbon dioxide since the Industrial Revolution. Although the overall concentration is still very low, concern about this increase arises from its possible effect on world climate.

Of the solar energy reaching the Earth, a considerable proportion is reflected back into space as heat. Carbon dioxide can interfere with this by preventing radiation reflected from the surface from escaping. It acts like a greenhouse. It has been postulated that, as the global atmospheric carbon dioxide concentration rises, the world's mean temperature will also rise. A rise of only 1° or 2°C could melt sufficient of the polar ice caps to flood many densely populated areas.

In the long term, say 1,000 years, most of the carbon dioxide produced by burning all the remaining fossil fuel would be taken up by the oceans as carbonates already present there and converted to bicarbonates. However, there may also be other, shorter-term, "sinks" that mop up carbon dioxide. Currently, the concentration of carbon dioxide in the atmosphere appears to be rising at only about half the rate at which the gas is being produced by human activity.

Acid rain

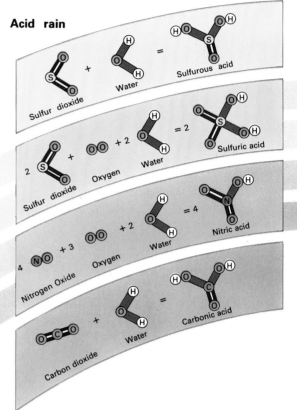

▲ Rain water is naturally acidic. Atmospheric gases, such as nitrogen oxide produced by lightning and sulfur dioxide from volcanos, dissolve in rain drops to give dilute solutions of various acids. But burning vast amounts of fossil fuels has caused the acidity of rain in some regions to increase so that it damages trees and lakes.

▼ Cleaning up the sulfur dioxide emission from power stations burning fossil fuels is not cheap. The more sulfur dioxide removed the more it will cost. Pretreating the fuel removes some sulfur, and the rest can be washed out of the flue gases or neutralized with limestone. Older power plants may have to be closed down.

The costs of control strategies

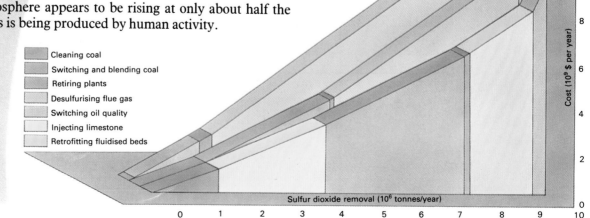

Cleaning coal
Switching and blending coal
Retiring plants
Desulfurising flue gas
Switching oil quality
Injecting limestone
Retrofitting fluidised beds

Cost (10⁹ $ per year)

Sulfur dioxide removal (10⁶ tonnes/year)

▼▶ The damage to trees by acid rain is written in their growth rings. In this section of a fir tree from West Germany the outer layers of growth, of the last 20 years, have been less vigorous than the previous 20. The cause is not due to acid rain falling on the tree itself, but to its effect on the soil. Whole areas of forest are now dying.

▲ Forest fires also add to the environmental burden of the atmosphere. Some, like this one on the Côte d'Azur, France, threaten mainly the tourist trade, but others cause damage to vast areas. It is partly the threat of such fires on a global scale that makes nuclear war unthinkable: the smoke produced could blot out the Sun's rays.

The fluorocarbons that cause concern in the ozone layer were originally developed as a safe refrigerant

Pollution from road transport

Much air pollution is localized. Examples of this are the smogs which used to occur in England and which still occur in Los Angeles. The London smogs were caused by climatic conditions which trapped a mass of cold air in the area. This then became heavily laden with particulate and gaseous material emitted from domestic and industrial burning of coal. These smogs killed several thousand people, and led to the Clean Air Act of 1956, which paved the way for much environmental legislation that has led to cleaner cities.

The Los Angeles smogs are also caused by trapped air masses combined with large-scale burning of fossil fuels, but there the automobile is the culprit. The smogs are photochemical. Sunshine provides the energy to convert nitrogen oxides and unburnt hydrocarbons in exhaust fumes into irritant compounds.

The automobile is implicated in another air pollution debate, related to lead. In the 1920s, Thomas Midgley (who later discovered the chlorofluorocarbons) found that an organic compound of lead, tetraethyl lead, could be used to upgrade the quality of gasoline. This compound is scavenged from the cylinders, where it decomposes, by an organobromine compound. Particulate inorganic lead then gets into the atmosphere through the exhaust. It does not travel far, but may occur in high concentrations in urban areas.

Lead is toxic to humans and may be particularly damaging to small children, when it can affect brain development. Because of the toxicity of lead, it is now used only rarely in paints, for example, although white lead was once a major pigment. Various attempts have been made to cope with the problem of lead emissions from automobiles, including the banning of lead additives from gasoline in the United States. Fitting filters to exhausts could trap much of the lead, but it would be essential that a system of recovering the filters and reprocessing the lead is adopted.

Smog formation

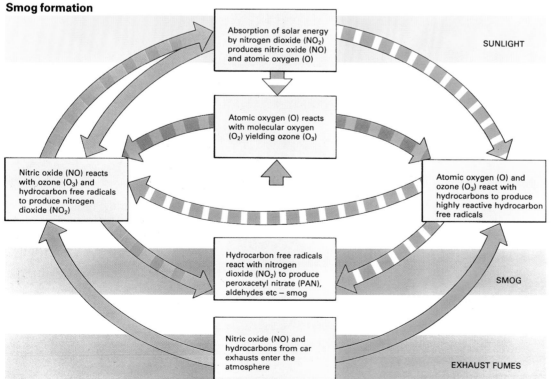

Absorption of solar energy by nitrogen dioxide (NO_2) produces nitric oxide (NO) and atomic oxygen (O)

Atomic oxygen (O) reacts with molecular oxygen (O_2) yielding ozone (O_3)

Nitric oxide (NO) reacts with ozone (O_3) and hydrocarbon free radicals to produce nitrogen dioxide (NO_2)

Atomic oxygen (O) and ozone (O_3) react with hydrocarbons to produce highly reactive hydrocarbon free radicals

Hydrocarbon free radicals react with nitrogen dioxide (NO_2) to produce peroxacetyl nitrate (PAN), aldehydes etc – smog

Nitric oxide (NO) and hydrocarbons from car exhausts enter the atmosphere

SUNLIGHT

SMOG

EXHAUST FUMES

▲ ◄ *Photochemical smog develops most days over Mexico City (above). This is caused by a combination of car exhaust fumes and strong sunlight. The exhaust fumes consist of a little incompletely burnt fuel and various oxide gases. Among these is nitric oxide, and this begins the sequence of events that produces smog. This gas reacts with oxygen to form nitrogen dioxide which can absorb sunlight and split off a reactive oxygen atom. This combines with the hydrocarbons of the unburnt fuel to produce the unpleasant aldehydes and peroxyacetyl nitrate that are the chief offenders of this form of chemical haze.*

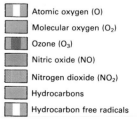

Atomic oxygen (O)

Molecular oxygen (O_2)

Ozone (O_3)

Nitric oxide (NO)

Nitrogen dioxide (NO_2)

Hydrocarbons

Hydrocarbon free radicals

Aerosols and the ozone layer

Arguments about pollutants are not necessarily more clear-cut when the pollutant does not occur naturally. During the 1970s, fears were expressed about the effects of chlorofluorocarbon emissions.

These hydrocarbon derivatives were discovered by the United States chemist Thomas Midgley (1889-1944), who was seeking a safe refrigerant gas in refrigerators. A new use developed in the 1960s, as propellants in disposable aerosol sprays.

In use, the propellant gas in an aerosol spray escapes into the atmosphere. In the case of the unreactive chlorofluorocarbons, it was postulated in 1974 that they would concentrate in the stratosphere, where ultraviolet radiation from the Sun would break them down, releasing chlorine free radicals (◀ page 94).

In the stratosphere there is a high level of ozone (O_3), formed by the interaction between molecules of oxygen and oxygen free radicals produced from oxygen molecules by ultraviolet radiation. The continual formation and breakdown of the ozone layer is important in preventing much of the Sun's ultraviolet radiation from reaching ground level, where it would harm living organisms. It was suggested that the free radical chlorine produced from aerosol propellants could initiate destruction of the ozone layer with catastrophic consequences.

Action has been taken in many countries to restrict propellant uses of the chlorofluorocarbons. What emerged from this concern was a much better knowledge of natural processes. It had been assumed, for example, that there were no natural sources of volatile organochlorine compounds. In fact, volatile methyl chloride is released into the atmosphere from the action of seawater on organic materials released by marine organisms.

Air pollution and industry

Air pollution on a local scale may also result from specific industrial activities. The production of soda by the Leblanc process in the 19th century led to large quantities of hydrogen chloride being liberated around manufacturing plants (◀ page 128). When this dissolved in rain and landed on the solid sulfide waste produced by the same process, it liberated hydrogen sulfide gas. This mixture of acidic and malodorous pollution led to the first industrial environmental legislation in England in the 1870s.

Since that time, gaseous emissions from factories have been controlled with increasing rigor. Nevertheless, as industrial activity expands into new areas (both technological and geographically), new pollution debates open up. In recent years, considerable effort has been expended on assessing the pollutant effects of fluoride emissions from giant aluminum smelters, for example.

With most cases of pollution from industrial processes, the pollutant can be kept from reaching the environment, but this costs money. A balance then has to be struck between the price which people are prepared to pay for goods and the cost of pollution prevention. Similarly, on automobiles, there is a cost to pay in environmental terms for removing lead additives from gasoline. To compensate for the decrease in engine performance means an increase in pollutant combustion gases or a higher energy cost in the manufacture of the fuel.

▲ *The volatile substances used as refrigerants and in aerosol cans may prove environmentally damaging. These simple compounds are chemically unreactive, uninflammable and safe to use. Dichlorodifluoromethane, one of the offenders, is the molecule shown above. When this molecule reaches the ozone layer it is broken up by strong ultraviolet rays from the Sun. Its chlorine atoms then attack the ozone and destroy it. However, ozone protects the Earth's surface from these same ultraviolet rays, which can kill lower forms of life and cause skin cancer in humans.*

Traces of lead, cadmium and mercury can reach water supplies from mines and industrial waste; all can form serious health hazards

Water pollution

For centuries, rivers have been used to dispose of human refuse. Alternatively, waste has been dumped on particular sites – archeologists have learned much from excavation of such places. Before large-scale industrialization, most of this waste consisted of substances which could be broken down readily by natural processes. Today, most sewage treatment is still based on the use of microorganisms to convert organic waste into materials which can be returned to the environment harmlessly.

Chemical pollution arises from three different sources: industrial production of chemicals, and agricultural and domestic use of these products. Pollution of the first kind can be caused at the earliest stage of the industrial process, the winning of raw materials. Water seeping through old mine workings and slag heaps can leach toxic metal ions which are then carried into rivers. It can also be caused at later stages in the manufacturing process, as with the escape of mercury from chloralkali plants which contaminated fish in Lake St Clair near Detroit in 1970. Another major form of pollution from raw materials may occur with spillages of oil into the sea, particularly from wrecked or damaged tankers transporting crude oil to the refinery. Oil slicks over a kilometer long may be washed ashore if not broken up by the action of the waves or by treatment with detergents.

Effluent disposal from manufacturing plants is increasingly controlled by legislation to minimize the amount of potentially harmful material which is released. For example, a metal plating works produces effluents containing both cyanide and heavy metal ions. This has to be treated by the manufacturer to destroy the cyanide and precipitate the heavy metals, before the waste can be poured away down the drain.

Similarly land storage of noxious wastes now usually takes place in specially designated areas to prevent rainwater leaching dangerous materials out of the dump and carrying them into aquifers or rivers.

▼ *This satellite picture of the Persian Gulf was taken in 1983 and shows the extent of oil pollution caused by the Iran/Iraq war. Iranian oil rigs at the northern end of the Gulf were damaged by Iraqi air raids and this resulted in millions of gallons of oil being dumped in the sea. The black areas show the heaviest contamination.*

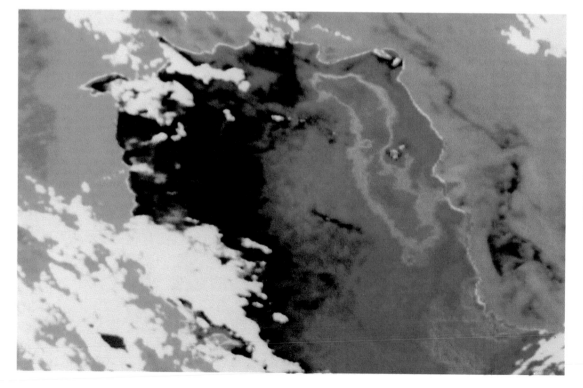

▲ *This buoy, anchored 5km off the coast of Long Island, New York, continuously monitors the environment. A set of instruments records and transmits data on the nature and movement of airborne pollutants. Some day a worldwide system of environment stations on land and sea will record all changes to the Earth's atmosphere and oceans.*

▲ The movement of oil
worldwide requires the use
of massive oil tankers, and
this has led to some huge
pollution problems. In 1967
the tanker Torrey Canyon
ran onto rocks off the Scilly
Isles and polluted the
British coast; in 1978 the
Amoco Cadiz repeated the
disaster on the French coast
(above). Inshore wild-life,
seabirds and tourism were
all badly affected. The
British used detergents to
disperse the oil, causing
further damage to the
environment. The French
(top) relied on mechanical
measures.

▶ Analytical chemists are
the front-line troops in the
battle against pollution.
They can measure the
minutest amounts of any
substance suspected of
polluting the environment.
The Rhine has been
accidentally polluted
several times in recent
years, resuting in massive
fish kills. Chemists keep a
daily check on its water.

▲ Because the oceans are so vast they can dilute and degrade almost anything we care to pour into them. Many coastal towns let both their sewage and chemical plants discharge effluent straight into the sea. Even so, this cheap and easy method of disposal creates problems for the creatures of the seashore, as well as making beaches unpleasant and unhealthy for visitors.

▶ While most things that we throw away are quickly degraded by microbes, some are not. These will accumulate in the environment, and while they may be harmless to humans they may affect other species. DDT, with its five chlorine atoms (blue), is a simple but very unreactive molecule. Only about 2 percent of the total DDT used is degraded in the environment each year. Even though banned, it will still take many years before the last traces disappear from the soil.

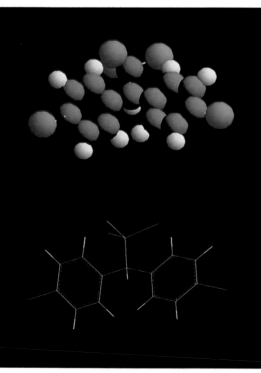

Chemicals in the food chain

Pollution from a single, clearly identified source, such as a chemical factory, is easier to regulate than the widespread pollution which can occur when chemicals are used agriculturally or domestically. The introduction of new detergents in the 1950s which contained substances that could not be degraded by microorganisms led to foaming rivers. This was overcome by replacing the offending substances with molecules which could be broken down by microorganisms.

One of the most persistent groups of chemicals to have been introduced in this century are the chlorinated pesticides, such as DDT (◀ page 150). The major use of these has been in agriculture.

DDT was introduced in the early 1940s and remained in widespread use until the early 1970s. During that time, it became dispersed throughout the environment to such an extent that even animals living in polar regions have DDT in their body fat.

Persistent pesticides remain on land which has been treated with them and are also carried into water by rainfall. The land-based material is taken up by creatures such as earthworms, from whence it gets into birds which eat the worms and so moves up the food chain. In water, it is taken up by microscopic organisms, which are then eaten by fish and so on. Every human being on Earth probably contains several parts per million of DDT.

Toxic metals and nitrates

In addition to the persistent organic pesticides, many pesticides contain toxic metals, such as arsenic. Indiscriminate use of these can also cause serious pollution. Mercury in fish can be caused not only by release of mercury from chloralkali plants, but also by leaching of mercurial fungicides from soils and from the effluents of paper mills where these fungicides are also used.

The other major agricultural input which can affect the environment is fertilizers. Excess amounts of nitrogenous fertilizers can dissolve in rain and be carried through to rivers and lakes. Here, the increased nitrogen content acts as a fertilizer for algae, causing explosive growth of these organisms and a serious upset to the ecological balance. The algae may also benefit from the fertilizing effect of phosphates, brought to them either from agriculture or from domestic use of phosphate-containing detergents.

Keeping drinking water clean

As population growth and urbanization have increased, there have been increasing pressures on water supply. In some countries, the water from a river is drunk several times before it reaches the sea. Water purification has recently come under suspicion as also having its pollutant effects.

The major chemical treatment for water to render it safe is chlorination. In addition to killing off potentially harmful organisms, the chlorine also reacts with organic matter in the water to produce chlorinated hydrocarbons. Recently, it has been suggested that these may be carcinogenic and that their levels in drinking water also need to be controled.

Credits

Key to abbreviations: ARPL Ann Ronan Picture Library; CD Chemical Designs Ltd; RHPL Robert Harding Picture Library; SPL Science Photo Library; b bottom; bl bottom left; br bottom right; c center; cl center left; cr center right; t top; tl top left; tr top right; l left; r right

1 CD 2-3 SPL/Jeremy Burgess 4-5 RHPL 7 SPL/ Powell Fowler & Perkins, "The Study of Elementary Particles" 8 ZEFA/T. Ives 9 SPL/Fred Espanak 10 John Hilleson Agency/R. Michaud 11, 12tl BPCC/Aldus Archive 12tc RHPL 12-13 CD 15 Tony Stone Worldwide 16tr Leo Mason 16-17 RHPL 17tl Colorsport 17tr Rex Features 18-19 Allsport 18b National Maritime Museum 19bl Biofotos/H. Angel 20 Mountain Camera/John Cleare 21 SPL/Jan Hinsch 22tl ZEFA/L. Vilotta 22cr Richard House 22b Bridgeman Art Library/Royal Society 23 © David Redfern/Stephen Morley 23r ZEFA/F. Damm 24 Michael Freeman 25t ARPL 25b SPL/Dr Gary Settles 26t John S. Shelton 26b SPL/Hank Morgan 27 ZEFA/Justitz 28 MOD (PE) RAE Farnborough/Crown Copyright 29 ARPL 30bl SPL/Claude Nuridsany & Marie Perennou 30cr David Lee Photography 30-31 Hutchison Library 31c SPL/Manfred Kage 31b Allsport 32t SPL/Andrew McClenaghan 32b Burndy Library 32-33 J. Allan Cash 33br © David Redfern/Andrew Putler 34t Bruce Bennett 34b Mountain Camera/Colin Monteath 35t Mountain Camera/John Cleare 35b Marin 36t Sonia Halliday & Laura Lushington Photographs 36-37 National Physical Laboratory/Crown Copyright 37cl SPL/James Bell 37r Bruce Coleman Ltd/Stephen J. Kraseman 38t Paul Brierley 38b John Watney Photo Library/Jerry Young 39 High Altitude Observatory and National Center for Atmospheric Research, Colorado 40cl Nils Abramson 40-41 John Watney 41r Perkin Elmer 42cr Royal Society 43t Tony Stone Worldwide 44 SPL/Martin Dohrn 45 SPL/Jeremy Burgess 46 SPL/Fermi National Accelerator Lab 47 French Railways Ltd 48 Paul Brierley 49t SPL/Alex Bartel 49b SPL/Vaughan Fleming 50b Michael Holford 50-51 SPL/Jack Finch 52t RHPL 52b Dr Kemshead 53t SPL/Lawrence Berkeley Lab 53b ARPL 54-55 ZEFA/T. Ives 55t British Library 55b SPL/James Stevenson 56t Paul Brierley 56b SPL/David Taylor 58 SPL/David Scharf 59t Paul Brierley 59b Minolta 60 RHPL 61 BBC Hulton Picture Library 62t Science Museum, London 62b Tim Woodcock 63 Tony Stone Worldwide 64t RHPL 64c National Maritime Museum 64bl SPL/Dr W.H. Ku 64br SPL/NASA 65tl Institution of Electrical Engineers 65tr ARPL 65bl SPL/NASA 65br SPL/Max Planck Institute for Radio Astronomy 66-67 Nikola Tesla Museum Belgrade 67inset Nikola Tesla Museum, M. Connolly Collection 67bl Independent Television News Ltd 68 David Redfern 69 ARPL 70t Bridgeman Art Library/Musée des Beaux Arts, Lille 70br SPL/Dr Mitsuo Chtsuki 71 London Pictures Service/Crown Copyright 72 University of Cambridge, Cavendish Laboratory 73 CD 75 Novosti Press Agency 77 Sally & Richard Greenhill 78t SPL/Manfred Kage 78b SPL/Prof. Erwin Mueller 79 Swanke Hayden Connell Architects/Dan Cornish 80 ZEFA/J. Behnke 81

SPL/Dr Jeremy Burgess 82,83,84 CD 85tl,bl Pye Unicam Ltd 85tc SPL/GECO UK 85r Science Museum 85bc New Methods Research Inc. 86 CD 87t SPL/Manfred Kage 87b,88,89,90,91t CD 91b ZEFA/P. Phillips 92l SPL/Sinclair Stammers 92r SPL/Andrew McClenaghan 93tl Anthony Blake 93r,bl Susan Griggs Agency 94t Action Press 94b SPL/Jeremy Burgess 95l Frank Spooner Pictures 95r SPL/H. Schneebeli 96t Brian & Sally Shuel 96b,97 CD 98 London Pictures Service/Crown Copyright 99t Lawrence Clarke 99bl,br Dr F.B. Pickering 100tr Leo Mason 100br Hoechst High Chem 100bl Frank Spooner Pictures 101tl Andrew Lawson 101br Michael Freeman 101bl Institute of Archaeology, University of London 102 ZEFA/W.H. Mueller 103tl Susan Griggs Agency 103bl Vautier de Nanxe 103br W.K. Young, National Meteorological Office 104tl SPL/Hank Morgan 104tr SPL/Swindon Silicon Systems/A. Sternberg 104br Art Directors 105t Bruce Coleman Ltd/Jane Burton 105b RHPL 106 SPL/Earth Satellite Corporation 107 RHPL 108SPL/US Naval Observatory 109t Sio Photo 109b,110,111,112t CD 112b RHPL 113l APCC/Aldus Archive 113r Michael Holford 114tl Wellcome Institute 114-115 Gesellschaft Liebig – Museum E.V. Giessen 114br, 115bl,115br ARPL 116bl SPL/GECO UK 116-117 ZEFA/Orion Press 117t RHPL 117c ZEFA/T. Horowitz 117b SPL/David Leah 118,119 CD 121 Farbdoppel 122t Hunting Geology & Geophysics Ltd 122b BASF 123t U.H. Mayer BFF 123b London Pictures Service/Crown Copyright 124 BASF 125,128-129,130-131 Bayer 133t Bell Labs 133b Suddeutscher Verlag 134,135 CD 136l ZEFA/T. Horowitz 136r BP Chemicals Ltd 137l ZEFA/Ung Werbestudio 137r CD 138bl ICI 138tl Hoechst 138cr London Pictures Service/Crown Copyright 138-139 Hoechst 140br Du Pont (UK) Ltd 141 BASF 141inset Shell Photo Service 142t BASF 142b Design Council 143t Brown Bros 143b CD 144tl TBA Industrial Projects Ltd 144tr BASF 144br Pirelli General 145t Holt Studios Ltd 145b South American Pictures/Tony Morrison 146-147 SPL/Sinclair Stammers 147tr Holt Studios Ltd 147br ICI Chemicals & Polymers 148 CD 148-149 Frank Spooner Pictures 149,150t Holt Studios Ltd 150b CD 151 ZEFA/K. Leeds 152 International Flavors & Fragrances Inc 153 CD 154t Distillers Company 154b Sounds Natural 155t SPL/Martin Dohrn 155b Distillers Company 156tl,cl,bl CD 156tr Novosti Press Agency 157 Photri Inc 158-159,159cl Explorer/P. Montbazet 159tr,cr,bl,br CD 160bl,br SPL/St Mary's Hospital Medical School 160-161 Glaxo Holdings 161bl SPL/CNRI 161tr,tl CD 162-163 Tony Stone Worldwide 162b ARPL 163l,r,164 CD 165r International Museum of Photography, George Eastman House 165l Science Museum 166,167 Kodak 168 Paul Brierley 169t Vautier de Nanxe 169b Michael Holford 170-171 Hamilton Kerr Institute/National Trust 172tl,tr Vauxhall Motors 172-173 Rex Features 173t Michael Freeman 174t Paul Brierley 174b Bruce Coleman Ltd/Wayne Lankinen 175 ARPL 176tl Bodleian Library, Oxford 176tl British Museum 176-177 SPL/NASA 177 Frank Spooner Pictures 178t,c Long Ashton Research Station 178br ZEFA 179t Paul Brierley 179bl John Watney 180 J. Allen Cash Ltd 181 Tony Stone Worldwide 182, Department of Defense,

Washington DC 183t ICI 183b CD 184tr Frank Spooner Pictures 184b Camera Press 184-185t Associated Press 184-185b Pixfeatures/Wakeman 185br Robert Hunt Library 186r Topham Picture Library 186l Popperfoto 187l SPL/NASA 187inset AIP Niels Bohr Library, William G. Myers Collection 187b Mansell Collection 188t University of Manchester, Physics Dept 188b University of Cambridge, Cavendish Laboratory 189 SPL/Lawrence Berkeley Laboratory 190 SPL/David Paker 191 Topham Picture Library 194t Niels Bohr Institute, AIP Niels Bohr Library 194b,195 Bibliothek und Archiv zur Geschichte der Max Planck Gesellschaft 196 SPL/Physics Dept, Imperial College 197 SPL/Lawrence Berkeley Lab 207 main picture SPL/Stanford Linear Accelerator Center 207tr CERN 207br SPL/Tasso Experiment, Desy/Oxford Nuclear Physics Lab 208,209 SPL/Lawrence Berkeley Lab 210tr SPL/Powell Fowler & Perkins, "The Study of Elementary Particles" 210-211 SPL/Irvine-Michigan-Brookhaven Proton Decay Experiment 212,214t SPL/David Parker 214-215 SPL/Patrice Loiez, CERN 215t SPL/Brookhaven National Lab 217 University of Manchester, Jodrell Bank 218 SPL/Irvine-Michigan-Brookhaven Proton Decay Experiment 219 Popperfoto 220tl SPL/C.T.R. Wilson 220tc SPL 220tr SPL/I. Joliot Curie & F. Joliot 220bl,br Popperfoto 221t SPC/Ohio Nuclear Corporation 221bl SPL/Martin Dohrn 222bl Susan Griggs Agency 222-223 British Museum 223tr Frank Spooner Pictures 224 Camera Press 225 SPL/Cea Orsay/CNRI 226t Greenpeace/Pereira 226b TRH/US Army 227 Chicago Historical Society 228-229SPL/US Army 229t Topham Picture Library 230-231 JET Joint Undertaking 231br,231bl Lawrence Livermore Lab 232 Anglo-Australian Telescope Board 233 Associated Press 234 ICI 235 Novosti/Gamma 236-237 Camera Press 236b,237 Rex Features 238 Associated Press 239 SKB 240 Associated Press 241tr Susan Griggs Agency 241bl Frank Spooner Pictures 241br Bruce Coleman Ltd/Melinda Berge 243t Frank Spooner Pictures 243bl SPL/Adam Hart Davis 243br Biofotos/Heather Angel 244-245 South American Pictures/Tony Morrison 245br CD 246bl TRH/Dept of Defense, Washington DC 246br SPL/US Dept of Energy 246-247 Susan Griggs Agency 247br Rex Features 248t ZEFA/Boutin 248b CD

Artists
Robert and Rhoda Burns, Simon Driver, Chris Forsey, Alan Hollingbery, Kevin Maddison, Colin Salmon, Mick Saunders

Indexer
Barbara James

Typesetting
Peter MacDonald/Ron Barrow

Production
Joanna Turner

Further Reading

General
Beiser, Arthur *Physics* (Benjamin/Cummins 4th ed)
Close, Frank, Martin, Michael, and Sutton, Christine *The Particle Explosion* (Oxford University Press)
Cotterill, Rodney *The Cambridge Guide to the Material World* (Cambridge University Press)
Kaye, G.W.C. and Laby, T.H. *Tables of Physical and Chemical Constants* Longman 14th ed
McGraw-Hill Concise Encyclopedia of Science and Technology 1984
Manahan, Stanley E. *General Applied Chemistry* (Willow Grant Press 2nd ed)
Marion, Jerry B. *Physics, the foundation of Modern Science* (John Wiley)
Moore, W.J. *Physical Chemistry* (Longman 5th ed)
Sherwood, M. *New Worlds in Chemistry* (Faber)

Specific topics
Westfall, Richard S. *Never at Rest: a biography of Newton* (Cambridge University Press)
Berg, Robert and Stork, David *The Physics of Sounds* (Prentice-Hall)

Walton, Alan J. *The Three Phases of Matter* (Oxford University Press)
Pendlebury, J.M. *Kinetic Theory* (Adam Hilger)
Dickerson, R.E., Gray, H.B. and Haight, G.P. Jnr. *Chemical Principles* (Benjamin/Cummins 3rd ed)
Sisler, H.H. *Electronic Structure, Properties and the Periodic Law* (Reinhold)
Murrell, J.N., Kettle, S.F.A., and Tedder, J.M. *The Chemical Bond* (J. Wiley 2nd ed)
Kirk-Othmer, *Concise Encyclopedia of Chemical Technology* (J. Wiley)
Williams, E.H. *Designing in Metals* (Iliffe Books)
Thompson, R. (ed) *The Modern Inorganic Chemicals Industry* (Chemical Society)
Waddams, A.L. *Chemicals from Petroleum* (Murray)
Brydson, J.A. *Plastics Materials* (Butterworth 4th ed)
McMillan, Frank M. *The Chain Straighteners* (Macmillan)
Hardie, W.F., and Davidson J. *A History of the Modern British Chemical Industry* (Pergamon)
Worthing, C.R. (ed) *The Pesticide Manual* (British Crop Protection Council 4th ed)

Pyke, Magnus *Food Science and Technology* (Murray 4th ed)
Sneader, Walter *Drug Discovery, the Evolution of Modern Medicines* (J. Wiley)
Sharp, D., and West, T.F. *The Chemical Industry* (Horwood)
Rhode, Robert B. and McCall, Floyd *Introduction to Photography* (Macmillan 4th ed)
Wittcoff, Harold A., and Reuben, Bryan G. *Industrial Organic Chemicals in Perspective* (Wiley)
Chemistry in the Economy (American Chemical Society)
Meyer, Rudolf *Explosives* (Verlag Chemie)
Herbert, Nick *Quantum Reality* (Hutchinson)
Hey, Tony and Walters, Patrick *The Quantum Universe* (Cambridge University Press)
Sutton, Christine *The Particle Connection* (Hutchinson)
Elsworth, Steve *Acid Rain* (Pluto Press)
Gribbin, John *Future Weather* (Penguin)
Harison, Roy H. *Pollution: Causes, Effects and Control* (Royal Society of Chemistry)
Lawless, Edward W. *Technology and Social Shock* (Rutgers University Press)

Glossary

Absolute zero
The temperature at which all substances have zero thermal energy.

Absorption
Any process by which a substance incorporates another into itself, or takes in radiant or sound energy.

Acid
A substance capable of providing hydrogen ions for chemical reaction.

Actinides
The 15 elements with atomic numbers 89-103, beginning with actinium.

Alcohols
A class of aliphatic compounds of general formula ROH, with a hydroxyl group bonded to a carbon atom.

Aldehydes
A class of highly reactive organic compounds of general formula RCHO, containing a carbonyl group.

Alicyclic compounds
A class of organic compounds in which carbon atoms are linked to form one or more rings.

Aliphatic compounds
A major class of organic compounds that includes all those with carbon atoms linked in straight or branched open chains.

Alkali
A water-soluble compound of the alkali metals (or ammonia). Alkalis neutralize acids to form salts, and turn red litmus paper blue.

Alkanes (Paraffins)
A homologous series of saturated hydrocarbons of general formula C_nH_{2n+2}, obtained from petroleum or natural gas.

Alkenes (Olefins)
A homologous series of unsaturated hydrocarbons having one or more double bonds between adjacent carbon atoms.

Alkynes (Acetylenes)
A homologous series of unsaturated hydrocarbons having one or more triple bonds between adjacent carbon atoms.

Allotrope
An element that occurs in varying forms, differing in their crystalline or molecular structure.

Alpha particle
Helium nuclei emitted from radioactive materials undergoing alpha disintegration.

Amides
A class of aliphatic compounds derived from carboxylic acids and ammonia.

Amines
A class of organic compounds derived from ammonia.

Analytic chemistry
Study of the compounds or elements comprising a chemical substance.

Aromatic compounds
A major class of organic compounds containing one or more planar rings of atoms whose electronic structure gives special stability.

Atom
Classically one of the minute, indivisible particles of which material objects are composed; in 20th-century science the name given to a relatively stable package of matter made up of at least two subatomic particles.

Atomic physics
Study of the physics of the atom.

Atomic weight
The mean mass of the atoms of an element weighted according to the relative abundance of its naturally occurring isotopes.

Baryons
In particle physics, a class of subatomic particles comprising the nucleons and hyperons.

Base
In chemistry, the complement of an acid, a substance that can accept hydrogen ions from an acid or that can donate an electron pair to an acid.

Beta ray
A stream of electrons or positrons emitted from radioactive nuclei undergoing beta disintegration.

Boiling point
The temperature at which the vapor pressure of a liquid becomes equal to the external pressure; that at which a liquid and its vapor are at equilibrium

Bond
The links that hold atoms together in molecules.

Boson
A behavioral classification of subatomic particles according to Bose–Einstein statistics.

Bubble chamber
Device used to observe the paths of subatomic particles, by reducing pressure as the particles pass through so that bubbles form along their paths.

Capacitance
The ratio of the electric charge on a conductor to its potential; or the ratio of the charge of a capacitor to the potential difference between its plates.

Carbonates
Salts of carbonic acid containing the CO_3^{2-} ion.

Catalysis
The changing of the rate of a chemical reaction by the addition of a substance that itself remains unchanged.

Chelate
A chemical complex formed from a polydentate ligand and a metal ion, thus making a ring.

Colloid
A system in which two or more substances are uniformly mixed so that one is extremely finely dispersed throughout the other.

Combustion
Rapid oxidation of fuel in which heat and usually light are produced, otherwise known as burning.

Condensation
The passage of a substance from gaseous to liquid or solid state.

Crystals
Homogeneous solid objects having naturally formed plane faces.

Cyanides
Highly toxic compounds containing the CN group.

Decomposition
A reaction in which a chemical compound is split up into its elements or simpler compounds.

Distillation
Process in which substances are vaporized and then condensed by cooling.

Elasticity
The ability of a body to resist tension, torsion, shearing or compression and to recover its original shape and size when the stress is removed.

Electric charge
An inherent property of matter, electrons carrying a negative charge and nuclei normally carrying a similar positive charge for each electron in the atom.

Electrolysis
The technique of producing a chemical reaction by passing a current through an electrolyte so that a substance is deposited at the cathode or anode.

Electromagnetism
The study of electric and magnetic fields, and their interaction with electric charges and currents.

Electron
A subatomic particle of negative charge, commonly in orbit around an atomic nucleus.

Electronics
Science dealing with semiconductors and devices where the motion of electrons is controlled.

Element
Simple substance composed of atoms of the same atomic number.

Elementary particle see subatomic particle

Energy
One of the fundamental modes of existence, equivalent to and interconvertible with matter.

Entropy
The degree of disorder of a physical system; the extent to which the energy in a system is available for doing work.

Equilibrium
The state in which a system – mechanical, electrical or thermodynamic – will remain if undisturbed.

Esters
Organic compounds formed by condensation of an acid with an alcohol, water being eliminated.

Ether
A medium postulated by 19th-century physicists to explain how light could be propagated as a wave motion through otherwise empty space

Ethers
Organic compounds of general formula E-O-R; the most important is diethyl ether.

Ethyl compounds
Organic compounds containing the ethyl group, C_2H_5.

Evaporation
The escape of molecules from the surface of a liquid into the vapor state.

Field
The area in which a body exerts its influence, electric, magnetic or gravitational.

Force
In mechanics, the physical quantity that, when it acts on a body, causes it to change its state of motion or tends to deform it.

Formula
A symbolic representation of the composition of a molecule.

Free radical
A molecule or atom that has one unpaired electron.

Frequency
The rate at which a wave motion completes its cycle, measured in hertz (Hz).

Gamma rays
High-energy photons emitted from atomic nuclei during radioactive decay.

Gas
One of the three states into which almost all matter can be classified. With its freely moving molecules, a gas will fill the available volume.

Gel
A colloid, usually with a semisolid base.

Gravitation
The force of attraction between all matter, one of the fundamental forces of nature.

Hadrons
A class of subatomic particles, including the baryons and mesons.

Halflife
The time taken for the activity of a radioactive sample to decrease to half its original value.

Halides
Binary compounds of halogens (highly reactive nonmetals), with oxidation number -1.

Heat
Form of energy that passes from one body to another because of a temperature difference between them.

Hydrate
A compound containing a definite proportion of water, known as water of crystallization.

Hydrocarbons
Organic compounds composed of carbon and hydrogen.

Hydrogen bonding
The formation of a weak bond between a hydrogen atom and another such electronegative atom.

Hydrogenation
A reaction in which hydrogen is added to a compound, converting unsaturated organic compounds into saturated ones.

Hydrolysis
A double decomposition of a chemical compound effected by water.

Hydroxides
Compounds containing the OH group or the ion OH^-, e.g. alcohols, phenols and carboxylic acids.

Impedance
The ratio of the alternating current voltage applied to an electric circuit to the current it produces.

Induction
The phenomenon in which an electric field is generated in an electric circuit when the number of magnetic field lines passing through it changes.

Inorganic chemistry
Major branch of chemistry comprising the study of all the elements and their compounds, except carbon compounds containing hydrogen (organic chemistry).

Interference
The interaction of two or more wave motions establishing a new pattern in the amplitude of the waves.

Ion
An atom or group of atoms that has become electrically charged by the gain or loss of electrons.

Ionic bonding
The transfer of electrons from one atom to another, causing electrostatic attraction to bond them together.

Ionization
The formation of ions.

Isomers
Chemical compounds with identical chemical composition and molecular formula, but differing in the arrangement of atoms in their molecules.

Isotopes
Atoms of an element with the same number of protons in the nucleus but different numbers of neutrons.

Kinetic theory
Statistical theory based on the idea that matter is made up of randomly moving atoms or molecules whose kinetic energy increases with temperature.

Lanthanides
The 14 elements with atomic numbers 58-71, following lanthanum in the Periodic Table.

Latent heat
The quantity of heat absorbed or released by a substance in an isothermal change of state, such as fusion or vaporization.

Leptons
In one division of subatomic particles, leptons include electrons, neutrinos and muons.

Ligand
An ion or molecule linked to a central metal ion by a coordinate bond to form a complex compound.

Liquid
One of the three states of matter, taking the shape of the container but of a fixed volume at a particular temperature.

Mass
A measure of the amount of matter in a body.

Mass spectroscopy
Technique in which electric and magnetic fields are used to deflect moving charged particles differentially according to their mass.

Matter
Material substance, with extension in space and time; the three physical states are solids, liquids and gases. Also regarded as a specialized form of energy.

Mechanics
Branch of applied mathematics dealing with the action of forces on bodies.

Mesons
Subatomic particles of the hadron group, including pions and kaons.

Metal
An element with high specific gravity, high opacity and reflectivity to light, malleable, ductile, and a good conductor of heat and electricity.

Methyl compounds
Organic compounds with the methyl group, CG_3.

Molecular weight
The sum of the atomic weights of all the atoms in a molecule.

Molecule
Entity composed of atoms linked by chemical bonds and acting as a unit; its composition is represented by its molecular formula.

Momentum
The product of the mass and linear velocity of a body.

Neutralization
A chemical reaction between compounds of opposite chemical character, giving an inactive product.

Neutron
An uncharged subatomic particle.

Nitrates
Salts of nitric acid, containing the nitrate ion No_3^-, almost all soluble in water.

Nitrites
Salts of nitrous acid, HNO_2.

Noble gases (inert or rare gases)
Chemically unreactive gases forming the elements in Group 18 of the Periodic Table, comprising helium, neon, argon, krypton, xenon and radon.

Nonmetal
A substance, particularly an element, showing none of the properties characteristic of metals.

Nuclear physics
Study of the properties and mathematical treatment of the atomic nucleus and subatomic particles.

Nucleon
Group of subatomic particles comprising the proton and the neutron.

Nucleus
The core of an atom, containing positively charged protons and electrically neutral neutrons.

Oil
Any substance that is insoluble in water, soluble in ether and greasy to the touch.

Orbit
The path followed by one celestial body revolving under the influence of graavity about another.

Orbital
The mathematical wave function describing the motion of an electron around the nucleus of an atom or several nuclei in a molecule.

Organic chemistry
Major branch of chemistry comprising the study of carbon compounds containing hydrogen.

Organometallic compounds
Class of compounds containing bonds from carbon atoms to metal atoms.

Oxidation
Large class of chemical reactions that, with reduction, are known as redox reactions; the two always go together, oxidation involving a loss and reduction a gain of electrons.

Oxides
Binary compounds of oxygen with the other elements.

Peptide
A compound containing two or more amino acids.

Periodic Table
A table of the elements in order of atomic number, arranged to illustrate periodic similarities and trends in physical and chemical properties.

Petrochemicals
Chemicals made from petroleum and natural gas.

pH
Measure of the acidity of an aqueous solution.

Phase
The proportion of a cycle already executed by an oscillating system, expressed as an angle.

Phenols
Class of aromatic compounds in which a hydroxide group is directly bonded to an aromatic ring system.

Phosphates
Derivatives of phosphoric acid, either phosphate esters or salts containing phosphate ions.

Photochemistry
Study of chemical reactions that produce light or are initiated by it.

Photon
The quantum of electromagnetic energy, often thought of as the particle associated with light or other electromagnetic radiation.

Plasma
Almost completely ionized gas, containing equal numbers of free electrons and positive ions.

Polymer
Substance composed of very large molecules built up by repeated linking of small molecules.

Pressure
The force per unit area acting on a surface.

Proton
Stable, positively charged subatomic particle found in the nucleus of all atoms.

Quantum mechanics
Theory of small-scale physical phenomena, such as the motions of electrons and nuclei within atoms.

Quantum theory
Theory developed at the beginning of the 20th century to account for certain phenomena that could not be explained by classical physics.

Quark
Fundamental subatomic particle, constituent of hadrons.

Radiation
The emission and propagation through space of electromagnetic radiation or subatomic particles.

Resistance
The ratio of the voltage applied to a conductor to the current flowing through it.

Resonance
The large response of an oscillatory system when driven near its natural frequency.

Salt
A compound formed by neutralization of an acid and a base.

Semiconductor
A material whose electrical conductivity varies with temperature and impurity, which can therefore be modfied for different electrical purposes.

Solution
A homogeneous molecular mixture of two or more substances, usually a solid and a liquid.

Specific gravity
Ratio of the density of a substance to that of a reference material, usually water at 4°C.

Specific heat
The heat required to raise the temperature of 1kg of a substance through 1 kelvin.

Spin
Intrinsic angular momentum of a nucleus or subatomic particle arising from its rotation about its axis.

Stereochemistry
The study of the arrangement of atoms in molecules, and of their properties.

Stereoisomers
Isomers with the same molecular structure but differing in the spatial arrangement of their atoms.

Subatomic particles (Elementary particles)
Small packets of matter-energy that are constituent of atoms or are produced in nuclear reactions or in interactions between other subatomic particles.

Sublimation
Transformation of a solid to the vapor state without its becoming liquid.

Sugars
Sweet, soluble carbohydrates comprising the monosaccharides and the disaccharides.

Sulfates
Salts of sulfuric acid, formed by reaction of the acid with metals, their oxides or carbonates, or by oxidation of sulfides or sulfites.

Superconductivity
A condition occurring in many metals, alloys, etc., which have zero electrical resistance at very low temperatures.

Superfluidity
The property whereby "superfluids" such as liquid helium exhibit apparently frictionless flow.

Suspension
System of macroscopic particles dispersed in a fluid in which settlement is hindered by intermolecular collisions and by the fluid's viscosity.

Temperature
The degree of hotness or coldness of a body, measured quantitatively by thermometers.

Thermodynamics
Division of physics concerned with the interconversion of heat, work and other forms of energy, and with the states of physical systems.

Transformer
A device for altering the voltage of an alternating current electricity supply.

Transition elements
Elements (all of them metals) occupying short groups in the Periodic Table.

Valence
The combining power of an element, expressed as the number of chemical bonds that one atom of the element forms in a given compound.

Vapor
The gaseous state of a substance, usually one that is solid or liquid at room temperature.

Velocity
The rate at which the position of a body changes in a given direction.

Viscosity
The property of a fluid by which it resists shape change or relative motion within itself.

Wave motion
The motion of a material or extended object, oscillating in such a way as to create an illusion of crests and troughs.

X-rays
Invisible elctromagnetic radiation with a wavelength between that of ultraviolet raditaion and gamma rays.

Index